Introduction to Water Resources and Environmental Issues

Second Edition

How much water does the world need to support growing human populations? What are the potential effects of climate change on the world's water resources? These questions and more are discussed in this thoroughly updated and expanded new edition. Written at the undergraduate level, this accessible textbook covers the fundamentals of water resources, water law, allocation, quality and quantity, and health issues, and provides examples of potential personal actions and solutions. There is a keener focus on climate change, as many of the predictions made in the first edition have now come to pass.

This new edition features improved artwork, more active learning prompts, more positive examples of beneficial changes, basic introductions to scientific approaches, and a discussion of emerging contaminants and LiDAR technology. It contains strong teaching features, with new "In Depth" and "Think About It" sections to encourage class discussion, and homework questions to test students' understanding.

Karrie Lynn Pennington has studied the interactions of land use, water quality, and quantity for over 30 years. The last 22 years were with the Natural Resources Conservation Service of the US Department of Agriculture (USDA) in the Mississippi Delta where, as in many places around the world, human actions changed the natural ecosystem into farmland, providing a classic example of land-use conversion for study.

She received her bachelor's degree in biology from the University of North Texas and completed her Master of Science in soils from the University of Idaho where she taught until moving to Tucson, Arizona. She taught for three more years at the University of Arizona (UA) before deciding to obtain her Ph.D. at UA. After completing her Ph.D. in soil and water science she moved to Mississippi, finishing her postdoctoral work with USDA's Agricultural Research Service. She is now retired.

Thomas V. Cech was born and raised on a farm near Clarkson, Nebraska. He graduated from Kearney State College with a Bachelor of Science degree in mathematics education, and later received a Master of Science degree in community and regional planning from the University of Nebraska, Lincoln. He was Executive Director of the Central Colorado Water Conservancy District in Greeley, taught undergraduate- and graduate-level water resources courses at the University of Northern Colorado and Colorado State University, and is now the Co-Director of the One World One Water (OWOW) Center at Metropolitan State University of Denver, Colorado.

"A comprehensive primer on our most precious, yet often neglected, resource. With systematic reviews of the critical challenges, this volume offers accessible treatment of thorny problems and provides useful summaries and additional readings. A must-read for students, scholars, and practitioners to safeguard water – the foundation of life."

Professor Christopher A. Scott, *University of Arizona*

"So glad to see this new edition! It's fully updated with contemporary trends and terminology. Particularly insightful are the 'Think About It' features embedded within each chapter, encouraging the reader to consider different facets of evolving water resource issues. Overall, it continues to be a well-organized, highly readable, and comprehensive treatment of a critically important resource."

Professor Ned Knight, *Linfield College*

"An encyclopedic journey into the connections between water issues and an array of topics, including human geography, public health, ecology, geology, history, and policy. It provides something for everyone – spectacular black-and-white photos, an introduction to virtually all aspects of water science, human interest stories, and great discussion questions."

Professor Kathy Jacobs, *University of Arizona*

Introduction to Water Resources and Environmental Issues

Second Edition

KARRIE LYNN PENNINGTON

THOMAS V. CECH
Metropolitan State University of Denver

Shaftesbury Road, Cambridge CB2 8EA, United Kingdom

One Liberty Plaza, 20th Floor, New York, NY 10006, USA

477 Williamstown Road, Port Melbourne, VIC 3207, Australia

314–321, 3rd Floor, Plot 3, Splendor Forum, Jasola District Centre, New Delhi – 110025, India

103 Penang Road, #05–06/07, Visioncrest Commercial, Singapore 238467

Cambridge University Press is part of Cambridge University Press & Assessment,
a department of the University of Cambridge.

We share the University's mission to contribute to society through the pursuit of
education, learning and research at the highest international levels of excellence.

www.cambridge.org
Information on this title: www.cambridge.org/9781108746847

DOI: 10.1017/9781108784221

First published 2022
Second edition 2022

A catalogue record for this publication is available from the British Library

ISBN   978-1-108-74684-7   Paperback

Additional resources for this publication at www.cambridge.org/pennington.

Contents

Preface

The stream of time moves forward, and mankind moves with it. Your generation must come to terms with the environment. You must face realities instead of taking refuge in ignorance and evasion of truth. Yours is a grave and sobering responsibility, but it is also a shining opportunity. You go out into a world where mankind is challenged, as it has never been challenged before, to prove its maturity and its mastery – not of nature, but of itself. Therein lies our hope and our destiny.

Rachel Carson, 1962 commencement address, Scripps College in California

We all live on one Earth, share one atmosphere, drink from the same waters. We are undeniably connected to our Earth and to each other. Global climate change affects all of us, in ways we do not recognize when we fail to make these connections and understand their significance.

We might not recognize that a global pandemic is a water issue, without considering that day-to-day actions that save lives require a source of plentiful, clean water. Washing hands, cleaning surroundings, and maintaining hydration are not possible without access to clean water. Without access to water, foods are not produced. Malnutrition leaves people more vulnerable to disease. The connections go on.

We might not recognize water as an issue of equality of justice, educational opportunity, medical care, or proper living conditions, but the connections are there for us to make if we will. Water supplies energy for development and is critical to many industrial processes that provide jobs. It makes agriculture possible in deserts. In areas where competition for water resources exists, we must consider the many people who depend on that same water for their fisheries and small farms; the family livelihoods that have existed for generations. How we choose to develop water can take those opportunities away forever, leading directly to more inequality. Unfortunately, the lack of inclusion of the many diverse peoples influenced by management of water resources has been apparent for many years. This generates inequality in the decision-making process which affects us all. Increasing diversity in the workplace, and in the policymaking process, are substantive changes which must occur.

The choices are not easy. We must draw upon creative problem solving and consider the needs of the Earth and humanity over development and profit. We will explore these issues more thoroughly in this text. It is our hope that students will follow the mandate presented to them by Rachel Carson.

Rachel Carson urged a generation to look at the interconnectivity of all of nature and realize that we are responsible for our part in challenging the distortion of truths which too often sacrifice our environment for commercial interests. Her warning about the misuse of chemicals in the 1950s and 60s mirrors today's warnings about increased greenhouse gases and global climate change. In this new edition, we examine more closely the connectivity of water resources issues and the associated environmental impacts of global climate change. We see the Earth's carrying capacity for humans already being strained by rising sea levels. We watch as the ferocity of storm events becomes greater and increase in destructive power. Extremes of weather – from drought to deluge – have countries facing environmental disasters ranging from deadly wildfires to more frequent floods. The human price for the neglect of our environment is already being paid, and without significant and timely action, the bill will only get higher.

Today our entire planet is in danger from global climate change. Our hydrologic cycle is being altered by choices from the individual to the global. We have known the dangers of the rapidly changing greenhouse gas concentrations in our atmosphere for over seven decades; with the solidifying of the fundamental science that was established in the 1950s and then well understood by the 1970s. Scientific evidence that has been agreed on by the majority of actively publishing climate scientists, 97 percent, assure us that we humans are the cause. Governments of the world have had decades to invoke actions to slow global warming, but efforts have been met with political roadblocks. Today we are seeing more action as most of the leading science organizations around the world have issued public statements expressing these concerns, including international and US science academies, the United Nations Intergovernmental Panel on Climate Change, and many more reputable scientific bodies around the world.

The Intergovernmental Panel on Climate Change (IPCC), formed in 1988 by the World Meteorological Organization (WMO) and the United Nations Environment Program (UNEP), was charged with providing regular assessments of the scientific basis of climate change, its impacts and future risks, and options for possible solutions. Their 2019 special report states that a 1.5 degree Celsius (2.7 degree Fahrenheit) change in global temperature could be reached in the next 12 to 20 years. This, the report states, will result in widespread damage to ecological, economic, and human environments.

The cost of protecting national economies is too often promoted above the need to protect our environment. This political cry against reducing human global warming activities does not bode well for future generations. Today we need to reframe the discussion in terms of the morality of harming the human population and our environment. Rachel Carson provides a roadmap: We must face realities of our environment instead of taking refuge in ignorance and the evasion of truth. Greta Thunberg, a young Swedish environmental activist on climate change, has provided inspiration and motivation to millions showing that you can make a difference with actions and words. She challenges: "*We know the solutions; all we have to do is wake up and change.*" We all have a grave and sobering responsibility, but also a shining opportunity to improve the health of our environment and all of humanity.

We will continue to examine issues, as described in the first edition preface, and solutions the students of today can choose to study and implement. Volunteering to talk about water to 4th graders on their natural resource field days was uplifting. I ended my talk about water by asking the students "what is the most important natural resource?" When they answered "water," I would say no – *you are* – because you have a brain and the ability to understand and change things that are wrong. Humans have endless creativity. We offer you the same hope that you will use that imagination and creativity to solve problems and find solutions.

In this Edition

- This updated text further examines water from global and historical perspectives – with its roles today and in the development of civilizations. These chapters have been updated to include issues of importance to Indigenous peoples, climate change influences on existing issues, and current global population issues involving the right to and the need for water.
- The text moves into areas of science, with the hydrologic cycle, water chemistry, and water quality providing the fundamentals. These areas have been expanded or updated to include new developments, and anthropomorphic influences on our hydrologic cycle.
- Subjects that require more understanding are presented as In Depth sections. These are intended to interest the curious student.

- The concepts of ecosystems are explored in chapters on watersheds, groundwater, lakes and ponds, rivers and streams, and wetlands. Sidebars and guest essays provide additional information and case studies.
- Human attempts to control natural systems are explored through the study of dams and structures. The importance of dams and reservoirs is considered as well as their consequences. This section includes updates on the global proliferation of dams since the first edition, in countries considered less developed, and water use in other energy production methods.
- The chapter on drinking water explores attempts to repair and restore damaged systems. Natural sources of pollution, as well as human-produced pollution, are discussed. Waterborne diseases and the complexities of their control are considered.
- Water law and water allocation are discussed. Who gets to use the water we have is a tremendously controversial subject, involving competing uses by cities, industry, endangered species, individuals, and the water source itself. Depleting groundwater or drying up a river is akin to killing the goose that laid golden eggs, but sometimes we don't think beyond right now.
- The roles of governments, and the various agencies they create, are presented primarily using the United States as an example. Case studies demonstrate positive and negative interactions of agencies all trying to do their jobs while competing for tax dollars.
- The final chapter summarizes the state of our Earth's water resources and encourages students to think about the future of water and humans as inseparable features. A final look at major issues – from global climate change to competing personal values – shows how much more there is to learn and the complexity of decisions that are still to be made.

1 Perspectives on Water and Environmental Issues

For many of us, water simply flows from a faucet, and we think little about it beyond this point of contact. We have lost a sense of respect for the wild river, for the complex workings of a wetland, for the intricate web of life that water supports ... We have been quick to assume rights to use water but slow to recognize obligations to preserve and protect it ... in short, we need a water ethic – a guide to right conduct in the face of complex systems we do not and cannot fully understand.

Sandra Postel, *Last Oasis* [1]

Chapter Outline

Aldo Leopold proposed his "land ethic" in his essay "A Sand County Almanac" published in 1949 just after his death. He believed, as Sandra Postel does and illustrates with water, that humans have a moral responsibility to protect the natural world. "A thing is right when it tends to preserve the integrity, stability, and beauty of the biotic community. It is wrong when it tends otherwise [2]."

1.1 Introduction

The study of water resources and environmental issues is a broad, fascinating field that can take many different directions. Some people are drawn toward the physical aspects of water and the environment – **rivers, lakes, wetlands, groundwater**, and associated **ecosystems**. Others enjoy learning about the historical aspects of water resources and hope to find opportunities to utilize past efforts to improve future activities. Still other students and practitioners focus on more human-related social or legal aspects of water resources management and the environment.

In this chapter, we'll start with an overview of water distribution across the globe. Then, ecosystems and watershed basics will be examined, followed by global water use and population issues. We'll finish with a discussion of the Earth's carrying capacity for humans and how global climate change influences these populations. Our goal is to whet your appetite – there's much to learn (and consider) as you read our work. We're pleased you're interested in water resources and environmental issues; now let's get started.

1.2 Distribution of Water on Earth

Water is found everywhere on Earth and is the only substance that can naturally occur as a liquid, solid, or gas. The Earth contains approximately 1.39 billion cubic kilometers (331 million cubic miles) of water. Examination of Figure 1.1 (Bar 1) shows that the majority, 96.5 percent, is stored in the oceans of the world. This water is **saline**. **Freshwater**, the water we use for almost everything, is only 2.5 percent of the total. Most freshwater is stored in **glaciers**, permanent snow, on **sea ice**, and in polar ice caps, or exists as groundwater (Bar 2). Surface water is still largely frozen, leaving human's major liquid water source in lakes, with smaller amounts in rivers, wetlands, and **soil**. Water in the atmosphere can be solid, liquid, or gas and although a very small amount, 0.04 percent, it is vitally important to global weather and climate conditions (Bar 3).

Globally, freshwater is abundant. However, it is not evenly distributed across the continents. Freshwater amounts vary season to season and year to year. Approximately two-thirds of the population – around 5 billion people – lives in locations that receive only one-fourth of the world's annual precipitation. Much of this does come seasonally as mountain snow or monsoon rains (see Figure 1.2). Much of India, for example, receives 90 percent of its annual rainfall, 300–650 millimeters (11.8–25.6 in), during the monsoonal season between June and September. The remaining eight months of the year are quite dry. India is home to 17 percent of the world population.

By contrast, the Amazon River Basin of South America has approximately 15 percent of the world's surface water runoff, with average annual rainfall of 2,300 mm (90 in), but only contains 0.4 percent of the Earth's population. Asia, on the other hand, has 69 percent of the world population and 36 percent of the Earth's surface water runoff, with annual average rainfall of 733 mm (28.9 in). Differences in water distribution, quantity, and population lead to water and crop shortages, wildlife and fisheries failures, and often disease, resulting in both ecosystem and human suffering.

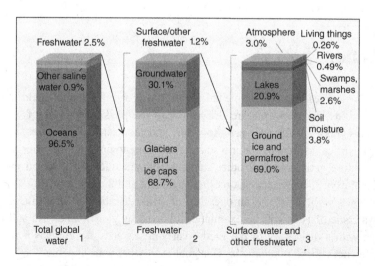

Fig 1.1 Distribution of Earth's water, numbers are rounded (NASA image of data from [3])

Fig 1.2 A monsoon over the Timor Sea near the Northern Territory capital city Darwin, Australia (Photograph Credit: Xian Yang via Getty images)

Think About It: Unequal Distribution of Water

Approximately two-thirds of the global population lives in regions that receive only one-fourth of the world's annual precipitation; this is a tremendous disparity worsened by global climate change. In today's world we are seeing countries, with much of their land at or near sea level like Bangladesh, suffer from climate change–induced extremes of weather. Floods are more frequent but often shorter in duration and so intense that they cause landslides. Landslides remove cropland, homes, and people, which destroys lives and livelihoods. Saltwater intrusion continues the destruction of native fisheries and cropland. Temperatures are becoming hotter and the normal monsoonal patterns have changed in Bangladesh. Droughts outside of the monsoonal period occur in areas of the country.

On a human scale, flooding in 2017 alone influenced about 4 million people. As homes, fisheries, and cropland were lost, people were forced to migrate into the slums of large cities unprepared to deal with them [4] [5]. What sorts of water management challenges does this create? What sorts of environmental and human challenges does this population/precipitation imbalance create? Can you think of alternatives to help solve these challenges? What types of resources do your alternatives require? How feasible are your alternatives when resources are limited?

1.2.1 Oceans

Oceans are not discussed in detail as part of this text since they are not freshwater and are an entire field of study on their own; however, oceans play a major role in our hydrologic cycle since they cover over 70 percent of the Earth's surface. Oceans too are unequally distributed. If you viewed the Earth from high above Great Britain, you would see almost 50 percent of the Earth covered by land. Conversely, if you viewed the Earth from a similar altitude above New Zealand, you would see a world covered almost 90 percent by oceans. This uneven distribution of land and oceans – caused by plate tectonics – greatly affects ocean currents, climate, and precipitation patterns around the world.

The depth of the world's oceans averages about 4,500 meters (14,800 feet), with the greatest depth measured at 11,035 meters (36,205 feet) in the Mariana Trench near the island of Guam in the western Pacific. In contrast, Mount Everest (the world's highest mountain, in the Himalayas of Nepal and Tibet) has an elevation of 8,850 meters (29,035 feet).

It is important to consider the ocean's ability to absorb the extra carbon dioxide (CO_2) humans are putting into our atmosphere, causing global climate change. Recent studies accurately place this number at one-third of the increased CO_2. This is a case of good news, bad news. The good news is that phytoplankton can convert some of the CO_2 to carbon and oxygen through photosynthesis. The carbon finds its way to the ocean depths where it remains sequestered primarily because there is little mixing of waters at these depths. The bad news is that the initial reaction of CO_2 in water produces acids increasing the acidity of the ocean. The increased acidity is harming coral reefs and fish populations that are unable to adapt. Therefore, increasing concentrations of CO_2 from human activities are partially mitigated by the oceans but not without damaging effects to the ocean and its inhabitants [6] [7].

The oceans are salty, or saline, and are composed of about 3.5 percent dissolved salts by weight. The dissolved salts are primarily sodium, calcium, magnesium, and chloride. Human tolerance for salt is less than 2 percent, which makes seawater undrinkable. When seawater is evaporated, more than 75 percent of the dissolved matter is precipitated as common table salt (NaCl). If all seawater were lost to evaporation, the remaining minerals would cover the seafloor with a layer of salt about 56 meters (183 feet) deep [8]. Most agricultural crops are not salt tolerant and most industrial processes cannot use saltwater. This has important implications to humans and ecosystems since nearly 97 percent of Earth's water is salty – difficult, expensive, or impossible to utilize for drinking water, agriculture, and industrial uses.

1.2.2 Glaciers, Permanent Snow, Sea Ice, and Polar Ice Caps

Glaciers are created when deep snow recrystallizes due to the weight of overlying snow and, over time, forms into dense ice sheets of freshwater. Glaciers form when the snow and ice become so thick and heavy that gravity causes the frozen mass to move. Presently, glaciers cover about 10 percent of the Earth's land surface while permanently snow-covered and frozen ground covers another 20 percent. This means that approximately 30 percent of the surface of the Earth is in the **cryosphere** (*cold* or *frozen sphere*).

Surprisingly, glaciers not only occur in polar regions, but also near the Equator. Mountain peaks in New Guinea, East Africa, and the Andes of South America contain glaciers at high elevations. In Tanzania, Mount Kilimanjaro (Swahili for *shining mountain*) is Africa's highest summit (5,895 meters, or 19,340 feet), and contains the massive Furtwängler Glacier on its Western Breach (see Figure 1.3). It's located only about 350 kilometers (220 miles) south of the Equator and is particularly susceptible to climate change. The glacier is named after Walter

Fig 1.3 Furtwängler Glacier from Uhuru Peak, 2001
(Photographer: Christopher J Bassett, https://creativecommons.org/licenses/by-sa/3.0/0)

Furtwängler, who, along with Siegfried Koenig in October 1912, achieved the fourth documented successful climb, and became the first to use skis to descend.

In Depth	**Kilimanjaro**

The glaciers of Kilimanjaro have been receding, and recently at an alarming rate. Ernest Hemingway described these ice fields as "wide as all the world, great, high, and unbelievably white in the sun." However, since 1912 – the year they were first extensively measured – the glaciers have lost 82 percent of their ice. Some claimed the glaciers would disappear completely by 2020. The year 2020 has passed and Furtwängler glacier is still here, but still shrinking. Scientists from the University of Massachusetts Kilimanjaro climate and glacier study concluded that the increasing fragmentation of the southern icefield will soon cause the loss of three glaciers, including Furtwängler [9]. Scientists debate the cause; one theory is that global warming is to blame for the loss of the 12,000-year-old mass of ice, while others point to the reduction in forest vegetation surrounding Mount Kilimanjaro. Trees are cut down or burned for agriculture production, sometimes inadvertently as honey collectors smoke bees out of hives. Less forest vegetation reduces evaporation from the forest canopy to the atmosphere, which reduces cloud cover and precipitation. The result is increased solar radiation and glacial evaporation [10]. Most likely, global warming and deforestation are working simultaneously to melt the glacier. This concept is called *destructive synergy* – multiple processes working together to produce worse or faster negative results than either one alone.

Snow and ice, appropriately named polar ice caps, perennially cover the polar regions of the Northern and Southern Hemispheres. In the Northern Hemisphere, much of the ice floats as a thin sheet of **sea ice**, frozen seawater, on the Arctic Ocean. By contrast, the Southern Hemisphere polar region consists of an extensive glacial system on the continent of Antarctica and sea ice beyond the coastline. Approximately two-thirds of the Earth's permanent cover of ice exists as sea ice. Surprisingly, sea ice only accounts for about 0.001 percent of the Earth's total volume of ice [11].

In Depth   NASA ICESAT – 2

Polar ice caps generate cold, dense water that creates deep ocean circulation. These ocean currents affect the world's climate by altering ocean temperatures in a process that fluctuates seasonally and over decades and centuries. It's a dynamic process that links polar ice caps with the oceans and the atmosphere. On September 15, 2018, NASA launched a new satellite called IceSat-2 (the Ice, Cloud, and Land Elevation Satellite-2) with the ability to measure Earth surfaces using laser pulses and precise timing to provide extremely accurate measurements. This allows scientists not only to monitor the elevation of ice sheets, glaciers, and sea ice in extraordinary detail but also to look at atmospheric conditions, vegetation, land ice, and inland water [12]. A lack of this kind of detailed data in the past has led to misunderstandings about how fast or if polar ice melting is significant. Analysis of new datasets provided by NASA will more clearly show the state of Earth's polar regions [13].

Indigenous People in Arctic Environment

Over the 42-year period covered in Figure 1.4, total area of sea ice loss is 1.75 square kilometers (676,000 square miles), an area approximately the size of the state of Alaska.

Discussions of ice melt and global climate change are important to the Indigenous people who live in these very cold Arctic environments –from Alaska to Siberia. Records indicate that Alaska's temperatures have risen at twice the global rate since the 1970s [14]. The 229 Alaskan tribes and the multitude of Siberian Russian peoples depend on sea ice stability for safe hunting and fishing to protect their food sources. The ice that has long preserved that food has melted, leaving

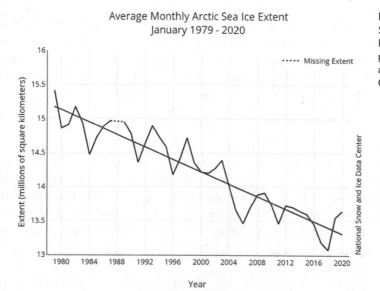

Average Monthly Arctic Sea Ice Extent
January 1979 - 2020

····· Missing Extent

Extent (millions of square kilometers)

National Snow and Ice Data Center

Year

Fig 1.4 Sea ice extent for 1979 to 2020 measured monthly in February shows a decline of 2.91 percent per decade National Snow and Ice Data Center, Boulder, Colorado [15]

rotten meat and hungry people. Equally important is the stability of permafrost, soil that remains below 32 degrees Fahrenheit all year. It is the foundation for all structures, roads, and livestock grazing in that region [16].

Climate change hits these Indigenous people, who are trying to maintain a way of life closer to that of their ancestors than city dwellers, partially because of their dependence on a stable environment. Their religious lives are often tied to their concepts of nature and living things, which are extremely personal. When your hunting grounds start to melt away and the soil you live on thaws and sloughs downhill or into the sea, it is as traumatic or worse than any earthquake or volcanic eruption; because you watch it coming in slow motion but are helpless to stop it. Whole villages have had to evacuate. In addition, there are no government programs to help rebuild because there is no solid foundation left to build on, no stable ice fields, and no permafrost. With increasing climate change, there will be no return to their former way of life.

> We humbly ask permission from all our relatives; our elders, our families, our children, the winged and the insects, the four-legged, the swimmers, and all the plant and animal nations, to speak. Our Mother has cried out to us. She is in pain. We are called to answer her cries.
>
> *Msit No'Kmaq* (All my relations!) [17]

Msit No'Kmaq means "All my relations" in the Algonquin language. This prayer from the Algonquin people expresses their belief in the value of all life and our obligation to respect and protect all living things and the planet that is our shared home.

1.2.3 Groundwater

Groundwater is water stored under the Earth's surface. It is replenished when precipitation falls on land surfaces and seeps down (percolates) through the soil and rock formations into an **aquifer** (a rock, sand, or gravel layer that can store and yield significant amounts of water). It's difficult to estimate global groundwater volumes, and it varies widely among sources. Groundwater represents about 30 percent of the total freshwater, with the remaining 70 percent found in polar ice caps, sea ice, permanent snow, glaciers, lakes, rivers, wetlands, and in the atmosphere. About 33 percent of the Earth's groundwater is found in the Asian continent, 23 percent in Africa, 18 percent in North America, 13 percent in South America, 6 percent in Europe, 5 percent in Australia, and the remaining 2 percent in other locations of the world [18]. The total volume of groundwater on Earth is small but is still 35 times greater than the volume of water in all the freshwater lakes and flowing rivers of the world [19].

The quality of the world's groundwater ranges from extremely salty (over 30,000 parts per million total dissolved solids) – particularly in some coastal areas – to relatively mineral-free groundwater in Iceland. Some groundwater sources contain high levels of naturally occurring minerals; this includes salts of manganese, sulfates, and chlorides that can create problems when used. Other locations may have significant amounts of nitrates, carcinogens, and other contaminates created by human activity. These too make water unsuitable for some uses, especially drinking water.

In Depth   **Units of Concentration in Solutions**

At the beginning of this chapter, salt concentration is given as a percentage. Later, we define salty water in terms of parts per million (ppm) total dissolved salts. Different units are often used to describe the same thing. This can be confusing, but is usually done either because of convention, the different methods used to determine the quantity, or the amount being measured. Each can be the correct unit to use. Scientists from several countries formed an international unit system to eliminate confusion, make certain that the same units are used for a specific purpose, and to ensure that experimental results are comparable. This is called the Système International (SI). Consider 1 percent to be 1 part salt in 100 parts of water, 1 part per million is 1 part salt in 1 million parts water and so on for parts per billion, trillion, etc.

Residents of Iceland have exceptionally high-quality freshwater. They obtain 95 percent of their drinking water from untreated groundwater through springs and wells. The capital city of Reykjavik has some 134,000 residents and obtains groundwater from reservoirs filled by wells which draw water from depths of 10–80 meters (33–262 feet). Low population densities and strong groundwater protection programs help maintain the high quality of groundwater supplies in the country. It is possible to maintain water quality if the motivation to do so is strong enough and the resources are available.

1.2.4 Rivers

Rivers, **streams**, **creeks**, and **brooks** all have the common feature of containing flowing water. These are the surface water transport system of the Earth's hydrologic cycle. There are over 4.8 million kilometers (3 million miles) of river channels in the United States alone, and yet rivers hold only 0.0001 percent of the world's water [20]. Rivers vary greatly in size, discharge, speed of current, aquatic populations, water quality, and temperature. However, the same river can also have these same variations. For example, the Amazon River of South America – flowing toward the Atlantic Ocean through the rainforests of Brazil – has much warmer temperatures, higher sediment loads, and lower oxygen levels than its faster-moving, colder, oxygenated headwaters in the Peruvian Andes Mountains (see Figure 1.5).

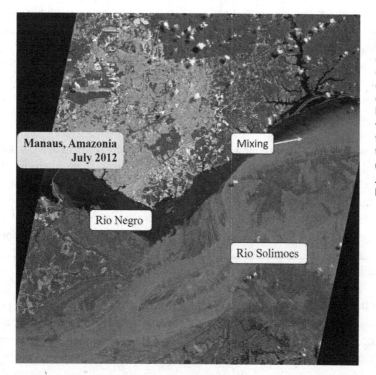

Fig 1.5 The Rio Solimoes, originating in the Peruvian Andes Mountains, has a pale, murky color caused by glacial silt and sand. The dark color of the Rio Negro is characteristic of clear waters that originate in rocky areas and carry little sediment. The pale and dark waters flow side-by-side as distinct flows before they eventually merge to help form the Amazon River (Image courtesy of NASA/GSFC/ JPL, MISR Team from Visible Earth; http://visibleearth.nasa.gov/ [21])

1.2.5 Lakes

A lake is any inland body of water, of appreciable size, that is found in a topographic depression. Most of the world's lakes (not to be confused with reservoirs constructed by humans) are located at high latitudes and in mountainous regions. Canada contains nearly half of the world's lakes, in part due to the ice sheets that carved depressions in the landscape some 10,000–12,000 years ago. Lakes are generally fed by surface water runoff and direct precipitation, and contain about 0.26 percent of the Earth's water.

Plant and animal life in lakes varies from nearly sterile pools in some desert regions to wonderfully rich communities of plants, fish, and insect life in more humid climates. For example, the Great Salt Lake, located in the dry desert region of Utah in the US, contains only small brine shrimp, fly grubs, and bacteria species. By contrast, a single lake in the tropical climate of Thailand supports nearly 40 species of plants, fish, and amphibians [22]. Lake history can be studied through sediment core analysis to show previous climatic changes, water quality changes, sediment levels over time, and biological productivity of the waterbody.

1.2.6 Wetlands

Wetlands can be generally defined as areas which are saturated by surface or groundwater, for long enough periods of time, to develop characteristic wetland soils that support primarily wetland vegetation. These may be called a swamp, marsh, fen, peatland, bog, moor, or estuary. Wetlands are generally located in flat spots, but can be found on mountains, in valleys, along a river, almost anywhere. The key features of wetlands are water (hydrology), wetland soils, and wetland plants. Wetlands do not exist at the polar ice caps or in fully arid regions but are found in all other climatic regions. Therefore, wetlands cover a wide range of precipitation and evaporation conditions and support a variety of plants and animals. One of a wetland's most important functions is providing habitat, food, and breeding or nesting areas to the diversity of life in wetlands; this includes phytoplankton, macrophytes, reptiles, amphibians, shorebirds, and waterfowl to mammals (see Figure 1.6). We will discuss their many other functions and their importance in mitigating the effects of global climate change in Chapter 9.

Fig 1.6 Wyoming, US, restored marsh, pond, willow habitat
(Photograph courtesy National Park Service)

In Depth   Water from Comets

An interesting hypothesis has been proposed in recent years suggesting that water-rich comet-like objects bombard the Earth depositing water vapor into the upper atmosphere. Scientists Louis Frank and John Sigwarth, of the University of Iowa, presented evidence from images taken by NASA's Polar satellite [23]. They believe that, over geologic time, meteors have put enough water into the Earth's hydrologic system to fill the oceans. It is a controversial idea and is an interesting area of scientific debate. More recently, in 2017, scientists studying meteor age and chemical composition have found that meteors 4,560 million years (Ma) old contributed enough carbon, oxygen, and hydrogen (the principle components of organic life) to the forming Earth to not be the limiting factor in the development of life on Earth [24].

No matter which field you choose to study, you will need to do research to learn about new ideas and progress made on an idea. Entire journal articles are not always available online to the public. Research university libraries have paid subscriptions to all kinds of journals as do many large public libraries that you may be able to use. Numerous sites are making journal articles available online to the public. Search for these in your studies. The Internet Archive Wayback Machine (anyone out there know who Mr. Peabody was?) is an excellent resource for finding articles of all types. It is available at https://archive.org/web/ (accessed 2020).

1.2.7 Atmosphere

An **atmosphere** is a gaseous envelope that surrounds a planet or other celestial body. By contrast, air is an invisible, odorless combination of gases and suspended particles that surrounds the Earth. For all practical purposes, air is the Earth's atmosphere. Water vapor moves into the air through evaporation from land and waterbodies. When the amount of water molecules evaporating (changing from a liquid to a gas) equals the amount condensing (changing from a gas to a liquid), the atmosphere is said to be **saturated**. This is the maximum amount of moisture that can be concentrated in the vapor (or gas) stage at a given temperature and pressure. When saturation is exceeded, and the air is cold enough, condensation occurs and water falls as precipitation (rain, snow, sleet, or hail). Although only 0.1 percent of the world's water is found in the atmosphere at any one time, it is critical to the replenishment of water supplies around the world as precipitation. Our atmosphere is also a key component of our hydrologic cycle and has a major role in global climate change.

1.3 Ecosystems, Biomes, and Watersheds

Water resources are found in a variety of forms, locations, and quantities throughout the world. Human life depends on adequate water supplies to survive and flourish, but so do plants, and terrestrial and aquatic wildlife. Trying to manage resources based on political boundaries can sometimes have severe consequences on the use and viability of water resources. Commercial development along one side of a border, such as the outer boundaries of a national park, can create land use changes that alter ecosystem boundaries within a protected area.

Similarly, the water development activities of one country could severely affect downstream land and water uses within an international watershed. For these reasons, it is advantageous to try to manage areas that are similar as a unit rather than by political boundaries. However, this is easier said than done, not only because of politics, but also because it is difficult to define the "correct" unit of land to consider. Three classification levels have been developed and could be useful in resource management. These are **ecosystems**, **biomes**, and **watersheds** (**catchments** or **river basins**).

1.3.1 Ecosystems

A general definition of an ecosystem is "a dynamic complex of plant, animal, and microorganism communities (the **biotic** environment) and their non-living (**abiotic**) environment interacting as a functional unit." Ecosystems are not static locations; at a minimum, energy and nutrients move in and out of such defined systems. An ecosystem can be of any size – a log, pond, field, lake, forest, or the Earth's biosphere (the portion of the Earth, which contains living organisms) – but always functions as a whole unit [25]. For example, the aquatic ecosystem of the Murray–Darling River Basin extends over 2,400 kilometers (1,500 miles) across southeastern Australia. By contrast, a single wetland covering less than half a hectare (about an acre) can also be specified as an ecosystem.

The term "ecosystem" is derived from the word "ecology" (from the Greek *oikos*, meaning "house" or "place to live"). The term ecology was first proposed by the German biologist Ernst Haeckel in 1869. It came into wider use about 1900. Ecology is the study of organisms, or groups of organisms, and their relationship to the environment [26]. In 1935, the concept of ecosystems as functioning units was proposed by Sir Arthur Tansley [27]. Eugene Odum's 1953 classic text [28] used ecosystems as the fundamental units for research and analysis bringing it into common use in modern ecology. Odum is considered the "father of the field of modern ecology" [29].

There are three important implications in the definition of an ecosystem. First, all parts of the Earth are parts of an ecosystem, from the smallest microbes to the largest plants and animal species. Usually, the smaller the ecosystem, the more interaction occurs between the living and non-living components in their localized environment. Second, all components of an ecosystem are not necessarily native to an area. Birds fly across the sky, plant seeds move in the air and into water, a large cat may decide to see what lives on the other side of the mountain – none of these are aware of their ecosystem boundaries. An ecosystem can become somewhat unstable or threatened through the introduction of new species or changing physical conditions of a region. Third, ecosystems generally do not have true geographic boundaries.

Designing an ecosystem research project involves many decisions. One could be determining the range of the project. For instance, the range of food sources for salmon in the northwest US might extend as far as the Sea of Japan in the western Pacific Ocean. However, it might be more useful for researchers studying a particular species of salmon to designate a given watershed as a geographic region for a salmon ecosystem; for example, an area extending from the ridge tops of a drainage basin in Idaho to the mouth of a river in Oregon. The map of global ecological land units could greatly improve this effort (see In Depth section).

Ecosystems can be delineated for a variety of purposes – to assess the environmental health of plants or animals, food sources, soil nutrients, flooding patterns, or other physical or chemical factors. The basic definition of an ecosystem (a dynamic complex of plant, animal, and microorganism communities and their non-living environment interacting as a functional unit) applies no matter what scale or reasons are used to define the ecosystem.

In Depth   **Global Ecological Land Units, GIS Coverage**

"A New Map of Global Ecological Land Units – An Ecophysiographic Stratification Approach" is the work of 35 authors, the American Association of Geographers (AAG), USGS, ESRI, and the Group on Earth Observations (GEO). The goal of this effort was to provide GIS coverage of global ecosystems that can be used for a variety of ecosystem research and management applications, such as assessments of climate change impacts to ecosystems, economic and non-economic valuation of ecosystem services, and conservation planning. The scale of the map allows use from the community level to the global. It is a groundbreaking achievement built on years of experience and research.

Four global raster datalayers were developed and acquired to represent the primary components of ecosystem structure (geology, landforms, soils, vegetation, climate, land use, wildlife, and hydrology), the bioclimate, landform, and lithology (study of general characteristics of rocks), and the land cover. The most accurate, current, globally comprehensive, and finest spatial and thematic resolution data available for each of the four inputs was used. A spatial combination of the four inputs was made into a single, new integrated raster dataset where every cell represents a combination of values from the bioclimate, landforms, lithology, and land cover datalayers.

This foundational global raster datalayer, called ecological facets (EFs), contains 47,650 unique combinations of the four inputs. Aggregation of the EFs resulted in 3,923 ELUs. This remarkable combination of effort owes much of its success to the tireless efforts of the primary author, Roger Sayre, a USGS Senior Scientist [30].

1.3.2 Biomes

A **biome** (pronounced bi-ome, plural biomes) is a significant regional ecosystem or community of plants and animals, such as a forest, grassland, or desert. The term "biome" is used most often for a specific community and is generally named after the dominant type of lifeform, such as a rainforest or a coral reef. Biomes can be found in numerous locations across the Earth. The grassland biome, for example, includes such regions as the tallgrass prairie of the United States and Canada, the llanos of Venezuela, the pampas of Argentina, the steppes of central Asia, the veldt (undulating plateaus) and savannas of Africa, and the grasslands of Australia. Such a grassland biome can also be part of a larger ecosystem (see Figure 1.7).

1.3.3 Watersheds

A **watershed** is the entire land area that drains water into a body of water such as a river, pond, lake, or ocean. The size of a watershed (also called a *river basin* or *catchment*) is determined by the topographic relief of a region. Smaller watersheds can be found within each larger watershed system. For example, the Waikato River – New Zealand's longest river on the North Island – contains smaller tributaries that encompass individual, smaller watersheds. All contribute to and are part of the larger Waikato River watershed which empties into the Tasman Sea. No matter where you live, you are in a watershed.

Watersheds can include a wide variety of ecosystems and biomes, and can cover multiple states, provinces, territories, or countries. Watersheds include both water (aquatic) and land (terrestrial) components. Watersheds, as study areas, lend themselves readily to geographical delineation since boundaries are easily established, i.e., their boundaries can be mapped by delineating the ridges or other highest boundaries of a region. A US Geological Survey topographic (topo) map can be used to determine the exact locations of watersheds in the United States. Worldwide, watershed delineation can be done for any body of water using local topographic maps.

As mentioned earlier, it can be difficult to draw accurate boundaries around an ecosystem since plants and animals do not necessarily confine themselves to an easily delineated region. Political lines, however, such as counties, townships, states, provinces, or international boundaries,

Fig 1.7 Wyoming grassland prairie, Hulett, Wyoming, US

are sometimes used to designate the boundary of an ecosystem. This presents another set of problems. Ecosystem boundaries are inherently inaccurate but do provide a physical region for scientific and political convenience. Unfortunately, this creates a lack of scientific precision. Ecosystem management based on watersheds is attractive for some but certainly not for all environmental management efforts. GIS maps of watersheds are available at several map scales in many countries. Check local, state, federal, or equivalent groups working in mapping watersheds such as Departments of Environmental Quality or their equivalent in the area of interest, United States Geological Service (USGS) [31] and World Resources Institute [32].

1.3.4 Human Effects on Ecosystems

We can see the adverse effects of human management on freshwater ecosystems. Dam construction, expansion of river water diversions, and poor water quality alter river regimes and cause degradation of wetlands. Modifications can fragment, change, or even destroy aquatic habitats, and alter wildlife migrating patterns. Results can be devastating for wildlife populations and entire species.

According to UNESCO (United Nations Educational, Scientific, and Cultural Organization):

- 63 percent of the world's largest 227 rivers are fragmented by dams, diversions, and canals, leading to ecosystem degradation.
  - Since 2015, dam building has expanded to at least 3,700 new dams in countries with emerging economies. This will fragment another 21 percent of free-flowing rivers.
- 64 to 70 percent of the world's wetlands have been drained between 1970 and 2000, with continued declines expected in the twenty-first century [33] [34].

The World Wildlife Federation reports a 60 percent decline in global wildlife populations in just four decades. Fish populations have fared no better. The extinction rate among vertebrates was highest for freshwater fish in the twentieth century [35] [36].

Loss of habitat and changes in resource use, including water, are major issues with land conversion. Displaced peoples can cause increased urban demands that may result in converting adjacent natural or agricultural land to accommodate the multiplying human needs. More often, land conversion is an economic issue; removing wetlands, grasslands, and forest to grow nonfood crops used in industry such as palm trees, corn, or soybeans for oil. Land conversion to urban or other uses can result in altered precipitation runoff patterns, siltation of rivers, lakes, and wetland complexes, loss of aquatic habitats, and a reduction in natural recharge. The introduction of exotic species often associated with urban development can lead to loss of native species, alterations in biodiversity, and reduced wildlife habitat or food production. Finally, pollution can alter the chemistry and ecology of waterbodies, contaminate water supply systems, and cause ecosystem degradation.

Think About It: Displaced People

There are nearly 2 million displaced persons in Darfur. Many are in large settlements such as this one, Abu Shouk in El Fasher, Northern Darfur (see Figure 1.8). Imagine the problems involved in getting water to all these refugees. Look at the land and ask yourself, where will food be grown? A 2020 update: Unfortunately, there are still about 1.6 million Sudanese who are not in relocation camps but are internally displaced, still in Sudan but not in their homes or villages. They cannot return to their villages because even though the conflict that started in 2003 is supposed to be ended, they are not safe from the violence of bandits and inter-ethnic conflict [37]. Internal displacement is a global concern in terms of humanity and resource needs. The Norwegian Refugee Council's 2020 global report lists 45.7 million internally displaced persons as of 2019 [38]. Who if anyone is responsible for helping internally displaced people during wars and after conflicts are settled? Is there a moral imperative here?

Fig 1.8 The water container queue at a well structure in the Abu Shouk camp in Sudan (Photograph credit: SALAH OMAR/AFP via Getty Images)

1.4 Global Water Use and Global Water Budget

Since the beginning of human settlements on Earth, water has been used for drinking, sanitation, and irrigation purposes. In prehistoric times, humans generally settled in areas of reliable water supplies. During times of drought, however, clans of humans were often forced to relocate to survive. Then, some 9,000 years ago, crude attempts were made to divert water from streams for crop irrigation. About 4,000 years ago, the first drinking water delivery systems were developed in what is today Iran, followed by the elaborate Roman aqueducts of 2,000 years ago.

Prior to the intervention of humans, the world's water supply remained in a natural state. Floods and drought were common in many regions of the prehistoric world. Natural global warming and cooling varied the Earth's climate, causing life in the various ecosystems to evolve, adapt, or disappear. As human population increased, the need for food supplies steadily grew. Simple irrigation methods were used in the Middle East, China, and India to meet these needs. With increased human use of water came pollution and water shortages.

The Earth's original supply of water is believed to have remained relatively unchanged for hundreds of millions of years. That means water that existed during the age of dinosaurs is still being used today. However, it is probably in a different location, and could be in a different form – as a solid, liquid, or gas. In the past 100 years, global water use has increased at a rate more than twice the Earth's population growth and has caused chronic water supply problems. In 2020, more than 4 billion people live with severe water scarcity for at least one month in the year, and half a billion live with scarcity all year long. Today, most countries in the Near East and North Africa, as well as Mexico, Pakistan, South Africa, and many parts of China and India are suffering from serious water shortages [39]. The World Economic Forum 2016 placed water crisis in the top five issues influencing global societal conditions [40]. See www.UNWater.org for more information on global water shortages, and strategies to alleviate future human and environmental distress.

Table 1.1 provides information regarding water availability and population by continent [41]. The water resources available in Asia are the highest in volume, but water available per capita there is lowest – because population totals in Asia are also the highest.

Increased demands for limited water supplies force millions of people (generally girls and women) in areas such as Mexico, China, Peru, Palestine, Afghanistan, and Niger, to spend a large part of their day transporting water in plastic bottles, buckets, and jugs for human consumption and cleaning (see Figure 1.9). The poor suffer the most. Daily chores include long walks to obtain water – too often at high prices or from contaminated streams or unclean taps. We'll discuss this tremendous human problem at greater length in later chapters.

The United Nations recommends that people should have at least 50 liters (13 gallons) of water each day for drinking, cooking, and sanitation. Water weighs 8 pounds a gallon. Imagine moving 104 pounds of water every day per person in a household (see Figure 1.10). This woman has access to a better tool, making the task easier, but definitely not easy.

1.4.1 Consumptive and Non-consumptive Water Uses

This is an especially important concept: water use can be divided into two types – consumptive and non-consumptive use. When water is fully consumed and not returned directly to a stream or other waterbody after use, it is considered a consumptive use. Non-consumptive use refers to water that is used, but then returned directly to a river, lake, ocean, or groundwater aquifer after use. A large

Table 1.1 Renewable water resources and availability by continent 2019–2020

Continent	Area million km²	Population million	Water Resources (km³/year)			Potential Availability 1000 (m³/year)	
			Average	Maximum	Minimum	Per 1 km² area	Per capita
Europe	11	741	2,900	3,410	2,254	277	3.9
North America	24	601	7,890	8,917	6,895	324	13.1
Africa	30	1,313	4,050	5,082	3,073	134	3.0
Asia	44	4,636	13,510	15,008	11,800	311	2.9
South America	18	430	12,030	14,350	10,320	672	28.0
Australia-Oceania	9	43	2,404	2,880	1,891	269	55.9
The World	135	7,800	42,785	44,751	39,775	317	5.5

Source: World Water Resources at the Beginning of the 21st Century
Note: Table 1.1 was updated with current populations and calculated water per capita using average water resource number divided by current 2019 or 2020 population. Created using Google search by continent.

Fig 1.9 Indian woman carrying water containers (Photograph by Galen R. Frysinger, Sheboygan, WI, US)

Fig 1.10 A woman in El Fasher, North Darfur, uses a Water Roller to more easily and efficiently carry water (UN Photo/Albert Gonzalez Farran)

portion of irrigation water is consumptively used by crops and lawns through plant transpiration. In addition, much of the consumptive use of water around the world is from evaporation off land and water surfaces [42].

Non-consumptive use of water can involve recreation, components of irrigation (surface water runoff that returns to a stream or shallow groundwater system), and portions of in-house water use for washing and other sanitation purposes that are recycled. Although recycled water quality is not at drinking water standards, it can save significant volumes of water. Any water quality problems depend on the extent of quality change and the intended use of the water. Improved water purification technologies such as microfiltration, reverse osmosis, and ultraviolet (UV) disinfection can raise the quality of recycled water to drinking water quality.

Think About It: Your Water Use

Determine your water use with a calculator from Home Water Works (www.home-water-works .org/). Could you do a better job conserving water? Why does this site also calculate your carbon footprint? Someone's carbon footprint is a measurement of the amount of carbon dioxide that their activities produce (see Cambridge dictionary: https://dictionary.cambridge.org/dictionary/english/carbon-footprint).

1.4.2 The Global Water Budget

Today, nearly 70 percent of all global water use is for agricultural irrigation. In some less-developed countries, the figure is closer to 95 percent. Irrigation water use is an extremely important component in feeding the world's population through food production or livestock crops, and maintaining industries using nonfood agricultural products. It is anticipated that 14 percent more freshwater will be needed to meet the world's growing demands for food in the next 30 years [43]. This looming problem becomes a challenge for everyone – how do we continue to meet human consumptive use water needs while protecting the limited water supplies for fragile ecosystems? This will be a challenging area of study for the near future.

> More than one-half of the world's major rivers are being seriously depleted and polluted, degrading and poisoning the surrounding ecosystems; thus, threatening the health and livelihood of people who depend upon them for irrigation, drinking, and industrial water [44]

Around the world, the demand for freshwater is escalating, primarily due to agricultural use, industrial development, and increased domestic use. This is due to population growth and increases in the number of countries able to provide public sanitation and increased water availability (see Figure 1.11). Water use has increased at twice the worldwide population growth rate, according to the United Nations Economic and Social Council. The US Water Resources Council estimated that, by the year 2020, water use in the United States would exceed available surface water resources by 13 percent [45]. The 2015 USGS water census showed surface water use in the US was 14 percent less than in 2010, owing primarily to changes in thermoelectric power use and increased irrigation efficiency [46]. However, severe droughts in 2015 decreased surface water availability in two key water-using states: Texas and California. Since the surface water was not available in the rivers, it could not be used, and would count as less surface water use [47].

Proposed solutions include increased use of groundwater supplies and implementation of improved water conservation programs. However, we will not be able to rely solely on groundwater because supplies are finite and cannot be expected to meet ever-increasing demands. Conservation programs are essential, but benefits are often limited or require time to produce measurable results.

Groundwater depletion is already a serious global issue, with estimated losses at 4,500 kilometers cubed (km^3) between 1900 and 2008. The yearly rate of groundwater depletion started increasing rapidly between 1951 and 2008 from 33 km^3/yr to 145 km^3/yr [48]. Groundwater use will become even less sustainable as increased pressure is placed on the resource in coming years. See Table 1.2 for a list of countries utilizing groundwater for more than 90 percent of their water needs.

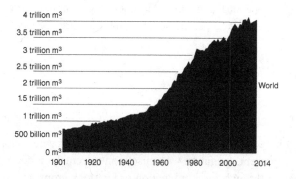

Fig 1.11 Global freshwater withdrawals for agriculture, industry, and domestic uses since 1900, measured in cubic meters (cm^3) per year
(Provided by Our World in Data [49])

Table 1.2 Countries and territories relying on groundwater for more than 90 percent of their water resources, 2002 [50]

Country	Region
Bahrain	Arabian Gulf Islands
Barbados	Atlantic Island, north of Venezuela
Egypt	Africa
Israel	Middle East
Jordan	Middle East
Kuwait	Middle East
Libyan Arab Jamahiriya	Africa
Malta	Islands in the Mediterranean Sea
Oman	Middle East
Qatar	Middle East
Saudi Arabia	Middle East
Turkmenistan	Middle East
United Arab Emirates	Middle East
Uzbekistan	Middle East
West Bank and Gaza Strip	Middle East
Yemen	Middle East

1.4.3 Drinking Water, Sanitation, Hygiene

Drinking water is the most basic human use for water. Humans can survive 8–10 days without food, but only 2 days without water. The United Nations Children's Fund (UNICEF) and the World Health Organization (WHO) monitor drinking water, sanitation, and hygiene worldwide in a program called WASH (water, sanitation, hygiene) as part of their sustainable development goals. Two billion people have no access to basic sanitation, no bathrooms, running water in a home, dishwashers, or clothes-washers, sinks, faucets, even soap and water, and in many cases not even maintained outhouses. Safe drinking water is particularly difficult without access to basic sanitation. Human fecal matter, loaded with viruses, bacteria, and parasites, is a major source of water contamination and unwashed hands a transport vehicle for disease.

However, between 1990 and 2002, global sanitation services increased from 49 percent to 58 percent. This increase was due largely to a global push for improved health through improved sanitation. In Latin America and the Caribbean, for example, 66 percent of the total population has access to sanitation linked to a sewage system. In Asia, that figure drops dramatically to 18 percent, and in Africa, it's a dismal 13 percent [51]. We'll discuss water and sanitation in detail in Chapter 11 and examine how some ancient sanitation systems were more advanced than portions of our modern world.

> *The 2030 agenda for Sustainable Development commits UN member states to take bold and, transformative steps to "shift the world onto a sustainable and resilient path", "realize the human rights of all", "end poverty in all its forms", and ensure "no one will be left behind" [52].*

There have been substantial increases in availability of safe drinking water globally. The inequalities in less-developed countries still exist and are especially dire in sub-Saharan Africa, hence the continued need to address this issue.

Unsafe drinking water is a daily problem faced by nearly 2 billion people around the world. In less-developed countries, 80 percent of all illnesses are water-related, resulting in 3.3 million deaths annually from diseases caused by *E. coli*, salmonella, and cholera bacterial infections, and from parasites like *Giardia* and *Cryptosporidium*. Between 1990 and 2000, more children died of these types of diseases than all the people killed in armed conflicts since World War II [53]. A 2019 UN news article reports that every year, 85,700 children under the age of 15 die from diarrhea and 72,000 more die from other illnesses associated with unsafe water, sanitation, and hygiene facilities [54].

In early 2005, the world was powerfully reminded of the tenuous nature of safe drinking water supplies. Soon after the December 2004 tsunami devastated southern Asia and parts of Africa and India, WHO warned that up to 150,000 (of the estimated 5 million people affected by this natural disaster) were at "extreme risk," primarily due to inadequate supplies of safe drinking water. Increased intensity of droughts and storms caused by global climate change can bring water insecurity to any part of the world as we have seen in reports in our daily news.

1.4.4 Water Quality and Quantity, Equally Important and Intertwined

Stresses on the global water budget are creating more water shortages. Water shortages in rivers, streams, and lakes can concentrate pollutants that are normally diluted by adequate flows. How can pollution concentration be eliminated? One method is dilution with additional scarce freshwater supplies, while a second is additional chemical treatment (which is often not available in less-developed regions of the world). Drought and increased irrigation demands can severely reduce river flows. The result of this can be seen along China's Huang He (Yellow River) and the Colorado River in Mexico and the US. These rivers are so overused that it is common for some stretches to completely dry up. This occurs near the Bohai Sea in China and the Gulf of California (Sea of Cortez) in Mexico. More and more, water management is practiced as a "zero sum game," where authorities provide water to one sector by taking it away from another.

It is one thing to talk about a country's water resources and express them relative to other countries in a global format, but it is essential to remember that water issues are not just about quantity. Water must be of good enough quality for its intended use, must be distributed close to where it is needed, and must be available in a timely manner. Poor quality water cannot be used as drinking water for human use. Water without a distribution system will not help people needing to travel long distances just to meet basic water needs. A great deal of rain falling outside the growing season will not help farmers. Unfortunately, water issues and solutions are not simple.

1.4.5 Agriculture

The second most basic human use of water is to produce food. Water serves an essential role in the production of food and fiber and can be supplied in a variety of ways. For example, some regions of the world receive adequate rainfall, and irrigation is unnecessary. However, if rain is not available when needed, irrigation may be the only way to save the crop. The timing of precipitation is more crucial than the annual amounts. In drier climates, on the other hand, irrigation is the only way to produce food and fiber crops such as wheat, cotton, and corn (maize) on a large scale (see Figure 1.12).

According to the Food and Agriculture Organization (FAO) of the United Nations, agriculture is the largest user of water resources around the world, accounting for 70 percent of all water withdrawals. This is distantly followed by industry: 20 percent; and domestic use: 10 percent. The water needed to produce daily human food requirements is in the range of 2,000 to 5,000 liters

Fig 1.12 Cotton that was irrigated with a center pivot system
(Getty image credit: dszc)

(525–1,320 gallons). Human absolute minimum needs (drinking water) are small – only 4 liters (1 gallon) per day. The FAO estimates that world food production must increase by 60 percent to feed the world's increasing population by 2030 [55].

Changing diets are also increasing demands on the world's freshwater supplies. It takes about 13,600 metric tons (15,000 tons US) of water to produce a ton of beef, but only 900 metric tons (1,000 tons US) of water to grow a ton of grain. Diets that rely on more beef and pork are exacerbating global freshwater supplies.

China is a good example of future water challenges. It contains 21 percent of the global population but has access to only 7 percent of the planet's freshwater. Millions of Chinese are migrating from rural communities to urban centers, increasing the use of indoor plumbing and choosing more water-consumptive use diets of beef and pork. The recent irrigation water shortage in eastern Shandong Province, which grows much of China's rice crop, was partially caused by the increasing water consumption of municipalities. Increased water use in cities led to shortages for farmers. An estimated 9 million Chinese suffered grain shortages because of this shift in water use. China is certainly not the only region experiencing shifting patterns of water use. Many people migrate to cities hoping for better lives, which also causes substantial shifts in water use.

1.4.6 Industrial Water Use

Industrial water demand is a primary component of increased demand on freshwater resources. For example, 300 liters (80 gallons) of water are required to produce 1 kilogram (2.2 pounds) of paper, and 215,000 liters (56,800 gallons) are needed to produce 1 metric ton (1.1 tons US) of steel. Countries that continue to industrialize rapidly, such as China, India, and Mexico, will continue to increase their industrial water needs to catch up with more-developed countries. Meanwhile, more-developed countries also continue to increase demands for the industrial use of water.

1.4.7 Recreation

We know that recreation is important to human health and well-being. Since ancient times, beaches, lakes, rivers, and pools have provided us with physical activity, rest, relaxation, pleasure, and exercise. Swimming is recognized as one of the most beneficial forms of exercise. Leisure and tourism are increasing significantly around the world with substantial economic benefits. However, these uses require good water quality. Recreational waters polluted with physical, microbial, or chemical hazards can lead to illness and disease in humans and animals as well as ecological damage to waterbodies and wildlife.

For example, implementation of the US Clean Water Act has made remarkable improvements in water quality of the Great Lakes. Local residents are very aware of this progress and guard their water. A recent request to allow BP Refineries to expand their Indiana facilities, to process Canadian crude oil, was met with protest petitions signed by tens of thousands of American and Canadian individuals. The US Environmental Protection Agency (USEPA) did not object to the request because the amounts of pollutants to be dumped into Lake Michigan were not considered harmful to uses of the lake. City councils and mayors also signed resolutions opposing the project, and protestors emphasized the message "do not pollute Lake Michigan." This wave of public protest resulted in BP agreeing not to increase the amount of pollutants added to Lake Michigan.

> See "Healthy recreational waters" at the World Health Organization website (www.who.int/features/2003) for additional information.

1.4.8 Hydropower

Hydropower generates electricity by using the kinetic energy of falling water to spin large electricity-generating turbines. We're seeing the demand for energy increase around the world, especially for electricity, because of demographic changes, improved standards of living, urban and industrial growth, and rising cultural expectations. Currently, 90 percent of the global population has access to electricity leaving about 80 million still without power. This is still a dramatic increase in availability due in part to the increased dam building discussed earlier in this chapter.

Hydropower currently provides 16 percent of the world's total electrical production, with China, Brazil, Canada, United States, and Russia being the largest producers [56]. Previously untapped hydropower potential in Latin America, Central Africa, India, and China are being used.

Dam construction for hydropower projects can cause displacement of millions of people and degradation of ecosystems (loss of wetlands, stream channel habitat and biodiversity, water temperature changes, and siltation behind dams in reservoirs). Decaying carbon sources in reservoirs are also releasing a greenhouse gas, methane, into the atmosphere contributing to global climate change (GCC) [57]. It seems that global economic improvements sometimes create new problems.

1.4.9 Virtual Water

Economists use the term **virtual water** to represent the amount of water required to produce food, energy, or other tradable products. When these items are exported around the world, water supplies from one region or country are "virtually traded" to the consumer in another region. Coffee consumed in Europe may have required as much as 20,000 liters (5,300 gallons) to produce 1 kilogram (2.2 pounds) of coffee in Colombia or Peru. A cotton T-shirt may have required 7,000 liters (1,850 gallons) of irrigation water – produced in Mississippi or China – but was sold and worn in London. Interestingly, Israel – which is a water-scarce country – discourages the export of virtual water by limiting the trade of oranges (a large water-use crop) to other countries. On the other hand, Japan and China import enormous quantities of food, and are large consumers of virtual water from exporting countries. Diets high in meat are responsible for the transport of large quantities of virtual water.

Energy production from petroleum is another good example of virtual water transfers when the country needing the energy is not the country using freshwater in extracting and processing the petroleum. Energy policy needs to be coordinated with water policy because the failure to treat undeniably connected environment issues makes policies and solutions less effective.

Commodities trade between regions and countries can be a stimulus for cooperation or can lead to dependency and potential conflict. The concept of virtual water has created new dialogue between politicians and researchers on the implications of this international water supply issue. Could water replace oil as a center of international conflict in the twenty-first century? The answer is yes, and it has. Water wars no longer mean only individuals feuding, but now include countries, states, and factions within countries and states or provinces fighting over the most essential of natural resources. We will discuss this in more detail later in Chapters 12 and 14.

1.5 Global Population Growth and Human Expansion

For thousands of years, our distant human ancestors lived precariously as hunters, gatherers, and scavengers. Life expectancy was short, with probably fewer than 10 million humans inhabiting the Earth at any one time prior to 5000 BCE. Exploitation of resources was localized. If water and food sources were depleted, human clans moved on or perished.

The world's population escalated, however, during the Industrial Revolution which began in the late 1700s because of improved living conditions and the use of fossil fuels. Before coal, oil, and natural gas were burned by the general population, fewer than 1 billion people inhabited the Earth (see Table 1.3). The 1 billion milestone was reached in 1804, and in a matter of 123 years – in 1927 – the 2-billionth person was born. It took from the beginning of human history until 1804 to reach a population of 1 billion, but only 123 years to double that number. This proliferation of people led to severe overcrowding in European cities. Severe societal pressures followed during the late 1800s and early 1900s. Europeans left in droves, as millions of people left their homelands for less crowded homes in the Americas, Australia, New Zealand, and other regions of the world.

The introduction of antibiotics and other improved health services further increased life expectancy. By 1959, the world's population reached its next milestone of 3 billion as advances in medicine, agriculture, and sanitation spread to less-developed countries. By the late 1960s, the world's population growth rate reached a new all-time peak of 2 percent annually.

It took only 15 years to add another billion people to the planet to reach the 4 billion mark in 1974. Growth rates were curbed somewhat by new birth control methods and the rising age of marriage, but so many people already existed that a true population explosion was occurring. Many

Table 1.3 Global population growth, CE 0–2100

Date	Population	Number of years
0	300,000	
1000	10.3 million	1,000
1804	1 billion	804
1927	2 billion	123
1959	3 billion	32
1974	4 billion	15
1987	5 billion	13
1999	6 billion	12
2011	7 billion	12
2020	7.8 billion	9
2030	8.5 billion	10
2050	9.7 billion	20
2100	10.9 billion	50

less-developed nations were at the center of these massive human increases. It took only 13 more years for the population to hit 5 billion in 1987.

The world's next population landmark was reached only 12 years later, in 1999 – 6 billion humans on Earth. Our population had doubled from 3 to 6 billion in only 40 years. This phenomenal increase occurred in less than the average human lifetime. Nearly 75 million additional people were added each year, a figure equal to nearly 25 percent of the entire US population. Currently, Europe and Africa carry 12 percent of the world's total population, with 9 percent in Latin America, 5 percent in North America, and about 60 percent in Asia (a proportion equal to that of the entire planet in 1800). By 2042, the world's population is expected to reach an amazing 9 billion people, with nearly all growth occurring in less-developed countries. This increase of 50 percent is projected to require only 43 years – an amazing and alarming situation (see Table 1.3) [58].

1.5.1 Settlement Patterns

It's important to understand the relationship between humans and the environment when considering the distribution and characteristics of present-day human settlements. People tend to live where topography, ecosystems, climate, natural resources, and economic development allow them to work, raise families, and experience productive lives. Population growth rates are generally influenced by education, religion, economic development, and urbanization. Mortality rates are directly affected by available medical care, food, shelter, safe drinking water supplies, and the age and sex of the local population. Table 1.4 is a listing of actual global vital statistics occurring in 2020, and offers an interesting look at how rapidly our population is growing.

1.5.2 Megacities

Megacities – massive, densely populated, urban centers of 10 million or more – are increasingly found in less-developed regions of the world. Globally, some 30 such megacities exist today (see Table 1.5 for the top 10). It's important to note that megacities, and other urban sprawl, can create

Table 1.4 World vital statistics by unit of time for 2020

Time unit	Births	Deaths	Increase in population
Year	146,474,696	57,920,669	88,753,981
Month	12,206,225	4,826,722	7,396,165
Day	401,300	158,686	243,161
Hour	16,720	6,611	10,131
Minute	278	110	169
Second	4	2	3

Table 1.5 Population of the world's largest cities, 2018 [59]

Rank	City	Country	Population
1	Tokyo	Japan	37,393,129
2	Delhi	India	30,290,936
3	Shanghai	China	27,058,479
4	São Paulo	Brazil	22,043,028
5	Ciudad de México (Mexico City)	Mexico	21,782,378
6	Dhaka	Bangladesh	21,005,860
7	Al-Qahirah (Cairo)	Egypt	20,900,603
8	Beijing	China	20,462,610
9	Mumbai (Bombay)	India	20,411,274
10	Kinki M.M.A. (Osaka)	Japan	19,165,340

environmental issues related to increased danger from flooding, increased contamination of surface and groundwater supplies, resource depletion, severe crowding, poor air quality, and general ecological degradation.

By 2030, over 60 percent of the world's population (nearly 5 billion people) will be living in urban areas. Today the figure is approximately 55 percent (~4.3 billion). In 1950, only 751 million people lived in urban areas. In 2000, some 777 million urban residents lived in slums [60]. In 2006, one-third of the world's urban dwellers (one billion people) lived in slums. Sub-Saharan Africa's slum population is almost two-thirds of its total urban population. The United Nations predicts that without significant change, the world's total slum population is approaching 1.4 billion (see Figure 1.13) [61].

1.5.3 Europe

Europe is one of the world's most densely populated and developed regions. It is home to approximately 748 million people and represents about 10 percent of the global population. People in the European countries are getting older, and nineteen of the twenty countries with the oldest population (median age 42) are located in Europe. Populations are expected to decline due to the aging population, low fertility rates, and out-migration.

Fig 1.13 Kibera slum houses in Nairobi, Kenya. These are not temporary homes. People can spend their whole lives living in poverty
(Photograph credit: YASUYOSHI CHIBA/AFP via Getty Images)

1.5.4 North America

North America has 23 countries with a total population of approximately 592 million – including 129 million in Mexico, 37 million in Canada, and 331 million in the United States, making up ~84 percent of the total population. The populations of the United States, Mexico, and Canada are expected to grow. A large percentage of the US population increase is anticipated to include immigrants from Mexico and other Latin American countries. Presently, some 1 million immigrants are attracted to the US annually, more than any other country. However, recent political leaders were not all as welcoming as in the past. Political policies are subject to change with new leadership.

Canada also anticipates an influx of immigrants, particularly from Asia and India. Canada's annual influx of immigrants is about 1 percent of its total population, or 300,000, with an expected increase to 350,000 by 2021 [62].

1.5.5 South America and the Caribbean

South America has a population of 430 million, or 5.5 percent of the world total. A key resource is its approximately 26 percent of the world's non-frozen freshwater resources. However, dense population centers caused by people leaving rural areas for urban life have created areas in which authorities are unable to safely manage untreated sewage. This trend creates significant water quality problems which can impact population growth if waterborne diseases spread [63]. Central America has a population of 51 million, and only 42 percent of the rural population and 87 percent of urban residents have direct access to drinking water. The Caribbean islands have a wide range of population densities and economic conditions. Water pollution and availability is also a serious problem on many of these islands [64].

1.5.6 Africa

Africa has a rapidly growing population of 1.4 billion, with many living in less-developed countries. Water problems are severe in many parts of the continent, and it has been estimated that 300 million people (nearly one-half of the continent's total population) are affected by water shortages.

Think About It: Weakened Immune Systems

In Africa, the HIV/AIDS epidemic has devastated many populations. It is not a waterborne disease; however, patients with HIV/AIDS, especially children, are more susceptible to and less able to combat waterborne diseases. As of 2018, ~40 million people are living with the disease worldwide, with 25.7 million of those found on the African continent.

What water-related illnesses could add to the high infant and mother mortality rates in less-developed countries? Is it possible to solve the problems related to water and health in these countries? Where would one begin to find a solution?

Expanding urban centers are creating severe drinking water and wastewater disposal issues throughout the African continent. Africa reached an urban population of 670 million in 2020, with about 50 percent still living in urban slums and lacking access to safe water and even basic sanitation, such as toilets, latrines, or faucets with soap and water. Progress is being made, but there is still much to do [65]. Alarmingly, Africa's urban population is predicted to nearly triple by 2050.

1.5.7 Arab States

More than 85 percent of the land area of the Arab States is classified as arid (annual rainfall less than 25 centimeters or 10 inches). Human activities, settlements, and economic growth have been controlled for centuries by limited water supplies. The Arab States region is home to some 420 million people living in 22 countries; including Morocco and Algeria in the west to Yemen and Oman in the east.

Life expectancy rates have increased in the past few decades throughout much of the region. Part of this increase is due to improved drinking water and sanitation systems. Safe drinking water is now provided to more than 90 percent of the regional population. However, nearly 25 percent of the population lives on less than US$2 per day. The region once had the highest population growth rate in the world (about 2 percent), compared to less-developed countries with a 1.4 percent growth rate. Although the growth rate has declined, its huge youth population will induce enormous population growth in future years (approximately 34 percent of the population is below the age of 15).

In 2020, about 372 million people live in the region, compared to 320 million in 2006 and only 150 million in 1980. Scarcity of water and cultivatable land will continue to be two major resource problems. The amount of land cultivated per capita will continue to decline due to population growth and **desertification** [66].

Desertification is the process caused by over-grazing and timber removal for fuel that turns productive land into non-productive desert. Desertification occurs mainly in semi-arid and arid regions (average annual rainfall of less than 60 centimeters or 15 inches) bordering on deserts. In the semi-arid region south of the Sahara Desert in Africa, called the *Sahel*, the desert moved 100 kilometers (60 miles) southwards between 1950 and 1975.

1.5.8 Asia and the Pacific Islands

The vast and culturally diverse countries of Asia and the Pacific Islands are home to 4.2 billion people – 60 percent of the world's population. This includes the rapidly industrializing nations of China (over 1.4 billion residents) and India (1.4 billion). The region also has an enormous population of youth. Over half of the world's young people, between the ages of 10 and 24, live in Asia and the Pacific region. Urbanization is rapidly expanding in many locations, with some 40 million people moving to urban areas annually. Too often, many people live in slum-like conditions with inadequate housing and no access to safe drinking water or adequate sanitation. In 2020, 18 of the projected 27 megacities (populations more than 10 million people, discussed earlier) are in Asia, and over half of these residents live in slums and other inadequate settlements. Human crowding will make millions vulnerable to natural disasters, disease, and other problems exacerbated by overcrowding.

Australia, the smallest continent, has a population of 25.4 million. Australia is also the driest continent, with an average annual precipitation of 20–30 centimeters (8–12 inches). The fertility rate overall is 2.1 live births per woman. The Indigenous Aboriginal and Torres Strait Islander rate is 1.8 live births per woman (anything below 2.1 is a declining population). Life expectancy is high at 83 years, and overall child deaths are low at ~3 per 1,000 births. Indigenous life expectancy has increased to ~70, and child death rate (6 per 1,000 live births) has fallen but is still 46 percent higher than overall rates [67]. Most of the population lives in urban settings with adequate water and sanitation services; however, Indigenous people tend to be in overcrowded, less than ideal urban settings with poor prenatal and birthing care. As with any country with urban population increases, Australia, which banned slums 100 years ago, is seeing an increase in substandard living conditions [68].

A unique water management partnership was created in Australia between the Wagga Wagga City Council and Charles Sturt University. Known as the *Global Water Smart City*, the project is a 10-year action plan to establish an international example of smart water use in an urban environment. Charles Sturt University (36,000 students) has five campuses spread across the state of New South Wales, along the Murrumbidgee River Basin in southeastern Australia. Areas of focus include effluent reuse and salinity management, water festivals, international seminars, exhibits, and other educational activities. Go to the Charles Sturt University website for additional information: https://news.csu.edu .au/latest-news/science/striving-for-international-excellence-in-urban-water-use (accessed 2020).

New Zealand, its neighbor to the east, is home to 4.8 million people, with 4.2 million in urban settings. Life expectancy and overall child deaths mirror Australia at 83 years and are low at ~3.4 per 1,000 live births [69]. New Zealand's annual rainfall is 600–1,600 mm (24–63 inches) but it has recorded more than 500 centimeters (200 inches) per year on the west coast of the South Island. About 85 percent of the population receives drinking water complying with government standards and 98 percent have water meeting bacteriological standards [70]. Other islands in the region add

~5 million more people. These vary in number and population demographics. The primary water issues are also varied, as with all islands. Water scarcity and floods, saltwater intrusion into drinking water, climate change impacts, and sinking land masses can all occur.

1.6 The Earth's Carrying Capacity for Humans, Is It a Set Number?

Carrying capacity is defined as the maximum life that our natural resources can support indefinitely. It is important to keep in mind that humans are not the only lifeform that our Earth needs to support. There is no set human carrying capacity for our Earth, largely because we, more than any other species, can change our environment. We can choose how to use our resources, how to help control populations, how to stop global climate change. The key word is **choose**. We can choose what we eat and how we produce it. We can choose to find a more equitable distribution of food and resources. We can choose to wear clothes longer, drive fewer miles, use less plastic. These are some of the things people in wealthy nations are able to choose.

We have been discussing world population and population growth by continent and said that the world's population growth is both amazing and alarming. The alarming part is the unequal distribution of growth, with the highest growth rates in some of the poorest countries that lack the infrastructure and resources to build an economy. About half of the human population (about 3 billion people) live in poverty, and 20 percent are severely undernourished. The rest live in relative comfort and health, with ample supplies of safe drinking water and adequate sanitation. Birth rates are higher in less-developed countries for many reasons, most relating to a lack of health education and healthcare, including availability of birth control measures. Globally, birth rates have declined, also for many reasons, such as the aging populations in many countries, increased use of family planning, and falling fertility rates. The economic viability of a region is tied to its population growth rates, ability to produce the goods it can sell, and its natural resources.

Think About It: Going to Town

As we have seen, urbanization was a dominant socioeconomic trend during the twentieth century. In 2020, 56 percent of the world's population lived in an urban setting – a first for the human species. How could this trend change demand for water supplies and associated environmental impacts? Is urban planning a wisely applied tool in most cities that you know about?

Urban residents require large quantities of food and water – historically provided from the surrounding countryside. Today, some cities must rely on water sources from great distances. Los Angeles, for example, obtains water from the Colorado River, which is over 320 kilometers (200 miles) to the east. Mexico City pumps water from over 150 kilometers (90 miles) away, and then must lift it 1 kilometer (0.5 mile) to its elevation of 2,100 meters (7,000 feet). Beijing is getting 70 percent of its water supply from the Yangtze River – some 1,500 kilometers (900 miles) away.

Food is also transported great distances to many urban centers – Tokyo is a good example. Residents still consume rice grown by Japanese farmers, but also utilize wheat imported from the Great Plains of Canada and the United States, and from Australia. Corn supplies for the Japanese consumer are shipped primarily from the Midwestern US. Soybeans come from the US and Brazil. Virtual water from Canada, Australia, Brazil, and the US is making its way in great quantities to Japan.

Increased water and food demands by countries such as India, China, and the United States create concern regarding the Earth's carrying capacity. Rapid population growth encourages farmers and the fishing industry to increase productivity, sometimes at the expense of fragile ecosystems. In some situations, these increased human needs have led to technological improvements, such as drip irrigation, improved seed varieties, or better harvesting techniques to meet increasing demands. In other situations, the environment has suffered terribly due to increased human impacts caused by food production efforts.

Think About It: What is *K*?

Carrying capacity is often symbolized as *K* by ecologists, and presents a vital question – Is it possible for humans to reach or go beyond this carrying capacity limit of a region or globally? Is water and food scarcity a looming disaster for humanity? Are we facing an impending global water and food crisis?

Eventually, the Earth's resources – particularly clean water supplies – will be inadequate to supply the needs of increasing population. Habitat degradation, polluted water resources, and inadequate food supplies may ultimately limit human numbers. The carrying capacity of a region, nation, or planet could be reached because of inadequate (or unusable) water sources.

Joel Cohen's book, *How Many People Can the Earth Support?* presents an excellent discussion of various methodologies for estimating the Earth's human carrying capacity – both past and present [71]. Cohen argues that little scientific consensus exists on the subject, but that food and water are generally regarded as the ultimate constraints on human population growth.

Think About It

Brian J. Skinner, a geologist at Yale University, states: "More than any other factor, availability of water determines the ultimate population capacity of a geographic province" [72]. Could the competition for adequate water supplies ultimately lead to war – which in turn could reduce population, which could reduce water demand? Humans must be capable of a better response than that of lower animals. Does water also determine the types and amount of food that can be produced?

Take a look at the *World Population Clock* at the US Census Bureau website (www.census.gov).

1.6.1 The Law of the Minimum and the Law of Tolerance

An organism is no stronger than the weakest link in its ecological chain. This concept was first expressed by Justus von Liebig in 1840 and is often referred to as "The Law of the Minimum" or "Liebig's Law." It is an appropriate analogy for human survival. The ability of a plant to survive in an ecosystem is determined by its ability to survive and reproduce. Extremes in water, light, nutrition, or temperature may be the limiting factors in plant growth and survival as well as for humans.

A good example of Liebig's Law can be seen west of the Mississippi River along the Great Plains of America. Rainfall gradually decreases as one heads west – average annual precipitation is roughly 100 centimeters (40 inches) along the Mississippi River Valley of Iowa, Missouri, and Arkansas, but only 30 centimeters (12 inches) or less along the eastern plains of Wyoming, Colorado, and New Mexico. Trees give way to grassland biomes as the amount of available water drops below the tolerance levels for trees. Likewise, the tall blue-stem grass (*Andropogon scoparius*), which needs an annual precipitation of 40 centimeters (16 inches), gives way to the shorter grama grass (*Bouteloua gracilis*), which requires less water.

In other situations, a lack of nitrogen or other nutrients may limit plant growth. Adding additional quantities of minerals, such as zinc or manganese, will not make up for a shortage of nitrogen. Liebig's Law is often shown as the staves (or slats of boards) in a wooden barrel, with each reaching a different height. When the barrel is filled with water, the shortest stave is the limiting factor on the amount of water stored in "Liebig's barrel." This same limiting factor affects plant growth if certain nutrients are missing.

In addition, an ecological maximum will limit the quantity and distribution of organisms in nature. This is known as Shelford's "Law of Tolerance"; it was developed by V.E. Shelford in 1913 and extends Liebig's Law of the Minimum to include animals [46]. The Law of Tolerance states that too much of a good thing can be as negative as too little. Shelford found that organisms might have a wide range of tolerance for one factor (such as temperature) but a narrow range for another, such as certain water quality parameters. The period of reproduction for animals is often a critical time, and environmental factors that affect seeds, eggs, embryos, seedlings, and larvae can affect populations and distribution. Brook trout eggs (*Salvelinus fontinalis*), for example, require a temperature range of 0–12 °C (32–54 °F), with an optimum reproductive temperature of 4 °C (39 °F). On the other hand, leopard frog (*Rana pipiens*) eggs will develop and hatch in the range 0–30 °C (32–86 °F), with the best temperature at 22 °C (72 °F). The lower temperature range of the trout eggs limits its range of reproduction. This is contrasted by organisms that have a wide tolerance range for several factors and are likely to be the most widely distributed.

Think About It: What Limits Humans?

Do humans have a Law of Tolerance – a point at which human populations will reach a peak? What roles do disease, reproductive rates, war, poverty, famine, and drought play in the overall human population of the world? Are freshwater resources the primary limiting factor when determining future world populations, or are other parameters more confining? Is there one minimum that outweighs other factors?

Summary Points

- Earth is a water planet; however, globally, its freshwater is not evenly **distributed**. Approximately **two-thirds** of the global population lives in regions that receive only **one-fourth** of the world's annual precipitation.

- This population/water disparity has striking consequences for quality of life especially in underdeveloped countries.

- Globally, 2.0 billion people lack access to improved sanitation facilities.

- Between 1959 and 1999, the **world's population doubled** – from 3 billion to 6 billion people. Current growth rates put the world population at 8 billion in 2024. It was 7.8 billion in early 2021, so we are already close to the 2024 prediction.

- **Global climate change** is a major water resource issue for many reasons. Its effect on stored ice is one issue. Although the total amount of water stored as ice is small, approximately 1.7 percent of the Earth's water is stored in **glaciers, permanent snow, sea ice**, and **polar ice caps**. This has the potential to become a major factor in the effects of **global climate change especially in terms of sea level rise, shifting ocean currents, and saltwater intrusion.**

- **Oceans** contain water too salty to drink by humans and other animals; however, it serves many other functions. Among these is its role in mediating global climate change by absorbing heat energy from the Sun and as a sink for CO_2.

- **Groundwater** comprises 30 percent of the Earth's freshwater supplies. Thirty-three percent of this is in the Asian continent and 23 percent is in the African continent. Groundwater is being overused and cannot be depended upon to satisfy expanding population needs.

- Freshwater is contained in **rivers, streams, lakes,** and **wetlands**. In addition to the river and stream hydrological functions of moving and storing water, they perform many functions also provided by lakes and wetlands from wildlife habitat, floodwater attenuation (holding floodwater), drinking water sources, to recreation.

- The last water reservoir is the **atmosphere. Water vapor** is not usually noticed day-to-day unless it is very humid, raining, snowing, or some other obvious precipitation event. It is a critical component of Earth's water cycle.

- **Ecosystems** are areas where all the biotic and abiotic features function together (i.e., as a system). The term **biome** usually refers to a single community within an ecosystem. **Watersheds** are all the land draining into a body of water. How the land in a watershed is cared for is a major factor in determining water quality in the receiving waters.

- The **global water budget** must include all human uses of water to maintain the health of our environment. Growing populations create strains on our water resources that will need to be addressed with careful planning, resource sharing, education, and technology.

- The Earth's ability to support its population is called its **carrying capacity**. Trying to determine the validity of a carrying capacity requires studying population numbers, distributions, lifestyles, life quality, etc. The fields of study are almost endless.

Questions for Analysis

1. a. What is the difference between a consumptive use and a non-consumptive use of water?
   b. Is using the phrase "consumptive use" accurate in terms of the hydrologic cycle?
   c. Why are these terms important?
2. What is "virtual water?"
3. Should water be a trade commodity? Justify your response.
4. The Earth's population reached 7.8 billion people in 2020;
   a. When is it expected to reach 10 billion?
   b. Where will most of the population increases occur?
   c. What is the significance of location on the effects of population growth?
5. Consider that approximately two-thirds of the global population lives in regions that receive only one-fourth of the world's annual precipitation.
   a. Will limited water supplies ultimately cause the Earth's carrying capacity to be reached?
   b. Will the disparities in precipitation location and amount play a role and, if so, how?
6. What are some differences between an ecosystem, biome, and watershed?
   a. Why are varying levels of classification needed?
7. What are some of the water-related environmental concerns specific to megacities?

Further Reading

Leopold, Aldo, 1949, *The Land Ethic in A Sand County Almanac*, New York: Oxford University Press.

Cohen, J., 1995, *How Many People Can the Earth Support?*, New York: W. W. Norton.

Jahren, Hope, 2020, *The Story of More, How We Got to Climate Change and Where We Go from Here*, New York: Vintage Books, a division of Penguin Random House LLC.

Odum, Eugene P., 1953, *Fundamentals of Ecology*, Philadelphia, Pennsylvania: W. B. Saunders.

Postel, Sandra, 1997, *Last Oasis: Facing Water Scarcity*, 2nd ed, New York: W. W. Norton.

Rich, Nathaniel, 2019, *Losing Earth, A Recent History*, New York: Farrar, Straus, and Giroux.

Westcoat, James L., Jr. and Gilbert ,F. White, 2003, *Water for Life*, Cambridge: Cambridge University Press.

References

1 Sandra Postel, 1997, *Last Oasis, Facing Water Scarcity*, 2nd ed, Worldwatch Publications, New York: W.W. Norton & Company

2 Aldo Leopold, 1949, *A Sand County Almanac*, Oxford University Press.

3 Peter H. Gleick, ed. et al., 1993, *Water in Crisis: A Guide to the World's Freshwater Resources*, New York: Oxford University Press, p. 514.

4 Tim McDonnell, 2019, "Climate change creates a new migration crisis for Bangladesh," *National Geographic*, January 24, www.nationalgeographic.com/environment/2019/01/climate-change-drives-migration-crisis-in-bangladesh-from-dhaka-sundabans/

5 Bangladesh Red Crescent Society, 2017, "Situation report: Monsoon floods in Bangladesh," August 16, https://reliefweb.int/report/bangladesh/monsoon-floods-bangladesh-situation-report-02-16-august-2017

6 D. K. Woolf, J. D. Shutler, L. Goddijn-Murphy et al., 2019, "Key uncertainties in the recent air-sea flux of CO_2," *Global Biogeochemical Cycles*, https://doi.org/10.1029/2018GB006041

7 European Space Agency, 2019, "Can oceans turn the tide on the climate crisis?", www.esa.int/Applications/Observing_the_Earth/Can_oceans_turn_the_tide_on_the_climate_crisis

8 Igor Shiklomanov, 1993, "Water in crisis: A guide to the world's freshwater resources," in *World Fresh Water Resources*, Peter H. Gleick, ed., New York: Oxford University Press Inc., p. 514.

9 University of Massachusetts Climate System Research Center, Kilimanjaro Climate and Glaciers Study, 2019, http://kiboice.blogspot.com/2019/

10 Lonnie G. Thompson, Ellen Mosley-Thompson, Mary E. Davis et al., 2002, "Kilimanjaro ice core records: Evidence of Holocene climate change in tropical Africa," *Science* 298, 18 October, 589–593.

11 Brian J. Skinner, 1969, *Earth Resources*, Englewood Cliffs, N.J.: Prentice-Hall, p. 232.

12 NASA's Goddard Space Flight Center, www.nasa.gov/content/goddard/icesat-2 (accessed 2020).

13 NASA EARTHDATA at National Snow and Ice Data Center (NSINC), https://nsidc.org/data/icesat-2 (accessed 2020).

14 P. Cochran, O. H. Huntington, C. Pungowiyi et al., 2013, "Indigenous frameworks for observing and responding to climate change in Alaska," *Climatic Change* 120, 557–567, doi:10.1007/s10584-013-0735-2

15 National Snow and Ice Data Center, https://nsidc.org/arcticseaicenews/ (accessed March 2020).

16 Neil MacFarquhar, 2019, "Russian land of permafrost and mammoths is melting," *New York Times Magazine*, www.nytimes.com/2019/08/04/world/europe/russia-siberia-yakutia-permafrost-global-warming.html

17 US Global Change Research Program, 2018, *Native Peoples–Native Homelands: Climate Change Workshop: Final Report: Circles of Wisdom*, Nancy G. Maynard, ed., NASA Goddard Space Flight Center, https://data.globalchange.gov/report/usgcrp-nativepeoples-workshop (available for download as PDF, accessed 2020).

18 James L. Westcoat, Jr. and Gilbert F. White, 2003, *Water for Life*, Cambridge: Cambridge University Press, p. 92.

19 Brian J. Skinner, James R. Craig, and David J. Vaughan, 2000, *Resources of the Earth*, 3rd ed, Englewood Cliffs, N.J.: Prentice-Hall, PTR, p. 202.

20 Luna B. Leopold, Kenneth S. Davis, and the Editors of LIFE, 1966, *Water*, New York: Time Inc., p. 42.

21 NASA/GSFC/JPL, MISR Team from Visible Earth: https://photojournal.jpl.nasa.gov/catalog/PIA02642

22 James L. Westcoat, Jr. and Gilbert F. White, 2003, *Water for Life*, Cambridge: Cambridge University Press, p. 92.

23 L. A. Frank and J.B. Sigwarth, 1993, "Atmospheric holes and small comets," *Reviews of Geophysics* 31, 1–28.

24 Adam R. Sarafian, Sune G. Nielsen, Horst R. Marschall et al., 2017, "Angrite meteorites record the onset and flux of water to the inner solar system," *Geochimica et Cosmochimica Acta* 212, 156–166, www.sciencedirect.com (accessed 2020, PDF available for download).

25 Government of British Columbia, Ministry of Forests, "Glossary of forestry terms," www.for.gov.bc.ca/hfd/library/documents/glossary/E.htm, December 2006.

26 Eugene P. Odum, 1953, *Fundamentals of Ecology*, Philadelphia, Pennsylvania: W.B. Saunders Company, pp. 3–4.

27 A. Tansley, 1935, "The use and abuse of vegetational concepts and terms," *Ecology* 16(3), 284–307.

28 E. Odum, 1953, *Fundamentals of Ecology*, Philadelphia: W. B. Saunders Company, p. 383.

29 James Hataway, 2018, "Eugene Odum: The Father of Modern Ecology," University of Georgia Today, https://news.uga.edu/the-father-of-modern-ecology/#:~:text=Eugene%20Odum%20is%20lionized%

20throughout%20science%20as%20the,ecology%2C%20which%20celebrates%20its%2010th%20anniversary%20this%20year

30 R. Sayre, J. Dangermond, C. Frye et al., 2014, *A New Map of Global Ecological Land Units – An Ecophysiographic Stratification Approa"ch*. Washington, DC: Association of American Geographers, https://pubs.er.usgs.gov/publication/70187380 (Available for download at www.researchgate.net/publication/269699546_A_New_Map_of_Global_Ecological_Land_Units_-_An_Ecophysiographic_Stratification_Approach/link/54a6bc5e0cf256bf8bb693f6/download).

31 USGS, "Water resources," www.usgs.gov/mission-areas/water-resources/maps

32 IUCN, IWMI, Rasmar Convention Bureau, and WRI, 2003, *Watersheds of The World*, Washington, DC: World Resources Institute.

33 United Nations World Water Assessment Program, *The United Nations Water Development Report*, January 2007, www.unesco.org/water/wwap/facts_figures/protecting_ecosystems.shtml

34 C. Zarfl, A. E. Lumsdon, J. Berlekamp et al., 2015, "A global boom in hydropower dam construction," *Aquatic Science*, 77, 161–170, https://doi.org/10.1007/s00027–014-0377-0

35 World Wildlife Federation (WWF), 2018, *Living Planet Report – 2018: Aiming Higher*, eds. M. Grooten, and R.E.A. Almond, Gland, Switzerland: WWF (downloaded from www.worldwildlife.org/pages/living-planet-report-2018, 2020).

36 N. M. Burkhead, 2012, "Extinction rates in North American freshwater fishes, 1900–2010," *Bioscience* 62, 798–808.

37 Catherine Wachiaya, 2020, "Sudan's internally displaced yearn for real peace to go home," United Nations Refugee Agency (UNHCR), www.unhcr.org/news/stories/2020/5/5ea866564/sudans-internally-displaced-yearn-real-peace-home.html

38 Norwegian Refugee Council (NRC), Internal Displacement Monitoring Center (IDMC), 2020, "Global Report on Internal Displacement (GRID)," www.internal-displacement.org/global-report/grid2020/ (accessed March 2020).

39 M. M. Mekonnen and A. Y. Hoekstra, 2016, "Four billion people facing severe water scarcity," *Science Advances*, 2, e1500323.

40 World Economic Forum, *The Global Risks Report 2016*, 11th ed, Geneva, Switzerland: World Economic Forum, www.weforum.org/reports/the-global-risks-report-2016

41 A. Shiklomanov and J. C. Rodda, eds., 2003, *World Water Resources at the Beginning of the 21st Century*, Cambridge: Cambridge University Press (data updated; the current populations, and calculated water per capita using average water resource number divided by current 2019 or 2020 population from Google search by continent).

42 Hannah Ritchie and Max Roser, 2020, "Water use and stress," https://ourworldindata.org/water-use-stress (accessed 2020).

43 UN-Water Thematic Initiatives, 2007, www.unwater.org/publications/coping-water-scarcity/ (accessed April 2020).

44 Ismail Serageldin, N. Borlaug, H. Kendall et al., 2000, "A report of the World Commission on Water for the 21st century," *Water International* 25(2), 284–302, www.researchgate.net/publication/299246060_A_report_of_the_World_Commission_on_Water_for_the_21st_century (accessed April 2020).

45 Water Resources Council, 1978, *The Nation's Water Resources, 1975–2000*, vol. 1, Washington, DC.

46 Ibid.

47 C.A. Dieter, M.A. Maupin, R.R. Caldwell et al., 2018, "Estimated use of water in the United States in 2015," *U.S. Geological Survey Circular*, 1441, 65, https://doi.org/10.3133/cir1441 [Supersedes USGS Open-File Report 2017–1131.] (available for download 2020 at https://pubs.er.usgs.gov/publication/cir1441).

48 Leonard F. Konikow, 2011, "Contribution of global groundwater depletion since 1900 to sea-level rise," *Geophysical Research Letters*, 38, L17401, doi:10.1029/2011GL048604, https://agupubs.onlinelibrary.wiley.com/doi/full/10.1029/2011GL048604 (accessed 2020).

49 Hannah Ritchie and Max Roser, 2020, "Water use and stress," https://ourworldindata.org/water-use-stress (accessed 2020).

50 United Nations Food and Agriculture Organization (FAO) Land and Water Development Division, 2002, "Unlocking the water potential of agriculture," Fact sheet, http://www.fao.org/3/bl040e/bl040e.pdf (accessed 2021).

51 United Nations World Water Assessment Program, 2007, "The United Nations Water Development Report," www.unesco.org/water/wwap/facts_figures/basic_needs.shtml, January (accessed 2020).

52 United Nations Children's Fund (UNICEF) and World Health Organization, 2019, "Progress on household drinking water, sanitation, and hygiene 2000–2017. Special focus on inequalities," www.unicef.org/reports/progress-on-drinking-water-sanitation-and-hygiene-2019 (accessed 2020).

53 Kofi Annan, 2003, "UN Secretary General address Environment Day," www.un.org/press/en/2003/sgsm8707.doc.htm (accessed 2020).

54 UN News, 2019, "More children killed by unsafe water, than bullets, says UNICEF chief," https://news.un.org/en/story/2019/03/1035171 (accessed 2020).

55 Food and Agriculture Organization of the United Nations, 2006, "No global water crisis – but many developing countries will face water scarcity," February, www.fao.org/english/newsroom/news/2003/15254-en.html

56 World Bank, 2015, "Electricity production from hydroelectric sources, % of total," https://data.worldbank.org/indicator/eg.elc.hyro.zs (accessed 2020).

57 Christina Nunez, 2019, "Hydropower explained," *National Geographic*, www.nationalgeographic.com/environment/global-warming/hydropower/ (accessed 2020).

58 UN Department of Social and Economic Affairs, 2019, *World Population Prospects 2019*, https://population.un.org/wpp/Publications/Files/WPP2019_Highlights.pdf (accessed 2020).

59 United Nations Department of Social and Economic Affairs, 2018, "World urbanization prospects 2018," https://population.un.org/wup/

60 United Nations Habitat, 2011, "The state of the world's cities," https://unhabitat.org/global-report-on-human-settlements-2011-cities-and-climate-change (accessed 2020).

61 United Nations Habitat, 2016, "2015–2016 slum almanac," https://unhabitat.org/slum-almanac-2015-2016 (accessed 2020).

62 Immigration.CA, 2020, "How many immigrants come to Canada each year?" www.immigration.ca/how-many-immigrants-come-to-canada-each-year (accessed 2020).

63 World Water Forum 8th Forum, 2018, "Pollution and deforestation obstacles to South American water," http://8.worldwaterforum.org/en/news/pollution-and-deforestation-obstacles-south-american-water (accessed 2020).

64 Caribbean Community Climate Change Center, 2020, "Tackling the Caribbean's climate-driven water resource problems," www.caribbeanclimate.bz/tackling-the-caribbeans-climate-driven-water-resource-problems/ (accessed 2020).

65 United Nations Habitat, 2016, "2015–2016 slum almanac," https://unhabitat.org/slum-almanac-2015-2016 (accessed 2020).

66 UN Population Fund (UNFPA), "Country profiles for population and reproductive health," www.unfpa.org/sites/default/files/pub-pdf/countryprofiles_2010_en.pdf (accessed 2020).

67 Australian Ministry of Health and Welfare, 2018, "Australia's health 2018," www.aihw.gov.au/reports/australias-health/australias-health-2018/contents/table-of-contents (accessed 2020).

68 Emma Baker, Laurence H. Lester, Rebecca Bentley, and Andrew Beer, 2016, "Poor housing quality: Prevalence and health effects," *Journal of Prevention & Intervention in the Community* 44(4), 219–232, doi: 10.1080/10852352.2016.1197714; https://theconversation.com/why-100-years-without-slum-housing-in-australia-is-coming-to-an-end-64153 (accessed 2020).

69 Worldometer, New Zealand Demographics, www.worldometers.info/demographics/new-zealand-demographics/#tfr (accessed 2020).

70 New Zealand Ministry of Health, 2019, *Annual Report on Drinking-water Quality*. Wellington, NZ: Ministry of Health, www.health.govt.nz/publication/annual-report-drinking-water-quality-2017-2018 (accessed 2020).

71 J. Cohen, 1995, *How Many People Can the Earth Support?*, New York: W.W. Norton & Co.

72 Brian J. Skinner, 1969, *Earth Resources*, Englewood Cliffs, New Jersey: Prentice-Hall, p. 130.

2 The Water Environment of Early Civilizations

Till taught by pain, Men really know not what good water's worth.

Lord Byron, *Don Juan*

Chapter Outline

2.1 Introduction

Water maintains all life on Earth. It is the tie that binds us together in a never-ending cycle. Humans need ample supplies for drinking and personal hygiene, food production, navigation, and power generation. Just as importantly, the health of our local, regional, national, and international environments – including rivers, lakes, ponds, wetlands, and groundwater – relies on adequate quantities and quality of water.

We've been altering our natural environment since prehistoric times when primitive societies relied on nature for their food, water, and shelter. Water attracts and supports a wide variety of plants and animals. Clans of humans often settled near lakes or rivers that provided readily available water and food supplies. Some human groups were nomadic, following herds of game and relocating during dry seasons or drought. Human impacts on the environment were probably minimal at first. However, acts as simple as drawing water from a stream, or killing game near a lake, can alter local ecosystems. Early humans soon learned to use fire as a deliberate action to modify and optimize food resources. This had locally important impacts on changing ecosystem structure including aquatic systems.

As human population increased and civilizations evolved, irrigation canal construction, diversion of water into aqueducts for delivery to faraway cities, and the disposal of human waste and refuse into local waterways had detrimental effects on the environment. These effects may have been subtle – such as a slight increase in water temperature caused by diminished surface water flows, or an increase in unwanted chemicals such as ammonia or nitrates in rivers, lakes, and groundwater aquifers. However, expanding populations caused localized environmental effects to become more pronounced. Water diversions from rivers and streams for human use altered local and regional ecosystems, and in some cases, caused the loss of entire plant and animal species. Human waste became more concentrated in larger settlements, and often entered streams and wetlands, fouling these water sources for human use downstream. Degradation occurred because growing populations used and consumed increasing amounts of water and food supplies with little regard for their sources.

Today we have made tremendous advances in understanding our environment and our role in its degradation. Where possible and with motivation for change, improvements are made daily; however, conditions today in some parts of the world have not improved significantly for centuries. Our Earth is suffering the effects of humankind's role in harnessing and exploiting our natural environment. Global climate change shows us that we need to make substantial changes in our use of natural resources now. Learning our history and learning from our past can help us make better decisions.

In Depth   Geologic Time

The history of the Earth is divided into eons, *eras*, *periods*, and *epochs*. The largest unit of time is called an *eon*, and there are four: the Hadean ("beneath the Earth"), Archean ("ancient"), Proterozoic ("early life"), and Phanerozoic ("visible life"). Some of the eons are divided into *eras*. The most recent eon, the Phanerozoic, is divided into three eras: the Paleozoic ("ancient life," a period of early land plants), Mesozoic ("middle life," which saw the rise of dinosaurs), and Cenozoic ("recent life," which saw mammals dominating after the extinction of dinosaurs).

Each of the three eras of the Phanerozoic Eon is divided into shorter units of time called *periods*, such as the Jurassic Period when dinosaurs grew to giant size. Periods lasted for tens of millions of years, so geologists divided these into *epochs*, such as the Holocene epoch which began 10,000 years ago (at the end of the last Ice Age and continues today). Importantly, humans have dominated and altered our environment so significantly that scientists are suggesting a new geologic epoch, the Anthropocene. The term is already in use informally to describe this time of human activity during discussions of global climate change [1].

2.2 Prehistoric Water Use

Our Earth is some 4.6 billion years old. About 3 billion years ago, the first forms of life appeared; these were primitive, single-celled creatures called *prokaryotes*, or primitive bacteria. The next step did not come for about 2 billion years. Then, single-celled seeds of life began to evolve into multicellular organisms. Oceans filled with a wide range of complex aquatic life. Some organisms moved to adjacent lands, perhaps to avoid being eaten by predators in the sea. A chain of new and even more complex land and water creatures evolved, only to be wiped out by a yet-unexplained mass extinction. The cycle of growth, evolution, and extinction has repeated many times during Earth's history.

The study of prehistoric fossils has shown that the Earth's climate has similarly experienced radical changes (see Figure 2.1). Regions of the world have transitioned from tropical forest to ice-frozen tundra, and back to warmer climate vegetation. By grouping climatic conditions and associated animal and vegetative species, the Earth's history can be classified into distinct geologic periods (see Figure 2.2).

For example, global warming during the Jurassic and Cretaceous periods (208 to 66 million years ago) produced an abundance of decayed plant and animal life which today provides fossil fuels

Fig 2.1 This trail scene is in Arizona Petrified National Park. It is named for the presence of beautiful crystals that can be found in the petrified logs. This trail offers one of the best opportunities to experience the petrified wood deposits. It is a reminder that what is desert today was once a forest. Climate changes have occurred throughout the Earth's history (Photograph courtesy of National Park Service)

for human consumption. By contrast, global cooling during the Pleistocene epoch (1.8 million to 10,000 years ago) allowed the formation of ice sheets in both the Northern and Southern Hemispheres. Entire species of plants and animals became extinct during these Ice Ages, the most recent one occurring at the beginning of the Holocene epoch some 10,000 years ago. Humans survived this most recent Ice Age, but the woolly mammoth did not.

In Depth   **Normal Geologic and Climatic Changes**

Imagine a prehistoric setting before human history began. Rivers, lakes, and our entire environment are in a natural state, dominated by natural phenomena, relatively unaffected by human activity. Plant and animal species are in a continual process of adapting to the Earth's changing climate. In some cases, a species will adapt and survive; in other situations, extinction occurs. Climate change is constant. Global cooling and warming cause ocean level changes and desertification. In addition, land changes were ongoing due to geologic activity – such as the upwards thrusting of entire mountain ranges, wind and water erosion, and movement of continents described by plate tectonics (Figure 2.3). Such changes caused entire continents, once covered with lush plant and animal life, to become glacier, desert, or rainforest and then repopulated and revegetated in later times. We must remember that these are normal geologic and climatic processes.

Early humans, called Neanderthals (*Homo neanderthalensis*), inhabited Europe and western Asia from about 230,000 to 30,000 years ago during the Pleistocene Epoch (see Figure 2.4). Stone tools were used, as well as bones, antlers, and wood for other purposes. These Stone Age people constructed complex shelters, used fire, and skinned animals. Neanderthals eventually became extinct, and were replaced by *Homo sapiens sapiens* (Cro-Magnon or modern humans). It is possible that competition for food and territory with *Homo sapiens sapiens* led to the Neanderthals' extinction.

Cenozoic Era *(Age of Recent Life)* – 66 MYA to present

> **Quaternary Period** – 1.8 million years ago to present
>> **Holocene Epoch** – 10,000 years ago to present
>
>> **Pleistocene Epoch** – 10,000 to 1.8 MYA, Great Ice Age
>
>> *Humans are believed to have developed in the Quaternary Period*

Mesozoic Era *(Age of Middle Life)* – 245 to 66 MYA

> **Cretaceous Period** – 66 to 144 MYA
>
> *dinosaurs dominate, seed-bearing plants appear*
>
> **Jurassic Period** – 144 to 208 MYA
>
> *large dinosaurs, flying reptiles, earliest known birds*
>
> **Triassic Period** – 245 to 208 MYA
> *first dinosaurs*

Paleozoic Era – 470 to 245 MYA

> *development of primitive fish, marine invertebrates primitive reptiles and land plants*

Phanerozoic Eon *(Greek visible life)* Eon – 570 MYA to present

Proterozoic Era

> *enrichment of the atmosphere with oxygen formation of eukaryotic, multicellular, life forms*

Archean Era – 3.8 to 2.5 BYA

> *formation of a low oxygen atmosphere, oceans, rocks, and unicellular life forms*

Hadean Era – 3.8 BYA

> *time from the start of the solar system to the development of stable Earth and Moon orbits and the oldest rocks*

Precambrian Eon – 4.6 billion years ago (BYA) to 570 million years ago (MYA)

> *represents 85 percent of Earth history*

Fig 2.2 Geologic timescales. BYA, billion years ago; MYA, million years ago. The Anthropocene would be first in the Quaternary Period. The exact timing of it is still a matter of scientific debate

Early prehistoric communities evolved near rivers and lakes in the Great Rift Valley of eastern Africa and in Europe, the Middle East, northern China, and India. Remnants of human occupation in Africa, from some 60,000 years ago, have been found at the present-day Kalambo Falls site in southern Africa. It is on the border between Tanzania and Zambia, along the Kalambo River at the southeastern corner of Lake Tanganyika. Evidence suggests that fire was used to clear forests in the area. Humans competed with animals, such as hyenas, for the use of caves and food [2]. Use of

Fig 2.3 This bridge is between continents in Reykjanes Peninsula, southwest Iceland across the Alfagja rift valley. On the midway point of the bridge, a plaque reads, "The Bridge Between Continents," and serves as a borderline between the Eurasian plate and North American plates. The two sides are symbolically connected and are marked: "Welcome to America" and "Welcome to Europe"

(Photograph from Nordic Visitor, Iceland, at https://iceland.nordicvisitor.com/travel-guide/attractions/reykjanes-peninsula/the-bridge-between-continents/)

Fig 2.4 First reconstruction of Neanderthal man (*Homo neanderthalensis*), 1888, by Hermann Schaaffhausen

fire may be the first anthropologic influence on ecosystems. For example, early human use of fire to burn grassland areas to drive animals during a hunt may have changed the natural ecological succession of such regions to brush land or forest [3].

During the hunter–gatherer stage of evolution, early people were directly dependent on nature, and competed with wild animals for food and shelter. Hunting tools gave humans an advantage over large predators, but humans were still sometimes the prey. This period of history saw the same controls on human population as occurs naturally with wildlife. The carrying capacity of the local and regional environment limited human population. Inadequate food or water supplies resulted in reduced numbers of humans, either through loss of life, adaptation of the clan size by choice, or by relocation to other regions for survival. This paradigm changed with the development of agriculture.

2.3 Water and Agriculture: The Basis of Civilization

Early attempts to master water and food needs occurred during the Neolithic, or New Stone Age, about 9000 BCE. In James Michener's novel *The Source* [4], a fictional cave-dwelling couple is described in the region of the Middle East, and their early attempt to grow food:

> "Ur's wife had found along the far banks of the wadi some young shoots of a vigorous emmer wheat and these she had chanced to place in proper soil along one edge of the great sloping rock, so that throughout the dry season enough moisture drained off the rock to keep the grain alive; and although its yield in edible wheat was disappointing, the grain lived as she had directed, and in the spring it reappeared where it was wanted."

Human knowledge progressed. Now, a major change occurred; the ability to grow crops, near a cave or other human dwelling, reduced the need for humans to wander after food. Over thousands of years, humans discovered that, if seeds were planted away from rocks and in areas where bits of new earth were added to old by wind or floodwater, crop yields improved. This became the basis for permanent communities, and was the start of civilization. Agriculture was the crucial single factor which allowed humans to settle in permanent communities. These later became the cities, states, and empires of ancient civilizations.

Farming developed in the Middle East between 8000 and 6000 BCE and started the transformation of tribal or family units into larger communities. Approximately 30 centimeters (12 inches) of water are necessary during the growing season to grow most grain crops. As Ur's wife discovered, seeds were the key to establishing crops and communities. During the initial history of agriculture, early farmers learned that seeds scattered on the land surface were often blown away by the wind or eaten by birds or rodents. With seed stocks limited, simple methods were developed to plant seeds below ground to encourage germination and growth.

Crude farming implements, such as a paddle-shaped digging stick, were used to create small holes in the ground for seedbeds. Later, the hand tool evolved into a human-pulled plow, called an *ard*, or scratch plow. Still later, around 3000 BCE, the ard was modified into an implement pulled first by humans, then by animals such as oxen. The plow was a major technological advancement but led to environmental damage such as the erosion of hillsides, the removal of topsoil, and the loss of native vegetation. Terracing (leveling a section of a hilly slope to reduce erosion) was an early attempt at soil conservation and is still used effectively today [5].

The initials BCE stand for "Before the Common Era." It replaces the "Before Christ" acronym BC. There is no difference between the two naming conventions except that *anno Domini*, Latin meaning "in the Year of our Lord," is replaced with the more neutral "Common Era," therefore CE replaced AD (*anno Domini*). This was done to try to make our reporting of time more universally applicable for people of many beliefs.

Think About It: Tools, How We Use Them Matters

The development of the plow was a great advancement in agriculture but came with a huge price. The benefits of plowing are immediate because great expanses of land can be planted into crops. However, pulverized soil and clearing of weeds changes the landscape. Erosion is a common problem, particularly on hilly slopes. Centuries of such tilling practices led, in some areas, to the complete loss of topsoil that covered an uplands region. Over time, the plow may have been one of humanity's most destructive forces on Earth, and yet one of its greatest tools for human survival [6]. What do you think? Are there better ways to farm? Can you think of examples of conservation methods related to water quality, water use, or aquatic life?

Farming allowed a shift from nomadic tribes of hunters and gatherers to small defensible communities, and then into larger urban centers. This transformation of human society primarily occurred along rivers – the Tigris and Euphrates Rivers in Mesopotamia (the "Fertile Crescent"), the Nile River in Egypt (see Figure 2.5), the Indus River Valley of modern Pakistan, and along the Huang He (Yellow River) in China. This was a cultural pattern that would change human society and their surrounding environments forever. Harvests of ample food supplies meant that less time was required for hunting and gathering. More time was now available for other activities of a progressive society; pottery replaced animal skins for food and water storage, homes were made of wood, stone, and even brick, and attention to the arts evolved [7].

Fig 2.5 This depiction of early Egyptian agriculture shows the use of tools and animals

The concept of living permanently in one place, called *sedentism*, was a huge change for humans. Instead of following game and living in transient camps, the new idea of living in a permanent location evolved. It resulted in the intensive use of land and water resources in a small area over many generations. This was a sharp contrast to the relatively minor environmental impacts created by hunter–gatherers who constantly relocated.

2.3.1 Mesopotamia

Agriculture evolved in Mesopotamia, the fertile river valleys located between the Tigris and Euphrates Rivers in present-day Iraq, eastern Syria, and southern Turkey. The two rivers of Mesopotamia were harnessed early for irrigation along their huge deltas which begin some distance north of Baghdad, Iraq. It stretches into a broad plain toward the Persian Gulf to the south. River water has been depositing sedimentary materials on the flat terrain of the delta since prehistoric times. Small dams were built for irrigation water storage in this region since the natural flows often arrived at the wrong time of year. Dams were vital to the permanent existence of life in Mesopotamia. Irrigation was extremely important throughout their written records, and irrigation was a constant topic, particularly in legal codes and religious myths [8].

The word "Mesopotamia" is derived from the Greek words meaning "between the rivers." This area is also known as the "Cradle of Civilization" and the "Fertile Crescent," since it was here that the first literate societies formed, and irrigation evolved. It should be noted that there never was a country called Mesopotamia; rather, it is a geographic region first denoted by Greek historians.

The climate of Mesopotamia was (and still is) harsh. The region has an annual rainfall of 15–20 centimeters (6–8 inches), and land seared by summer temperatures of 50 °C (120 °F) in the shade. Even so, with the anticipation of melting snow waters from the Armenian Mountains to the north, which includes Mount Ararat (elevation 5,137 meters, or 16,854 feet) in modern-day southern Turkey, native vegetation was cleared to create tracts of land to raise crops and domesticated animals for human consumption. The length of these two river valleys is extensive with some 1,850 kilometers (1,150 miles) for the Tigris and 2,865 kilometers (1,780 miles) for the Euphrates. The river valleys are composed of extensive sand and fertile silt deposits washed down from the mountains over the ages. Even though the climate is quite dry, the lush soils of the flat river valleys, and the water from the rivers, provided ample opportunities for irrigated agriculture to flourish.

Early settlement in this arid region began around 7000 BCE. Irrigators constructed small breaches (cuts) through riverbanks to allow water to flow onto the fertile floodplains [9]. These first crude attempts at irrigation in the harsh desert climate were later refined with the construction of an elaborate network of irrigation canals, dikes, and reservoirs. As early as 3000 BCE, every major city in the region was the center of a canal system. Such larger irrigation projects required more cooperation than the prior, smaller efforts, and led to cooperative agreements, expanding economies, and a social order previously unknown.

Civilization evolved as government and laws formed and refined to accommodate more elaborate social and economic systems. Calendars were created to keep track of rainy seasons, floods, and planting times. Crops became more diverse, and included barley, peas, lentils, wheat, dates, onions, garlic, lettuce, leeks, and mustard. Animals were domesticated, and included cattle, sheep, goats, and pigs. Oxen and donkeys were used for transportation, while fishing and hunting provided for an adequate food supply [10].

Think About It: Water Use, Who Decides?

It was vital that ancient rulers developed adequate food supplies for their residents. Around 1760 BCE, King Hammurabi personally directed his governors to dig and dredge canals in Mesopotamia on a regular basis. He knew that irrigation held the key to his power and, therefore, applied enormous resources to the development and maintenance of ancient water systems. Laws enacted under the Rule of King Hammurabi included provisions and penalties for neglect of irrigation canals. Queen Semiramis, Assyrian Ruler, 800 BCE, had this epitaph engraved on her tomb: "I constrained the mighty (Tigris) river to flow according to my will and led its water to fertilize lands that had before been barren and without inhabitants."

Today "constraining the river" also means diverting its flow for human uses and preventing it from flooding. Removing river water from its natural floodplain changed ecosystems that depended on the river to interact with its floodplains. What changes in technology can you think of that impact rivers today? Do you think the water needs of larger human settlements are more important than the protection of an ecosystem? Is there an absolute right or wrong here? Are there compromises that help all water users and ecosystems remain sustainable?

In Depth   **The Hanging Gardens of Babylon**

Humans have a physiological and cultural need for large quantities of water, for fountains, ponds, and other water features. It's believed the Hanging Gardens of Babylon were in Nineveh. It was a beautiful city of wide streets, large public squares, parks, and gardens located along the Euphrates River about 80 kilometers (50 miles) to the south of modern-day Baghdad. These gardens may have been only a myth. King Nebuchadnezzar II supposedly constructed them for one of his wives who was homesick for the landscape of her childhood home of Media in northwest Iran. The gardens were reported to be in a building of vaulted terraces, 37 square meters (400 square feet) and 23 meters (75 feet) above the ground. Since the region received very sparse rainfall, slaves worked in shifts to pull a chain of buckets that lifted water from the Euphrates River to irrigate fruit trees, herbs, vines, and flowers. A bas-relief of the hanging gardens is found today in the British Museum in London [11].

> The approach to the Garden sloped like a hillside and the several parts of the structure rose from one another tier on tier . . . On all this, the earth had been piled . . . and was thickly planted with trees of every kind that, by their great size and other charm, gave pleasure to the beholder . . . The water machines (raised) the water in great abundance from the river, although no one outside could see it.
>
> Diodorus Siculus, Greek historian (90 BCE–21 BCE)

Away from Mesopotamian cities, hanging gardens were not a feature seen on the desolate desert landscape. The government required farmers to clean irrigation canals to remove silt and other sediments deposited by irrigation water. Silt and mud settled out rapidly and had to be removed to keep irrigation canals flowing. The excavated material was piled along the sides of canals, sometimes to a height of 9 meters (30 feet) or more. This created a barrier to water drainage, and sometimes caused the Tigris or Euphrates Rivers to alter their courses during a flood, with devastating results [12].

Mesopotamians held a great distinction between tame and wild, civilization and wilderness. It was important that animals be domesticated, such as donkeys, water buffalo, cows, pigs, sheep, and goats. An animal was hunted if it could not be domesticated, sometimes to eradication, as occurred with lions of the region. The goal of Mesopotamian culture was to bring order out of nature's chaos. However, the civilization ultimately failed because it could not successfully maintain order.

Extreme crowding occurred in cities such as Babylon, which contained narrow streets and small houses. Thousands of individual cooking fires filled the overhead skies with smoke. Garbage piles were common, dirt floors were the norm, and human waste was left to accumulate in the streets. Flies and rodents were everywhere, particularly in areas that held domesticated animals, such as sheep, cattle, goats, and pigs. Drinking water supplies from rivers and groundwater wells became polluted by human and other animal waste. Infant mortality was high [13].

As the lush river valleys filled with agricultural fields and human settlements became more and more crowded, newcomers had to settle on less desirable ground away from the fertile valleys. Often, this ground was hilly, rugged terrain, complete with rough ground, rocks, thickets, and highly erodible soils. Land first had to be cleared of rocks, thickets, and trees before crops could be planted. Deforestation was a major negative impact of agriculture in these formerly tree-covered uplands regions. Tree cutting, both to clear land for cultivation and for firewood, desiccated large regions and caused hillside erosion.

Limited precipitation was conserved. Terraces, land leveling, and other crude methods of capturing water were implemented. To maintain soil fertility, some cultivated lands were allowed to rest, or to be on sabbatical, one year out of seven – called a *shmitah* by the Israelites. Upland regions with sloped lands allowed some rainfall to run off the land, but some runoff collected as surface water in ponds, marshes, and other wetland complexes. Groundwater levels may also have been elevated in some localized regions. Often, rainfall in these waterlogged locations created infestations of malaria mosquitoes [14].

Altogether, the uplands of the Middle East provide a telling example of how societies have tended to destroy their environment. Many of the areas that once supported a thriving agriculture are now largely unproductive. The face of the land itself is a more revealing document than all the written records [15].

2.3.2 Egypt

Egypt Is … the Gift of the River (Herodotus, 484–425 BCE)

To the south of Mesopotamia, agriculture developed and flourished along the long, narrow valley of the Nile River in Egypt. This occurred at approximately the same time as along the Tigris and Euphrates Rivers around 6000 BCE. The Nile operated with clock-like regularity. In late June, the lower Nile River swelled with tropical rains and melted snow from the upstream mountains of Ethiopia to the south. By late September, the entire Nile River Valley was a lake of sediment-filled water. About a month later, floodwaters receded and left behind rich deposits of mud and silt across the valley. Irrigation systems were passive systems, built high enough along the riverbanks to divert water only during the high flow periods. Without the Nile River, Egypt would have been nothing but an eastern extension of the Sahara Desert.

Egypt is still extremely dry. A miserly 2.5 centimeters (1 inch) of average annual precipitation is the norm in the capital city of Cairo. To the south of the city, a rain shower or two may occur every two or three years. Temperatures range from above 50 °C (120 °F) to below freezing. Earlier, around 3000 BCE, an elaborate system of dikes, canals, and reservoirs helped divert, store, and later deliver floodwaters from the Nile River to cropland. Fields were designed as large, flat basins to capture and hold floodwaters. Irrigation water flowed from one basin to another, and then another, controlled by simple floodgates. (Rice is cultivated with a similar method today.) Silt in the floodwaters served as fertilizer to enhance crop growth. Settlement along the Nile River coincided with the decline of Saharan vegetation in northern Africa some 5,500 years ago. Climate and vegetation change in the Sahara region led to an out-migration of inhabitants to the more fertile banks of the Nile, and probably laid the groundwork for the rise of the Egyptian Empire.

Nilometers are a unique aspect of irrigation along the Nile River. These devices provided measurements of flood flows (see Figure 2.6). Nilometers provided valuable information regarding the quantity of floodwaters available for irrigation and provided historic trends. Between June and September, torrential rainstorms pound the uplands of the Nile River in Ethiopia. These seasonal rains cause flooding along the lower Nile River to the north. Regularly, the Nile bursts its banks and inundates large regions of the Nile River Valley. As floodwaters retreat, fertile black silt and mud remains as new deposits to the cropland. The valley is enriched, year after year, with this natural fertilizer.

The simplest nilometers were vertical stone columns submerged in the Nile, with intervals etched in the stone, like a measuring stick. The height of water on the column indicated the relative volume of water in the river. A second nilometer design was constructed adjacent to a stairway leading

Fig 2.6 Egyptian nilometer on Elephantine Island at Aswan, Egypt. This nilometer was still in use, measuring the height of the Nile River, into the nineteenth century. Here women are getting water at low stages (Photograph by Sepia Times/Universal Images Group via Getty Images)

down to the riverbank. The stonewall of the stairway was etched with a measurement scale indicating the height of the Nile. A third, more elaborate measuring system included a ditch or pipe that led away from the river to a well or tank. Floodwaters filled the conveying ditch or pipeline, and then spilled into the well or tank. The water elevation was measured and provided more easily monitored depth of flood-waters. These types of nilometers were located within a temple, where only priests or rulers had access.

The annual flooding of the Nile River was critical to the Egyptian civilization. Below-average monsoonal moisture created little flooding downstream and resulted in famine for ancient Egyptians. On the other hand, heavier than normal rains in Ethiopia caused too much flooding, with equally disastrous results. Fortunately, the Nile's floodwaters were generally predictable, and most years' flows were adequate to produce a crop. Nilometers helped monitor these flows, and Egyptian rulers utilized the valuable information.

The productive Nile River valley of Egypt historically supported a human population of somewhere between 1.5 and 2.5 million. In addition, waterfowl and other wildlife were abundant in marshes and wetlands (Figure 2.7). The present human population of the region is about 84 million. This 42-fold increase in people has sorely stretched the soil and water resource base of the valley. Rather than being an exporter of food, as was the case during the Roman Empire, Egypt today imports 40 percent of its food and 60 percent of its wheat [16]. Valley soils are also suffering as salinity and soil erosion increase. Today, once-abundant wildlife species are gone due to draining of marshes for cropland and hunting.

Ancient Egyptian attitudes toward nature reflected the Nile River's dependable cycles. Ra, the Sun god, was the primary deity because his actions were regular, and all nature responded. Osiris was associated with the growing and dying of crops and other vegetation closely related to the Sun and the Nile River. Egyptians considered the natural world friendly to their civilization. This was a contrast to the chaotic perspective of nature that Mesopotamians held toward the environment [17].

Fig 2.7 Ancient hieroglyphs of Egyptian wetland wildlife (Egyptian Museum in Cairo, Egypt)

2.3.3 India

The Indus River Valley of present-day Afghanistan, Pakistan, and northwestern India is located east of Mesopotamia, and is considered the birthplace of the Indian civilization. Around 5000 BCE, irrigators diverted snowmelt waters from the adjacent Karakoram, Hindu Kush, and Himalayan Mountains to the fertile floodplains of the Indus River. Their drainage and water storage systems are considered the most sophisticated in the ancient world. A well-organized society developed around extensive agricultural surpluses of vegetables, grains, and domestic animals. Extensive trade developed with Mesopotamia, and ancient methods of irrigation were exchanged and probably shared between residents of the two regions.

The Indus Valley contained some of the most advanced cities of its time. Around 2600 BCE, the city of Mohenjo-Daro had a population of many thousands of people. It was a busy place where people lived and worked, used public and private wells for their water supply, and had a grid system of streets. It even had wastewater drainage ditches down the center of streets. For some unknown reason, the citizens of Mohenjo-Daro started to leave about 1700 BCE, and eventually the city was abandoned. The empty city was later buried under layers of dust and sand [18].

Timber cutting was widespread in the Indus Valley, and supplied firewood to bake bricks. These were used for building materials in ancient cities such as Harappa, which flourished between 2600 and 1700 BCE. (Many Indus cities were well-planned communities, including wide streets, public baths, wells, and reservoirs.) Similar to Mesopotamia, the need for firewood produced widespread deforestation, and resulted in soil erosion, flooding, and loss of vegetative cover [19].

2.3.4 China

The fourth great river cultural region of ancient times, the Huang He (Yellow River) Valley of northern China, began around 5000 BCE. The area was more temperate and forested than today but raging floods of the Huang He are still legendary. Sections of the Great Wall of China acted as levees to constrain part of the Yellow River. Here, many large marshes and swamps were drained to grow crops. The Huang He often turned violent with floodwaters, both from rainstorms and melted snow water of the Kunlun Mountains in Qinghai Province, and made the river unnavigable for months at a time. During the dry season of the winter months, the Huang He was (and still is) a slow-moving river laden with silt. However, summer storms turn it violent. Since 200 BCE, it has flooded some 1,500 times, and the river has changed course over nine times [20]. Great canals were constructed to obtain water for irrigation, transportation, and communication. Travelers from Mesopotamia may have provided information regarding irrigation and other farming methods to the Chinese.

2.4 Ancient Drinking Water and Sanitation Systems

In prehistoric times, humans lived in primitive structures, such as caves, wooded shelters, or cliff overhangs, located near rivers, lakes, or springs. Such water sources provided adequate drinking water for small, prehistoric clans of humans. As these groups of hunters and gatherers moved from one region to another, they devised utensils from animal skins to transport and store water. This allowed nomadic tribes to follow reindeer, bison, mammoths, and horses to obtain food and skins for clothing and utensils. As time progressed, shallow groundwater wells were dug to provide new sources of drinking water.

Fig 2.8 One of the three greatest water projects of ancient China is a *qanat* system, the Turfan water system located in the Turfan Depression, Xinjiang, China. It was started during the Han Dynasty (206 BCE–24 CE). Tunnels remaining today are a protected area of the People's Republic of China
(Digital photograph produced by Colegota, available at http://en.wikipedia.org/wiki/Irrigation)

Later, as human populations increased and settlements developed, the concept of moving water through canals, channels, and aqueducts evolved. By 2500 BCE, great urban centers formed along the Indus River and its tributaries of modern-day Pakistan. The most important of these were Mohenjo-Daro on the lower Indus River, and Harappa on the Ravi River on the upper portion of the Indus Valley. Both were laid out on a gridwork pattern of streets and had extensive wastewater systems that included bathrooms linked to sewers. This had a profound effect on the public health of these cities which each had as many as 40,000 residents.

Around 2000 BCE, the concept of transporting water great distances, both for irrigation and drinking water supplies, was expanded in present-day Iran. Since the region receives only 15–25 centimeters (6–10 inches) of average annual precipitation, water development was critical. The system is called a *qanat* (pronounced ka-nat), from the Semitic word meaning "to dig" and continues in use in the Middle East, China, and other regions of the world. China's famous qanat system of tunnels can be visited today (Figure 2.8).

A qanat system consists of numerous vertical shafts that join into one horizontal shaft. This transports groundwater from the side of a hill, or foothills of a mountain, to lower elevations by gravity. The vertical shafts were generally 60 meters (200 feet) apart. Some horizontal shafts were dug over 120 meters (400 feet) below the land surface for 30 kilometers (20 miles). There are tens of thousands of qanat systems in Iran, and hundreds of thousands of kilometers of tunnels [21]. Benefits of qanat systems include relatively little loss of water supplies due to evaporation, and access to relatively clean water supplies.

Qanats are a fascinating water delivery system that span from ancient times to the present. Marcus Vitruvius Pollio (ca. 70 BCE–23 BCE), the Roman architect and engineer, wrote of the qanat system in his work *De Architectura* about 25 BCE. He explained that the terrain of the prospective qanat was first surveyed, and then a well was dug at a foothill of a mountain or other large hillside. These locations often contained sands and gravels holding significant quantities of groundwater. Two diggers, called *muqanni*, dug a test well. If groundwater was found, a horizontal tunnel was then constructed at a depth equal to the bottom of the well. This horizontal tunnel sloped gently so that groundwater could be delivered to fields and villages several kilometers from the well.

Around the sixth century BCE, the ancient city of Athens had "public wells of great depth, covered with stone slabs, with small apertures, the necks of which are well furrowed by the ropes which for centuries have drawn the dripping buckets form the cool depths" [21]. Moreover, Greeks and Romans developed large systems of **aqueducts** that greatly improved the lifestyles in their cities. As far back as 343 BCE to 225 CE, engineers constructed water supply systems that delivered 492,000 cubic meters (130 million gallons) per day [22]. Vitruvius wrote in *De Architectura* [23]:

> "There are three methods of conducting water, in channels through masonry conduits, or in lead pipes, or in pipes of baked clay. If in conduits, let the masonry be as solid as possible, and let the bed of the channel have a gradient of not less than a quarter of an inch for every hundred feet, and let the masonry structure be arched over, so that the sun may not strike the water at all. When it has reached the city, build a reservoir with a distribution tank in three compartments connected with the reservoir to receive the water ..."

By 97 CE, Julius Frontinus, the Roman water commissioner and surveyor, distributed water throughout Rome (a city of over 1 million residents at its peak in ancient times) through lead pipes to public fountains, baths, and a few private customers [24]. "A day at the baths" provides a virtual tour of a Roman bath, see *The Baths of Caracalla*, at the Public Broadcasting System (PBS) website (www.pbs.org/wgbh/nova/lostempires/roman) (accessed 2021).

Very few ancient Greek waterworks remain, but some Roman facilities are still in use today. Three of ancient Rome's eleven aqueducts continue to deliver water supplies to that city's water system. Tiers of aqueduct arches, made of massive granite blocks, span above the Old Town area of Segovia, Spain, the remains of a Roman water delivery system from the second century CE. Ancient Rome's public baths, pools, open water channels, gardens, country estates, tunnels, archways supporting aqueducts, and the volume of freshwater transported are unprecedented in human history (see Figure 2.9).

Ancient drinking water and sanitation systems evolved from non-existent to nearly modern standards as far back as the beginning of the Common Era. However, the past 2,000 years have seen drinking water quality in some regions of the world sink to near ancient levels, while water systems in other regions of the world are quite elaborate and risk-free. The disposal of wastewater also ranges between the two extremes. We will explore these health issues in later chapters.

Today, sand blows past the crumbling columns and eroding walls of Leptis Magna, a World Heritage Site on the Mediterranean coast of Libya. This abandoned settlement in North Africa was one of ancient Rome's most important shipping ports. Around 2,000 years ago, Leptis Magna was a thriving seaport, due in part to the tremendous work performed by water supply engineers. Aqueducts fed storage reservoirs, fountains, and public baths. However, in the fifth and sixth centuries CE, the city was pillaged by the Vandals and Berbers (Indigenous tribes) of the area, and residents fled. (The term "vandalism" comes from their harsh conquest of Rome in 455 CE.) As the waterworks of Leptis Magna were abandoned, desert soon reclaimed the city. It was not until the twentieth century that portions of the city were excavated to reveal its former splendor (see Figure 2.10) [25].

One of the greatest feats of Roman engineering was the aqueduct Pont du Gard over the River Gardon in southern France. It is the highest (165 feet, 49 meters) of the Roman aqueducts and is the best preserved. It was built to serve the city of Nîmes which had no source of water for drinking, agriculture, or other basic needs (see Figure 2.11).

Fig 2.9 Roman baths at Bath, England. Imagine how important these baths were in Roman life (Photograph by David Iliff)

Fig 2.10 Theater at Leptis Magna, a Roman seaport on the Mediterranean Sea, at present-day Alkhuns, Libya (Credit: Rachele Rossi, via Getty Images)

Fig 2.11 The Roman aqueduct at Pont du Gard is considered by The World Heritage Society as a masterpiece of creative genius, a unique example of Roman civilization, architectural, and technical skill. Aqueducts were used to span rivers and other obstacles, but were not the Roman's first choice for water-delivery systems; buried pipes were easier to build. As a working aqueduct, it enabled a water channel to flow across the Gardon River. It was conceived as the principal piece in a 50 km (~31 mile)-long aqueduct, 90 percent of which was underground. It helped supply water to the town of Nîmes, known as *Nemausus* in Roman times.

Although aqueducts use gravity to move water, the engineering feats of the Romans are evident in that the vertical drop from the highlands source to Nîmes is only 56 feet, yet that was enough to move water over 30 miles. It is estimated that the aqueduct supplied the city with around 200,000,000 liters (44,000,000 imperial gallons) of water a day, and water took nearly 27 hours to flow from the source to the city [26]
(Credit: PASCAL GUYOT/AFP via Getty Images)

2.5 Water, Humans, and the Environment

For centuries, individuals have tried to understand the relationship between water and the natural environment. Thales (pronounced Thay-leez) (636–546 BCE) was a Greek philosopher who reasoned that all things in the Universe were made of water. He considered that the rains from heaven, flow in rivers, and waters of the oceans were components of a rudimentary hydrologic cycle. Two centuries later, one of the earliest ecologists/botanists, another Greek philosopher, Theophrastus (372–287 BCE), wrote *Historia Plantarum* (History of Plants); a 10-volume encyclopedia of the plant kingdom, and is considered the beginning of a scientific understanding of plants. The relationship between water, humans, and the environment has been at times harmonious, but too often one of great conflict.

Did ancient humans live in harmony with the environment, and cause little degradation to our water environment? The answer is "No." Sadly, the destructive capacity of humans on our natural environment is not a recent phenomenon. For example, many regions of the Middle East have experienced severe erosion problems caused by deforestation which occurred thousands of years ago. Evidence also shows that human settlement radically altered many island environments.

This also led, in some cases, to the extermination of most large animals once native to the islands [27]. For example, Easter Island (well known for its massive, stone-carved statues) is a small, hilly, and now treeless island located in the Pacific Ocean off the coast of Chile. It was completely deforested of palms, trees, and shrubs when Polynesians first settled there, between 400 CE and 1500 CE. The reason for the loss of vegetation is uncertain. It may have been used for firewood or building material, or perhaps climate change in the region was the cause. Regardless, the result has been heavy soil erosion and severely altered vegetation.

The Anasazi settlements in the desert regions of the southwest United States had to maintain harmony with nature because resources were so scarce. The loss of balance contributed to the abandonment of these desert villages. It is believed that these desert communities were abandoned around 1300 CE due to a combination of factors which may have included prolonged drought, intertribal warfare, salt buildup in irrigable lands, over-population, poor sanitation, endemic disease associated with poor hygiene and crowded conditions, and the loss of food supply from over-hunting (see Figures 2.12 and 2.13).

More recently, in the Middle Ages (400–1400 CE), humans' need for warm clothing led to the beginning of widespread decimation of beaver throughout Europe, and later in North America. Medieval castles were cold, and fur-lined robes, cloaks, and bed covers were more a necessity than a luxury. The need for warm animal pelts gave rise to the business of beaver trapping. In Europe, declining beaver populations pushed trappers deeper into the Russian interior. Records survive from the 1300s when from July to September of 1384 CE, there were 382,982 pelts (beaver, squirrel, marten, ermine, and sable) imported to England from the Baltic region of western Russia. Scotland's

Fig 2.12 Mesa Verde cliff-dwelling ruins at Mesa Verde, Colorado. Native Americans used the upper areas for grain storage while the dwellings were built into the cliff faces. The aquifer dripped out of the cliff faces at different points, allowing pueblo builders to collect water for use in the dwellings (Photograph by Karrie Pennington)

Fig 2.13 This is a view of Mesa Verde taken from a neighboring hillside. The planting season required much of the tribe to travel to the land above the dwellings where they planted their crops. This was not an easy life. Notice that the trees today are all dead. These pinyon pines were killed by a combination of extreme drought years in the early 2000s and the ips beetle (Ips confusus), also called a bark beetle. Normal pines close beetle borings with sap, but water-stressed trees cannot produce sufficient sap. A die-off like this one would have been devastating to early cliff-dwellers. A normal pinyon lifespan is 800 years
(Photograph by Karrie Pennington)

Fig 2.14 Canadian beaver (*Castor canadensis*) came close to extinction from over-hunting, but numbers have recovered as the demand for beaver pelts has decreased. The lack of economic gain saved this species from extinction. Are there other non-monetary reasons to save a species?
(Credit: GeoStock Getty images)

beaver trade ended in 1350 CE because of a lack of animals. By 1388 CE, France's supply of beaver pelts dwindled to small numbers, and by the mid-1500s, most of Europe's remaining beaver population survived only in regions of Siberia and Scandinavia [28].

North American beaver met a similar fate, albeit a bit later in history. Settlers in the New World found abundant beaver (*Castor canadensis*) (Figure 2.14) and began trapping to supply the European fashion market. Beaver skins provided early settlers with financial capital and caused a flourish for European markets. As mountain men and French voyageurs pushed westward into the Rocky Mountains of western Canada and the United States, pristine beaver ponds were raided. Trapped animals were killed for their pelts, and beaver numbers plummeted.

In the course of just 400 years, the fate of significant numbers of Canadian and American beaver was sealed. Today, it is estimated that the beaver population in the United States is about 5 percent of historic numbers [29]. Fortunately, in Canada, the beaver has made a remarkable comeback after being close to extinction in the mid-nineteenth century. Some 100,000 pelts were shipped to Europe each year at the peak of the fur trade. However, Europeans changed their fashion preferences to silk hats rather than beaver fur as they used other materials for warmth. Today, beaver populations are alive and well in Canadian and some American waterways thanks to significant wildlife conservation efforts [30].

The Earth's water resources provide habitat for a great diversity of life – an estimated 4,000 species of mammals, 9,040 species of birds, and about 19,000 species of fish. In 1995, the United Nations Environment Program's Global Biodiversity Assessment estimated the world's total number of described species at approximately 1.75 million. It has been estimated that throughout the history of life on Earth, some 1 billion species of living organisms have existed at one time or another [31]. It could be estimated, on a very gross basis from fossil records, that about four species per year went extinct out of approximately 10 million total species. Given that rate, today one would expect about one species extinction every 4,000 years, and among birds one every 2,000 years. These are estimates based on broad assumptions. In fact, the current extinction rate is much higher than this hypothetical prediction [32].

Natural selection can evolve over eons, or in a matter of decades. In Europe, the peppered moth, *Biston betularia*, has changed in color, from a pale gray or whitish color to a darker-colored form, in less than 50 years. This occurred between 1850 and 1898 in highly industrialized regions – such as the Ruhr Valley of Germany and the Midlands of England. The darker color provided better camouflage in the soot-darkened tree trunks of the region [33]. Human activity, in the pursuit of water resources development, has also caused some plants and animals to adapt. Sometimes a species is unable to cope with rapid environmental change and is lost.

Think About It: Species Extinction

What other modern factors contribute to a high species extinction rate? How is global climate change contributing to species loss? Is the loss of a species, for example, of a small but once plentiful fish, important in the large scheme of things? Provide an example.

2.6 Historical Perspective: Humans and Environmental Change

Historians often group world development into nine Eras. The arrows indicate how the human population influenced their environment. A vertical arrow pointing up is neutral, and an arrow rotating clockwise until it points vertically down indicates increasing influence.

• **The first Era, 13,000,000 BCE to 200,000 BCE**, included the first people on Earth and the development of primitive tool use.

↑ **Environmental change – humans reacted to their environment rather than altered it.**

• **Era 2 was about 200,000 BCE to 10,000 BCE.** *Homo sapiens* survived this period by developing delicate tools, controlled fire use, hunting, and gathering. They also communicated between groups, and shared knowledge so that, collectively, the basis for living together began.

Primitive art developed, perhaps as part of communication and learning. Humans in large family-sized groups populated most areas of the world by the end of this period and developed the skills necessary to survive varied climates and conditions.

↑ **Environmental change – humans were still controlled by their environment but learned to adapt to it.**

• **Era 3, 10,000 BCE to 1000 BCE**, saw the development of farming, and tools were used to produce food. This led to an agrarian society, with a culture of beliefs and organizations that was a complex interaction of weather, climate, landscape animals, and tools. Farming appeared on all major landmasses except Australia within 8,000 years – a very short time. Early deforestation and land erosion caused problems with food shortages and neighbors. Religion, writing, and horse-drawn vehicles were big developments in this era.

↘ **Environmental change – humans learned to alter their environment.**

Think About It: Farm Development

What environmental conditions existed in some regions of the world that helped develop farming systems? What was the role of water?

• **Era 4, 1000 BCE to 500 CE.** Human use of the Earth's resources intensified, extracting food, energy, using forests, developing animal herds, and growing population centers. Erosion, famine, species extinction, and conflict were consequences. Empires developed, while religions and cultural ideas also evolved.

↘ **Environmental change – humans continued to alter their environment.**

• **Era 5500 CE to 1500 CE** saw China, Europe, and India as the largest world economies. Genghis Khan formed the Mongol Empire – terrifying, but unifying Eurasia with expanding commercial and cultural exchange. The Aztecs and Incas developed civilizations in the Americas, and the savanna and forest areas of Africa became centralized states. Communication and travel between areas spread disease. Societies in Afro-Eurasia and the Americas developed water management systems: aqueducts, canals, and underground channels, devices for lifting water, and methods of irrigating crops and draining land (see Chapter 1). This produced bigger yields and allowed farming and urban populations to grow, which permanently altered the landscape. Deforestation for agriculture and city development led to wood shortages for heat, construction, and metal smelting, and caused intense soil erosion and soil degradation and increased flooding.

↘ **Environmental change – humans continued to alter their environment.**

• **Era 6, 1500 to 1750 CE,** was the age of globalization. Ocean travel linked the Americas, Africa, and Eurasia, finally bringing cultural and commercial exchanges between the major groups of the civilized world. Unfortunately, for Indigenous populations of the Americas and Island lands, disease came with the travelers, literally killing much of these populations in an event called "The Big Dying." A major ecological shift in population types ensued.

Change was rapid and drastic. Science developed as a disciplined body of knowledge, producing new and better tools for industrial advances in almost every area of human life. Environmentally, **humans had taken major control**. Mining, shipbuilding, and the sugar industry all used the Earth's resources to excess.

↓ **Environmental change – humans continued to alter their environment more extensively, creating greater environmental harm.**

• **Era 7, 1750 to 1914 CE**, was the age of revolution in politics, industry, and agriculture. Population growth, communication, fossil-fuel use, industrialization, democracy, and colonial empires were major factors of change. With these came increased deforestation, mining wastes, and water and air pollution on a major scale, invasive species, spreading plant disease, and famine with huge loss of life.

↓ **Environmental change – humans significantly harm their environment.**

• **Era 8, 1914 to 1945 CE**. This was the time of manned flight. World Wars I and II killed 45 million people, and modern weaponry became more effective. Genocide, or our knowledge of genocide, became horrifically common. In contrast, the spread of literacy and radio communication, and the cure and control of many terrible diseases, improved human existence. A world of change – both good and evil – developed. **Humans took almost total control of the biosphere with the ability to destroy it with nuclear power.** Individual environmental impact became significant with the acquisition of "things" by millions of people. Waste production started a huge industry.

↓ **Environmental destruction – humans radically altered their environment.**

• **Era 9, 1945 to present**. Increasing human effects on the Earth are our most significant "contribution" to our times. Huge population growth, infrastructure development, the chemical industry – good and bad – continued the spread of disease, massive worldwide pollution, overuse of water resources, species extinction at an increasing rate, air pollution, continuous increases in atmospheric CO_2, ocean warming and sea levels rising, melting tundra, extreme droughts, extreme flooding, and increasingly violent storm events are all signs of the times. Anthropologic changes and, perhaps more importantly, the unintended consequences of these changes, have caused scientists to consider creating a new geologic time unit called the Anthropocene. Although the arrow (below) is tilted fully downward, advancing technologies, increased global awareness of the problems, education, and above all the global incentive to find solutions can slow and over time begin to halt the destruction.

↓ **Environmental change – there are efforts to minimize environmental damage; however, global climate change is a reality and there is much that students, like yourselves, can contribute.**

Think About It: Development

Reread Era 9 above. The study of history shows that our environment can be radically altered, sometimes forever, if proper limits are not used. Are there any global responsibilities of more-developed countries to encourage and assist that new development be done in environmentally protective ways? Do we need to take Aldo Leopold's "land ethic" and Sandra Postel's "water ethic" to a global "environmental ethic"? Consider global climate change, how well is the environmental movement doing? Are countries taking this issue seriously enough? What needs to be done to make certain progress is made rapidly enough to halt increasing global climate change? Can we do better? Must we do better?

2.6.1 Ownership vs. Stewardship

Does human dominion over plants, animals, and water resources mean *ownership* or *stewardship?* Some believe that dominion means that you have a responsibility to care for and protect rather than exploit the natural world. Should natural resources, without exception, be available for the convenience of humans, and do we have the right to harvest them for the purposes of humankind? Alternatively, do we have a moral responsibility to protect these resources for future generations? Some resource managers of forests, grasslands, or fisheries are too often only concerned with protecting the resource for future economic gains. Others argue that plants and animals have value for purposes other than human consumption. These conservationists strive to preserve nature for its natural beauty, or for nature's own purposes, regardless of any utility to humans [34].

In 1918, Leon Trotsky, the Russian revolutionary, stated, "The proper goal of communism is the domination of nature by technology and the domination of technology by planning, so that raw materials of nature will yield to mankind all that it needs and more besides." This was the view that Nature had no intrinsic value and was only a resource to be used as humans saw fit. Environmental disasters followed for decades, under Communist rule, as it did in many other countries, as industrialization ignored its influence on the natural world.

Many believe that human population growth and increased resource consumption are root causes of increased resources exploitation and biological extinction around the world. The need for food, water, and clothing has led to the development of irrigation, domestication of animals, and deforestation. Human activities – many associated with water resources development and use – have increased the strain on ecosystems around the world. This has led to between one-third and one-half of the Earth's land surfaces being altered by human action [35]. What level of conservation would you chose to practice? What do you think is a sustainable level of anthropological land use?

Think About It: Is the Value of Animal and Plant Life Equal to That of Humans?

Consider two examples:

(1) In 1974, a tragic accident occurred when a ship sank off the eastern coast of the United States. Six people and a 36-kilogram (80-pound) Labrador retriever abandoned ship. Space in their lifeboat was limited. Regardless, the ship's captain kept his dog in the cramped lifeboat while forcing three others to swim in the frigid waters of the Atlantic. Nine hours later, two of the swimmers perished, but the others, including the Labrador, were rescued.

The following year, the captain was required to appear in a federal court for the death of his crew members. He argued that he simply couldn't bring himself to sacrifice his dog and feared the lifeboat would capsize if he took on the other three crew members. The Court did not agree and charged him with manslaughter. Do you agree with the Court's ruling? Why or why not?

(2) Should trees have legal standing as entities deserving protection? In 1972, the US Supreme Court was faced with a decision of historic proportions regarding the rights of trees. The lawsuit involved a request by Walt Disney Enterprises for a permit, from the US Forest Service, to develop a ski resort in the wilderness area of the Mineral King Valley in California.

   The Sierra Club argued against the project, but ultimately, the US Supreme Court ruled in *Sierra Club v. Morton* (1972) that the Sierra Club did not have "standing" or was not directly affected by the proposal. The debate centered on allowing trees to have standing in a court of law. Why not provide legal standing to natural beings themselves? The Supreme Court was split on the ruling (with a narrow majority vote of 4–3).

Justice William O. Douglas made his dissent for the minority as follows [36].

> *The critical question of "standing" would be simplified and put neatly in focus if we fashioned a federal rule that allowed environmental issues to be litigated before federal agencies or federal courts in the name of the inanimate object about to be despoiled, defaced, or invaded by roads and bulldozers and where injury is the subject of public outrage. Contemporary public concern for protecting nature's ecological equilibrium should lead to the conferral of standing upon environmental objects to sue for their own preservation.*
>
> This legal concept is being put to the test with rivers in New Zealand. In 2017, the country granted legal personhood to the Whanganui River on the North Island. Two years later, in 2019, the Te Awa Tupua River received the same legal status as a person in New Zealand. In 2018, Bangladesh became the first country to grant all of its rivers the same legally protected status as humans. Today, communities, industries, developers, and others cannot do anything in these two countries that could injure their uniquely protected rivers.

The above discussion provides a direct relationship to water resources issues. Should humans have the right to divert and use surface water even if it negatively alters the habitat for downstream plants and animals? What would happen if an entire species is lost due to human water use? Should humans have the right to pollute surface water or groundwater, even minimally, if that use of water is necessary for human life? Since ancient times, water and the environment have been altered and used to serve the needs of humans. Is there a point where the needs of the environment outweigh the needs of humans? There certainly is a point where the resource can no longer support its use. Is that too far to go before taking positive action? These are reoccurring themes, introduced in Chapter 1, and will continue to be considered in several upcoming chapters.

Summary Points

- The development of **agriculture**, especially **irrigated agriculture**, is a vital component of civilization.

- **Drinking water and wastewater systems** evolved as human settlements grew and in number.

- **Groundwater wells** and **qanats** were early water delivery systems that are still in use today.

- Early humans did not necessarily live in harmony with nature because human activity caused severe soil erosion and deforestation even during ancient times.

- Climatic conditions, with their associated vegetative and animal communities, are used to classify Earth's history into specific **geologic periods**.

- Human influence has so profoundly altered the climate and associated ecological communities that a new epoch, the Anthropocene, has been proposed.

Questions for Analysis

1. Specifically, what factors made irrigated agriculture essential for civilizations to flourish in ancient times? Explain your reasoning.
2. Are the same factors still a major part of modern agricultural production? Explain your reasoning.
3. Choose a distribution system for drinking water or a wastewater disposal system which developed as human settlements grew. Describe the system, how it was constructed, and where it was developed.
4. Are qanats still in use today? Is there a need for such ancient structures, given changing climate patterns?
5. Give examples of how early humans caused ecosystem damage in their daily lives.
6. Would you agree that modern society does a better job of protecting our water environment than early humans? Why or why not?

Further Reading

Hillel, Daniel, 1994, *Rivers of Eden: The Struggle for Water and the Quest for Peace in the Middle East*, Oxford: Oxford University Press.

Hughes, J. Donald, 1975, *Ecology in Ancient Civilizations*, Albuquerque, N.M.: University of New Mexico Press.

Michener, James A., 1965, *The Source*, New York: Random House.

Outwater, Alice, 1996, *Water: A Natural History*, New York: Basic Books.

Smith, Norman, 1972, *A History of Dams*, Secaucus, N.J.: Citadel Press.

Stiling, Peter, 1999, *Ecology: Theories and Applications*, 3rd ed, Upper Saddle River, N.J.: Prentice-Hall.

White, Gilbert F., David J. Bradley, and Anne U. White, 1972, *Drawers of Water: Domestic Water Use in East Africa*, Chicago, Ill.: University of Chicago Press.

Wulff, H. E., 1968, "The qanats of Iran," *Scientific American*, April 1968.

Zalasiewicz, Jan, Colin N. Waters, Mark Williams, and Colin Peter Summerhayes, eds., 2019, *The Anthropocene as a Geological Time Unit: A Guide to the Scientific Evidence and Current Debate*, New York, NY: Cambridge University Press.

References

1　Jan Zalasiewicz, Colin N. Waters, Mark Williams, and Colin Peter Summerhayes, eds., 2019, *The Anthropocene as a Geological Time Unit: A Guide to the Scientific Evidence and Current Debate*, New York, NY: Cambridge University Press.

2　Hillary Mayell, 2005, "Neanderthals, hyenas fought for caves, food, study says," *National Geographic News*, May 3.

3　J. Donald Hughes, 1975, *Ecology in Ancient Civilizations*, Albuquerque, N.M.: University of New Mexico Press, p. 22.

4　James A. Michener, 1965, *The Source*, New York: Random House, p. 93.

5　J. Donald Hughes, 1975, *Ecology in Ancient Civilizations*, Albuquerque, N.M.: University of New Mexico Press, pp. 25–26.

6　Daniel Hillel, 1994, *Rivers of Eden: The Struggle for Water and the Quest for Peace in the Middle East*, Oxford: Oxford University Press, pp. 44–45.

7　World History Encyclopedia, www.worldhistory.org (accessed 2021).

8　Norman Smith, 1972, *A History of Dams, Secaucus*, N.J.: The Citadel Press, pp. 7–8.

9　World History Encyclopedia, www.worldhistory.org (accessed 2021).

10　Luna B. Leopold and Kenneth S. Davis, 1966, *Water*, New York: Time Inc., p. 122.

11　The Public Broadcasting Service (PBS), 2006, NOVA Online, "The hanging gardens of Babylon," April www.pbs.org/wgbh/nova/sunken/wonders/ans24.html

12　J. Donald Hughes, 1975, *Ecology in Ancient Civilizations*, Albuquerque, N.M.: University of New Mexico Press, pp. 34–35.

13　J. Donald Hughes, 1975, *Ecology in Ancient Civilizations*, Albuquerque, N.M.: University of New Mexico Press, pp. 30–31.

14　Daniel Hillel, 1994, *Rivers of Eden: The Struggle for Water and the Quest for Peace in the Middle East*, Oxford: Oxford University Press, pp. 46–47.

15　Ibid.

16　Joe Wisenthal, 2011, "Egypt's food problem in a nutshell," *Business Insider*, www.businessinsider.com/egypts-food-problem-in-a-nutshell-2011-1 (accessed 2020).

17　J. Donald Hughes, 1975, *Ecology in Ancient Civilizations*, Albuquerque, N.M.: University of New Mexico Press, pp. 37–38.

18　The British Museum, 2006, "Indus Valley," August www.ancientindia.co.uk/indus/index.html

19　J. Donald Hughes, 1975, *Ecology in Ancient Civilizations*, Albuquerque, N.M.: University of New Mexico Press, pp. 35–36.

20　Britannica, Editors of Encyclopaedia, "Huang He floods," *Encyclopedia Britannica*, www.britannica.com/event/Huang-He-floods (accessed 2021).

21　H. C. Butler, 1902, *The Story of Athens*, New York: Century Croft, pp. 74–77, as cited in C. F. Tolman, 1937, *Ground Water*, New York: McGraw Hill, p. 593.

22 H. E. Wulff, 1968, "The qanats of Iran," *Scientific American*, April, p. 94.

23 M. Cartwright, "Aqueduct," *History Encyclopedia*, www.ancient.eu/aqueduct/ (accessed 2021).

24 *Vitruvius: The Ten Books on Architecture*, 1914, translated by Morris Hicky Morgan, Cambridge, Mass.: Harvard University Press.

25 Gilbert F. White, David J. Bradley, and Anne U. White, 1972, *Drawers of Water: Domestic Water Use in East Africa*, Chicago, Ill.: University of Chicago Press, p. 3.

26 Donald Langmead and Christine Garnaut, eds., 2001, *Encyclopedia of Architectural and Engineering Feats*, Santa Barbara, CA: ABC_CLIO, pp. 254–256.

27 J. Diamond, 1986, "Archaeology: The environmentalist myth," *Nature* 324, 19–20, https://doi.org/10.1038/324019a0

28 Alice Outwater, 1996, *Water: A Natural History*, New York: Basic Books, pp. 5–6.

29 Alice Outwater, 1996, *Water: A Natural History*, New York: Basic Books, p. 32.

30 Government of Canada, 2006, "Canadian heritage," August. www.pch.gc.ca/progs/cpsc-ccsp/sc-cs/o1_e.cfm

31 Peter Stiling, 1999, *Ecology: Theories and Applications*, 3rd ed, Upper Saddle River, N.J.: Prentice-Hall, p. 30.

32 Peter Stiling, 1999, *Ecology: Theories and Applications*, 3rd ed, Upper Saddle River, N.J.: Prentice-Hall, p. 85.

33 Peter Stiling, 1999, *Ecology: Theories and Applications*, 3rd ed, Upper Saddle River, N.J.: Prentice-Hall, pp. 66–67.

34 Carl F. Jordan, 1995, *Conservation: Replacing Quantity and Quality as a Goal for Global Management*, New York: John Wiley and Sons, p. 7.

35 Peter Stiling, 1999, *Ecology: Theories and Applications*, 3rd ed, Upper Saddle River, N.J.: Prentice-Hall, p. 3.

36 *Sierra Club v. Morton*, Secretary of the Interior, 405 U.S. 727 (1972).

3 The Hydrologic Cycle

Water in one form or another circulates around us, all the time, and everywhere.

Evelyn C. Pielou, *Fresh Water* [1]

Chapter Outline

3.1 Introduction

Water literally surrounds us. It is in the air we breathe as an invisible vapor and in tiny liquid droplets. Water as a liquid fills streams, bayous, rivers, wetlands, groundwater aquifers, and vast oceans. It is stored in the soil beneath our feet. Frozen water forms ice caps at the Earth's poles, icebergs, and mountain glaciers. About 8 percent of the Earth's surface is composed of soils with permafrost within the top 2 meters (7 feet). In June 2020, Verkhoyansk, a small town in the Sakha Republic of Russia known for its extreme weather range of –90 °F (–67.8 °C) to a record 98.96 °F (37.2 °C), hit a new record high of 100.4 °F (38 °C). As mentioned in Chapter 1, increased temperatures are melting permafrost with disastrous results. NASA scientist Patrick Taylor explains that our planet's average temperature increase over the past 40 years has been 1.44 °F, but north of the Arctic Circle the increase is 3.5 °F [2]. Global climate change will alter our understanding of the distribution of water on Earth.

Ours is a water planet. In 1972, the Apollo 17 astronauts photographed Earth from space in full color. We saw for the first time a beautiful sphere of blue water, white clouds, and the browns and greens of our world floating in space. It was, and still is, a breath-taking image. We learned in elementary school that frozen and fluid water covers about 75 percent of the Earth's surface. That photograph from space made this simple fact come to life.

The concepts regarding water's existence "all the time and everywhere" are explained by considering the three phases (stages) of water – liquid, solid, and gas, in the context of the hydrologic cycle. We're all familiar with this concept.

3.2 The Hydrologic Cycle

The hydrologic cycle describes the continuous movement of water on the Earth's surface, through the Earth's soils and geologic materials, and into the atmosphere. It is the process of how water moves on, in, and above the Earth (see Figure 3.1). All water on Earth is constantly reused, recycled, and purified in this process. This dynamic water system is powered by energy from the Sun, which drives a continuous exchange of water molecules between the atmosphere, oceans, and the land. The major components of the hydrologic cycle are evaporation (vaporization), transpiration, condensation, precipitation, runoff, infiltration, percolation, and storage [3].

Since the **hydrologic cycle** is circular, we can start our discussion anywhere – so we'll start with the atmosphere. The most recognizable chemical formula is that of water, H_2O, a strongly bonded combination of two atoms of hydrogen and one of oxygen. The cycling of water is linked with energy exchanges between the atmosphere, oceans, and land. The hydrologic cycle is important in determining both weather and weather variability. We know that water moves easily between its three phases and there are energy transfers during the phase changes (see Figure 3.2). Through a phase change, the energy input is used to overcome the intermolecular attraction (IMF) of the water molecules. Until the phase change is complete, the temperature of the reaction is constant (see Figure 3.3).

There is another phase change pair to consider, **sublimation** and **deposition**. This is a rather rare solid to gas (vapor) change that usually occurs at high elevations often with a wind to move the air. It requires a latent heat of sublimation (L_s) of approximately 620 to 680 cal/gram at –33 °C to go in either direction for the same reasons as the other phase changes getting the molecules moving and then bringing them together again. Sublimation explains the gradual shrinking of snowbanks even though the air temperature has remained below freezing for an extended period.

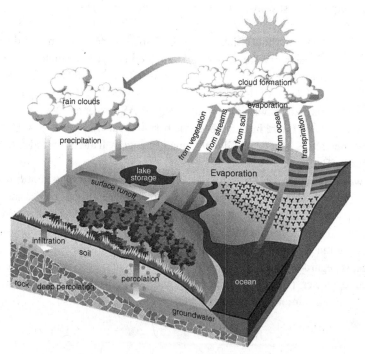

Fig 3.1 The Natural Hydrologic Cycle. The slight modification in placement of the word evaporation is to emphasize that evaporation occurs from vegetation, streams, lakes, soil, and oceans
(Image courtesy of Federal Interagency Stream Restoration Working Group (FISRWG)) [4]

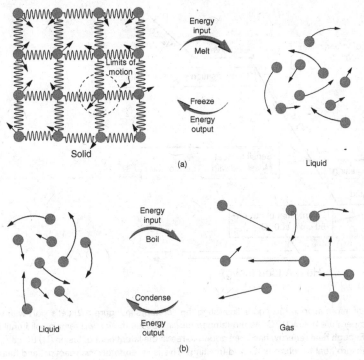

Fig 3.2 Phase Transitions: (a) Energy is required to partially overcome the intermolecular forces (IMF) between molecules in a solid to form a liquid. This energy is called latent heat. Latent means hidden. It is called hidden because it cannot be measured with a thermometer since it is reacting at the molecular level to increase the kinetic energy (energy of motion) of the molecules. It also takes energy to move the molecules back closer together and to keep them close, energy that must be released for freezing to take place. (b) Molecules are separated by large distances when going from liquid to vapor, requiring significant energy to overcome molecular attraction. The same energy must be removed for condensation to take place (Image courtesy of Boundless Physics, now The Libre Texts Project. Image originally from OpenStax College, College Physics. September 18, 2013.

In our discussion of the hydrologic cycle, we use the term evaporation instead of vaporization. However, they are equivalent. Evaporation is influenced by several factors. The boiling point 100 °C is where the liquid and vapor forms of water are at equilibrium (see Figure 3.3). The temperature assumes standard atmospheric pressure. Atmospheric pressure, near the Earth's surface at sea level, is 1 kilogram per square centimeter (14.7 pounds per square inch). This pressure is produced by the density, mass per unit volume of the air pressing down. Atmospheric pressure has the effect of pushing down the water molecules, under the force of gravity, effectively slowing their escape to the air. The higher the atmospheric pressure, the slower the rate of evaporation.

Think About It: Will My Water Ever Boil?

Have you ever been hiking at high elevations and desperate for a hot meal? Did you notice that it takes longer to cook even though you had boiling water? High elevations have lower atmospheric pressure, which allows water to boil at lower temperatures. For instance, if you live in a mile-high city like Denver, Colorado (elevation 1,609 meters, or 5,280 feet), water boils at 95 °C (203 °F). Cooking food takes longer at this lower temperature due to the higher elevation. By contrast, cooking the same food in New Orleans would take less time because of its sea level elevation. There, water boils at 100 °C (212 °F). Do you think the boiling temperature of water is substantially different on the top of Mt. Everest compared to Denver or New Orleans? Explain. How does vapor pressure influence evaporation?

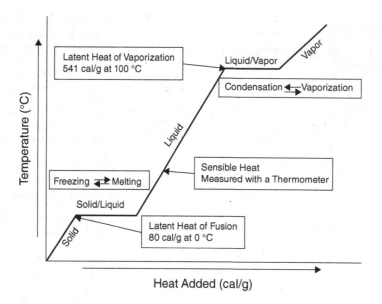

Fig 3.3 There is a lot of information in this figure illustrating the concepts in Figure 3.2. Let's start with water as a solid (ice) at some temperature below 0 °C. As the temperature increases the ice will remain solid until it reaches 0 °C (freezing point). Enough heat (energy) has been added to reach the latent heat of fusion (L$_f$) 80 cal/gram. Water in its solid phase must absorb heat to return to a liquid (melting). This is an endothermic reaction, and heat is absorbed. When this reaction is reversed, liquid to a solid (freezing), that same 80 cal/gram is released making this an exothermic reaction. The horizonal line indicates that there is no measurable (sensible) temperature change until all the ice has melted or all the liquid has returned to solid.

As the water boils, we can use a thermometer to measure sensible heat until another phase change. Vaporization of a liquid to a vapor occurs at 100 °C (boiling point) when the latent heat of vaporization (L$_v$) 541 cal/gram has been reached. Again, the horizonal line indicates that there is no temperature change until all the water becomes vapor

Substances dissolved in water, such as salts, alter water's properties. For example, salts increase the boiling temperature of water and decrease its freezing temperature. This occurs, on a simple level, because salts dilute water. Salt molecules, in a given volume of water, reduce the number of water molecules in that volume. This physical process is significant if, for instance, you add salt to ice in an old fashion ice cream freezer to make the ice cream freeze faster, because the salt causes the freezing temperature to be colder. The effects of salinity, however, in terms of the hydrologic cycle, are generally too small to be significant.

Evaporation is also dependent upon the size of the water surface, or its surface area since evaporation is a surface phenomenon. The larger the surface area, the more molecules available to evaporate, the higher the evaporation rate. Therefore, evaporation rates are different for oceans and lakes versus ponds and puddles. A larger surface area also allows more air circulation. This removes water vapor molecules as they form rather than having them return to a liquid state. Therefore, surface area and wind simultaneously influence evaporative rates.

Evaporation is also influenced by how much water is already in the air. Anyone who has ever lived in a humid climate knows that evaporation is slower when there is a high relative humidity. Plants, animals, and people all feel hotter because evaporation uses heat. This is called an **endothermic** or cooling process. High relative humidity slows the normal cooling effects of evaporation. This makes us uncomfortable because our sweat does not evaporate. Plant and animal cooling mechanisms are similarly slower.

To summarize, evaporation rates are influenced by the amount of heat, surface area of a waterbody, atmospheric pressure, wind conditions, and relative humidity. Higher heat, wind

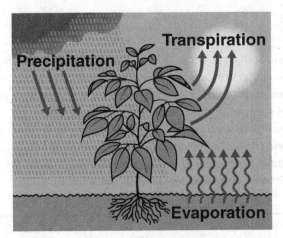

Fig 3.4 Evapotranspiration is the simultaneous transpiration and evaporation from plant and soil surfaces. (Image courtesy of USGS) Evapotranspiration rates are controlled by factors like evaporation. There must be heat energy from the Sun, available water, and a humidity gradient to allow diffusion and air movement to remove the water vapor. (A gradient is simply a change in one thing relative to another.) A humidity gradient, that allows water to evaporate, occurs when the humidity at the plant surface is higher than the humidity in the surrounding air

turbulence, and surface area increase evaporative rates. Higher atmospheric pressure and relative humidity decrease it. Research has shown that nearly 90 percent of all atmospheric moisture originates as evaporation from oceans, seas, and other bodies of water (rivers, lakes, wetlands) [5].

The remaining 10 percent of water vapor found in the Earth's atmosphere is released by plants through **evapotranspiration** (also referred to as ET). This water loss from plants, caused by evaporation, is called transpiration. The reason scientists combine evaporation and transpiration is that it is difficult to separate the two processes (see Figure 3.4). **Transpiration** brings water to the plant surface where evaporation can occur. Plants use water as an internal transportation and cooling system. Roots take water and nutrients from the soil via **osmosis**, creating an area of high water potential. Water evaporating at the leaf surface creates an area of low water potential. Water moves from areas of high potential (the roots) to areas of low potential (at the leaf surface). Water moves up through the plant vascular tissue, to the leaves, and out through small pores (stomata) on the underside of the leaf, which are connected to the vascular tissue. Simultaneously, water is also evaporating from the warming soil surface and from water drops on the leaves and stems.

In Depth	Water Potential

The process of water flowing from areas of high potential to areas of low potential is called *passive movement*. Imagine the classic example of water at the top of a waterfall. Water at the top of the falls has a high potential energy, or ability to flow. After the water flows over the falls to the bottom, its potential energy, or ability to flow, has been greatly decreased. *Water potential* is measured as [energy/volume], or pressure, in units of bars, atmospheres (atm), or megapascals (MPa): 1 bar is about 1 atm or 0.1 MPa.

Water potential has four parts, making up its total potential (ψ_w). These are osmotic potential (ψ_o), pressure potential (ψ_p), matric potential (ψ_m), and gravitational potential (ψ_g). *Osmotic potential* is the ability of a solution to draw water in when it is separated from another solution by a semipermeable membrane. Osmotic potential is dependent on the concentration of solutes, usually salts, in solution (water flows from a dilute solution into a concentrated one). *Pressure potential* is a measure of how pressure or tension influences water flow (water flows from high pressure to low). *Matric potential* refers to water bound in a thin layer (one or two molecules) to a dry surface. This creates a decrease in the free energy of water (any solids present could decrease the rate of water flow). *Gravitational potential* is the effect of gravity on water flow since water flows downhill.

3.3 Scientific Debate

How water moves through plants has been a subject of debate, experimentation, and discussion for many years. The concepts of cohesion, atmospheric pressure, and capillarity do not individually explain how water reaches the tops of tall trees. The cohesion–tension theory, first proposed in 1894 by H. H. Dixon and J. Joly [6] suggests that transpiration "pulls" water up trees. Dixon argued that, as a water molecule evaporates, it pulls another one along, and so forth up the tree. This theory was accepted by many scientists, and challenged by others; however, the basic principles still hold even though the theory itself was found insufficient to fully explain how water moves in trees. Let's look a bit at the idea of scientific debates which commonly occurs in science.

In Depth   **Scientific Debate**

Scientific debate is common. Debate helps clarify old ideas, generate new ones, and provide improved ways of looking at a problem. Debate challenges the scientific community to re-examine their ideas. Scientists need proof, data, facts. You can promote a hypothesis if you can scientifically back up your ideas with data, and other scientists can repeat your results. This is the basis of the scientific method. The following are the basic steps in the scientific method.

1. The first step is something we all do. Observe something that makes us want to know more.
2. Scientists then develop a testable question. What are we trying to find or discover?
3. Then the research into the scientific literature begins. What is already known about this idea? Is the scientific community asking the same question? Is there debate? What new or additional support are needed for either side of the debate?
4. Now a hypothesis is proposed. This is important; by definition, a hypothesis is a proposed explanation based on a limited amount of evidence as a place to start examining further. It is not just a guess; it is a well-thought-out guess!
5. Testing the hypothesis through experimentation (testing methods) appropriate to the field of study comes next.
6. Analysis of the data provides conclusions based on that data. Another important concept, the conclusion, can be that the hypothesis was right, wrong, or inconclusive. The conclusions are not opinions, they must be supported by the facts, the data.
7. Next the process requires publication of what has been done. The whole process in steps 1–6 is usually prepared as a paper that is submitted for review by other scientists in the field of study (peer review). The reviewers decide if the process has been done well enough to recommend publication or needs improvement. This can require many steps and is not a trivial matter.
8. After publication, the debate can begin (evaluation of the work). Can other scientists repeat the process and get the same results (reproducibility)? Has anything new been added to support or question previous work? The scientific community is free to evaluate published papers and question any part of them. This helps to ensure the integrity of the process. This scientific method is what makes the difference between just an idea and an idea with factual data to support it. That does not mean that new research and advances in techniques will not change how data are interpreted or create entirely new ideas. That is the work of science to continue to explore, learn, develop concepts, and solutions.

Scientific debate can become just as personal as any disagreement because scientists hold no moral high ground on always being nice about their differences. An example is an idea presented in Chapter 1. University of Iowa professor of physics, Louis A. Frank, and his co-author, senior research scientist, John B. Sigwarth, received personal as well as professional criticism for publishing their findings regarding water entering the Earth's atmosphere from meteors [7]. Prior to about 1986, scientists generally believed that all water on Earth today was present when the Earth was formed. The hypothesis was simple – water is a constant volume of about 1.4 billion cubic kilometers (335.9 million cubic miles) cycling through different phases, but neither gaining nor losing volume.

Today, other studies have supported Frank and Sigwarth's hypothesis. Small comets (more like "snowballs" the size of small houses) enter Earth's atmosphere every few seconds. These 20 to 40 ton "snowballs" fall at a rate of about one every three seconds, and would provide about 2.5 centimeters (1 inch) of water across the Earth's surface every 20,000 years. That may not seem like a lot of water, but if this occurred for the entire age of the Earth (some 4.6 billion years) it could account for much of Earth's water [8]. What was considered a radical idea in 1986 has gained acceptance from new studies, broadening the field of study and examination of data (facts) considering new information. Science is not static, methods get more sophisticated, minds look at ideas from different perspectives, scientists collaborate, and change happens.

The amount of water in the atmosphere is not great (only 12,900 cubic kilometers, or 3,100 cubic miles), and comprises 0.001 percent of all water found on Earth. If all that moisture fell as rain, it would only cover the Earth's surface to a depth of 2.5 centimeters (1 inch) [9]. Precipitation generally equals the amount of global evaporation. However, precipitation usually exceeds evaporation over the continents, while evaporation exceeds precipitation over the oceans. This discrepancy is corrected by surface water flows from the continents. These outflows and precipitation replenish the oceans with renewable water lost to evaporation.

Evaporation, transpiration, evapotranspiration, sublimation, volcanic emissions, and perhaps meteorites account for all water in the atmosphere. Evaporation from oceans is the primary source of moisture for the atmosphere, about 90 percent. Transpiration from plants is also a significant component. Plants can lose up to 98 percent of the water they take in through evapotranspiration. Large oak trees can transpire as much as 150,000 liters (40,000 gallons) of water per year. An area of 0.5 hectare (1 acre) of corn can transpire 15,000 liters (4,000 gallons) in one day.

Think About It: Trees and the Hydrologic Cycle

Evapotranspiration from trees is another area of debate. Some scientists believe that reafforestation (planting trees in areas that were converted from forest or in depleted forest) reduces available water supply by leaf interception, ET, and reduced runoff going into streams. These scientists and land managers are often concerned with the many uses of water in a local watershed or catchment. Others believe that trees play a different role improving water availability on both regional and global scales.

Is it possible that the water vapor reinvigorates the hydrologic cycle as added moisture in the air? Could that vapor increase local precipitation as well as global precipitation as it travels on air currents? What else can be in the transpired water that influences cloud formation? Should tree placement and variety be considered in reforestation efforts? Researching both sides of this issue would be an excellent class, team, or individual activity.

After water vapor enters the lower atmosphere, warm air currents lift the moisture upward. This vertical movement of moisture, in large air masses, is a major factor in the occurrence of precipitation. As moisture increases in altitude, air temperature decreases and water droplets and aerosols cool. At a given temperature and altitude, **condensation** occurs when water vapor changes from a gas to a liquid, and forms water droplets. When condensation is complete around a water droplet, the latent heat is released. As more droplets form, the latent heat increases causing instability in the forming cloud (see Figure 3.5). These water droplets can join forming clouds that grow heavy with moisture, producing precipitation (rain, snow, sleet, hail, and freezing rain). This is the primary method for water vapor to move from the atmosphere back to the surface of the Earth.

When precipitation falls on a land surface, it can take several paths. First, it could immediately evaporate and return directly back to the atmosphere, as a vapor, to continue the hydrologic cycle. Second, some precipitation could seep into the ground and become soil moisture to be consumed by plants (such as trees, shrubs, crops, or pastures – see Figure 3.6). This layer of soil, just below the land surface, is called the **vadose zone**. It can be filled with both air and water, although in very dry climatic locations it would be primarily filled with air.

The layer of geologic material beneath the vadose zone can be saturated with water and is called the **saturated zone**. Water found in this location is appropriately called **groundwater**. The

Fig 3.5 Latent heat is important in the development of thunderstorms and hurricanes. Clouds form as the warm air rises and water vapor begins to condense around particles. When the water vapor condenses to form clouds, it releases latent heat into the atmosphere. The latent heat then warms the surrounding air, causing the cloud to expand. This increases the cloud height and, depending on how unstable the atmosphere is, thunderstorms can form from these growing clouds. Thunderstorms release enormous amounts of latent heat which adds to the instability in the atmosphere and causes some thunderstorms to become severe. In a hurricane, the latent heat is released within its clouds, increasing expansion and instability. This causes the intensity of the storm to increase
(Photograph by Karrie Pennington)

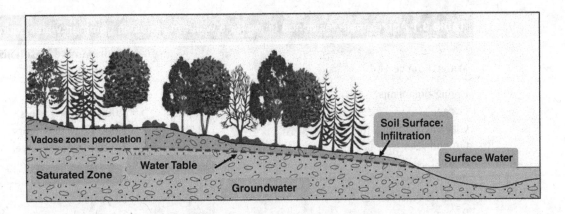

Fig 3.6 Zones of water in the environment. Soil is the storehouse for plant-available water. After a rain event, the water that remains in the soil after gravitational movement is designated as field capacity which varies with soil texture. Sandy soils with large pores lose most of the precipitation to gravity whereas clay soils with small pores hold onto more water. When the water content of the soil is no longer available to plants, the soil is said to be at its wilting point. The water held between the field capacity and the wilting point is plant available water.

Loamy soils (approximately equal mixtures of sand, silt, and clay) have a more balanced texture and hold the most plant-available water. The rooting structure of the plant determines where it picks up most of its water. For many plants, this is in the top 12 inches or so of soil

(Image modified from USGS)

area where the vadose zone meets the saturated zone is called the **groundwater table** or simply the **water table**. The boundary between these two zones rises and falls as precipitation events, infiltration, and percolation cause groundwater levels to increase or decrease. In Chapter 5, on watersheds, we discuss groundwater recharge from surface water, lakes, rivers, and streams.

The third potential path of precipitation, after it falls on the land surface, could be into rivers, lakes, ponds, wetlands, and other bodies of water. Almost all of this **surface water runoff** eventually flows into an ocean, unless it evaporates back to the atmosphere, or seeps into the ground.

Snowmelt is also an important process in the hydrologic cycle. Snow that falls at high elevations, or in mid-latitude areas with colder seasonal temperatures, serves as a storage reservoir of frozen water. When warmer spring temperatures arrive, snowmelt is discharged into rivulets, brooks, streams, and rivers to enhance river flows, and can replenish water levels in ponds, wetlands, lakes, reservoirs, and groundwater.

Eventually, precipitation returns to the atmosphere to continue the hydrologic cycle, but the time involved can vary. Precipitation can reside for long periods as groundwater, in oceans, glaciers, or permanent snowfields. These locations can hold water, in solid or liquid form, for years and even centuries before it evaporates. The length of time water remains as surface or groundwater, before returning to the atmosphere, is called its **residence time (see Table 3.1)**.

Notice that some of these values are given as ranges and all are approximations based on the formula Time = Volume/Inflow or Outflow. Since volumes, inflows, and outflows are all estimates and a state of equilibrium is assumed, this is to be expected. Imagine calculating the actual total volume of the ocean (ocean water does not stop moving around, precipitation and evaporation are always happening somewhere, and the ocean floor resembles mountain ranges) or try calculating the volume of water discharging from all the rivers in the world into the ocean. (Consider that there are droughts, dams, more dams, rain, floods, people taking water, etc.). There are research projects, usually involving computer modeling, using as much real-time data as

Table 3.1 Approximate residence times for a water molecule in common water sources

Water Source	Approximate Residence Time (water molecules)
Living organisms	1 week
Ocean	3,000 to 4,000 years
Glaciers, icebergs, permafrost	1,000 to 10,000 years
Groundwater	2 weeks to 10,000 years
Soils	2 weeks to 1 year
Atmosphere	10 days
Lakes & reservoirs	10 years
Rivers	2 weeks
Wetlands	1 to 10 years

possible, to calculate many of these values. The estimates in this table provide a framework to help visualize residence time in the hydrologic cycle.

3.4 Weather, Climate

Weather is the state of the atmosphere at a given time and location. This can mean wet or dry conditions, cloudy or clear skies, windy or calm, hot or cold temperatures. The given state of weather can remain almost constant for days or may change radically in a few hours or even minutes. Weather can include such events as thunderstorms, blizzards, windstorms, tornadoes, heat waves, hurricanes, and brutal cold spells. Floods and drought can also be part of local and regional weather patterns but can be quite unpredictable. By contrast, **climate** is the state of the weather that can be expected months, years, and even centuries into the future. The climate of a region is created by the interaction of the atmosphere, topography, solar radiation, and the oceans.

Global climate change is the most serious moral and environmental issue facing society today. We are seeing extreme events occurring globally on an almost daily basis. Australia endured drought, massive wildfires, and flooding storms all in 2020 [10]. A plague of locusts so thick that you could not see through them devastated agriculture in East Africa and South Asia in June 2020 leaving 5 million people in danger of starvation [11]. Houston, Texas, has seen five 500-year floods in five years. Remember, a 500-year flood has a one-in-500 chance of occurring in any given year so, five in a row is extremely rare.

As if starvation, loss of homes, and death by tornado, hurricane, drowning, fire, etc. where not enough, there are more humanitarian damages. The COVID-19 global pandemic was made much more likely by limiting wildlife diversity, and destroying habitats that create a niche for smaller animals such as rodents, rats, and bats, that are known to transmit disease [12].

It is not hard to see that global climate change, with all of its disastrous consequences, is hardest on poor countries and poor people everywhere without the resources of richer countries and people. Globally, these inequalities of life circumstances have caused people to rise up in protest, whether in support of stopping global climate change or over issues of race and justice. It seems that, in 2020, the peoples of the world were looking for answers and action.

David Wallace-Wells of *New York Magazine* sums up our state of events very well in the following: "Climate change will continue, and those records – high temperatures, historic rainfall, drought, and wind speed and all the rest – will continue to fall. From here, literally everything that follows, climate-wise, will be literally unprecedented" [13].

Yes, climate change occurs naturally; however, it is the rate of change as influenced by human activity that is the issue. Sunlight is the controlling factor, and how much sunlight the Earth receives depends on its orbit and celestial orientation. The Earth has cycled through ice ages (**glaciations**) and warmer periods (interglacial) over the last million years. (The change interval is believed to be about 100,000 years.) During the last Ice Age (approximately 70,000 to 11,500 years ago), ice covered much of North America and Europe. We are now living in the Holocene epoch, an interglacial. The warmest period of time in this interglacial is called the hypsithermal interval, and it lasted about 3–4,000 years. The rate of temperature change in this interval was about 1 degree per 500 years [14].

Today, the US National Oceanic and Atmospheric Administration (NOAA) reports that the combined land and ocean temperature has increased at an average rate of 0.07 °C (0.13 °F) per decade since 1880; however, the average rate of increase since 1981 (0.18 °C/0.32 °F) is more than twice as great [15]. For comparison to the rate of change in this warming period (0.32 °F/10 y)*10 = (3.2 °F/100 y)*5 = 16 °F in 500 years. That is 16 times faster than the previous natural cycle.

The two most abundant naturally occurring greenhouse gases include water vapor and carbon dioxide (CO_2) – with methane (CH_4), ozone (O_3), and nitrous oxide (N_2O) occurring in smaller amounts. These **non-condensable** (not easily condensed by cooling) gases allow sunlight, relatively short wavelength ultraviolet (UV) and visible light (heat energy), to reach the Earth. The long wavelength infrared light (heat energy called **thermal IR**) is reradiated back to the atmosphere. Greenhouse gases capture this energy so that less heat leaves the atmosphere, and results in warming. This is called the **greenhouse effect**. It is climatologically especially important because it keeps the Earth's surface temperatures about 32 °C (58 °F) warmer than it would be without them. This temperature range has allowed human civilizations to live for millennia. Changing the balance in this natural greenhouse effect is resulting in global climate change. The increase in greenhouse gases due to the burning of large amounts of fossil fuels since the Industrial Revolution (~1790) has severely altered that balance.

Water vapor is an important greenhouse gas because it is the most abundant. In addition, because it combines with sunlight, water vapor becomes a heat source accounting for, on average, about 60 percent of the heat effect. Heat radiated from Earth's surface is absorbed by water vapor molecules which radiate that heat in all directions. Some of the heat returns to warm the Earth's surface. It is not hard to imagine a warm water blanket if you have ever experienced a southern summer. However, water vapor does not control heat, heat controls water vapor.

The temperature of the Earth's atmosphere controls the maximum amount of water vapor it can contain. Warm air holds more water vapor than cold air. High temperatures caused by the increase in greenhouse gases cause water to evaporate and increases water vapor in the atmosphere (humidity), which in turn increases the temperature. This is what scientists refer to as a **positive feedback mechanism**. Basically, one thing leads to another in a loop. Without an increase in the greenhouse gases causing an increased temperature, the amount of water vapor in the atmosphere would not have changed.

A phrase we have used before is that "nature is seldom simple." Professor Adam Sobel, Columbia University, gives an appropriate analogy, "Saying water vapor is a more important greenhouse gas than CO_2 is like saying the amplifier in a sound system is more important than the volume dial for producing the sound. It's true, in a literal sense, but very misleading. CO_2 and other long-lived greenhouse gases are the volume dial on the climate, and the water vapor amplifies the warming that they produce" [16].

Another consideration with water vapor is cloud formation. Clouds reflect heat energy creating a cooling effect; however, at the same time there is more liquid water in the air, increasing warming. A cloudy winter day tends to be warmer than a cloud-free day. Is it possible that these two effects cancel each other out? These are important areas of research in the field of climate change [17].

Carbon dioxide is released through respiration and is also an end product of combustion. The burning of fossil fuels (including oil, coal, wood, and natural gas) releases CO_2. Prior to the Industrial Revolution of the late 1800s and early 1900s, the amount of CO_2 in the atmosphere was relatively stable at 280 parts per million (ppm). On June 27, 2020, the level was 49 percent higher at 418 ppm. This is not natural variation.

Polar mesospheric clouds are the focus of a two-year mission of NASA's AIM satellite launched in 2007 and continuing in 2020. Polar mesospheric clouds are the Earth's highest and form an icy membrane 80 kilometers (50 miles) above the Earth at the edge of space. Mesospheric ("night-shining") clouds are visible from the ground with the naked eye, but historically only at high latitudes (poleward of 55°). In recent years, they have become brighter, more frequent, and visible at lower latitudes (40° N). This is a potential sign of climate change. Understanding the mesosphere is another step in understanding the role our atmosphere plays in creating a planet suitable to sustain us, humankind [18].

Methane is 80 times more effective at trapping heat than CO_2, but it has a shorter lifespan in the atmosphere. Its role in global climate change is nonetheless significant. A major contributor is the use, extraction, and processing of fossil fuels natural gas, oil, and coal. It is also produced through anaerobic processes that occur in rice production, wetlands, and reservoirs with large loads of organic material fueling anaerobic processes. Agricultural sources of methane are from cattle and other ruminants that release significant amounts of methane as part of their digestive mechanism. Livestock waste management, rice cultivation, and burning of agricultural wastes are additional sources. Methane concentration was relatively constant, changing from 1.52 ppmv (parts per million by volume) in 1978 (when the technology to measure it became available) to ~1.77 ppmv in 1990 ~2.5 times preindustrial values. Whether or not the increase is due to increased releases of methane, or to changes in hydroxyl radicals (OH) that remove methane from the atmosphere, it is another area of research and debate [19].

Ozone is formed in the stratosphere (the layer that is 14–22 kilometers [8–12 miles] above Earth's surface) when oxygen reacts with UV light. This creates an ozone layer that protects the Earth from harmful radiation by reflecting UV light. Some of the ozone moves lower into the troposphere (Earth's lower atmosphere, 8–15 kilometers [5–9 miles] above the surface, where clouds form and precipitation occurs).

Twentieth-century transportation exhaust emissions, wood burning, and industrial emissions have released large amounts of nitrogen and carbon into the atmosphere which react with UV light to produce ozone in the troposphere. Nitrous oxide from agricultural fertilizers, some industrial processes (fossil-fuel-fired power plants, nylon production, and nitric acid production) and vehicle emissions is one of these molecules. Since 1900, the amount of ozone in the troposphere has more than doubled over what can occur naturally. Ozone is part of the photochemical smog holding heat in the atmosphere. So, what is protective in the stratosphere can also become a pollutant that is harmful to human health, and acts as a greenhouse gas.

Synthetics with no natural source include chlorofluorocarbons (CFCs), carbon tetrafluoride (CF_4), sulfur hexafluoride (SF_6), and hydrofluorocarbons (HFCs). These are greenhouse gases of varying importance. The long-lived (about 100 years) CFCs were recognized as a problem, and a highly successful public interest campaign was launched to stop their use in refrigerants, aerosol propellants, and cleaning solvents. Fortunately, CFC levels are remaining stable or declining. This is particularly important because, in addition to being greenhouse gases, CFCs react in the cold stratosphere to destroy ozone. This can create holes in the ozone layer, which allow harmful UV radiation to enter the atmosphere [20].

Think About It: Hydrologic Cycle and the Anthropocene

The authors hope that making the causes and consequences of the water crisis visible in diagrams will be an important step towards the goal of a sustainable relationship with water, ecosystems, and society (Figure 3.7) [21]. **Green water footprint** is water from precipitation that is stored in the root zone of the soil and evaporated, transpired, or incorporated by plants. **Blue water footprint** is water that has been sourced from surface or groundwater resources and is either evaporated, incorporated into a product, or taken from one body of water and returned to another, or returned at a different time. **Gray water footprint** is the amount of fresh water required to assimilate pollutants to meet specific water quality standards [22]. (Figure 3.7 modeled after Figure 1 from reference [25].)

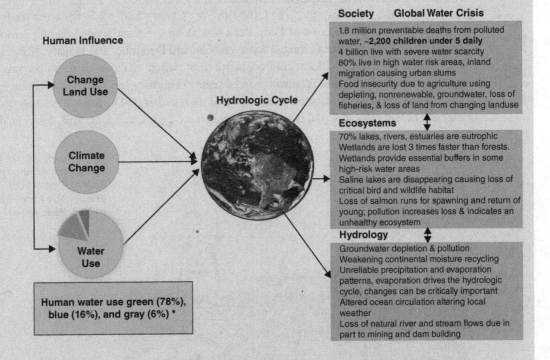

Fig 3.7 *A water cycle for the Anthropocene* was created to help clearly show the tremendous influence humans have on the natural hydrologic cycle (seen in Figure 3.1)

Do you think this diagram fairly portrays human actions and their effects on society, ecology, and hydrology? Explain your thoughts. Is climate change really a water resource issue? Can you provide everyday water use activities influenced by climate change not on the diagram? Go to the Water Footprint website (https://waterfootprint.org/en/water-footprint/personal-water-footprint/) where you will find much to ponder. Is the concept of a water footprint practical? What scale (from global to personal) would you apply it to? Explain your thoughts. Calculate your own personal water footprint. Are you surprised by any specific part of your own water use?

3.4.1 El Niño and La Niña

According to NOAA [23] **El Niño** is "an oscillation of the ocean–atmosphere system, called **ENSO (El Niño Southern Oscillation)**, in the tropical Pacific that has important consequences for weather around the globe." It is a natural rhythm of the ocean–atmosphere system that can interact with other climatic systems to increase rainfall in some regions of the world, while contributing to drought in other locations. To provide forecasting data, NOAA operates a network of buoys in the Pacific Ocean to measure air and ocean temperatures, wind, and ocean currents (see Figure 3.8). The buoys transmit data in real time to researchers and forecasters around the world.

You can go to the National Oceanic and Atmospheric Administration, Global Tropical Moored Buoy Array (www.pmel.noaa.gov/tao/drupal/disdel/) for real-time data from approximately 70 moored ocean buoys. These data help detect and predict El Niño and La Niña events.

El Niño is Spanish for "The Little Boy" or "Christ Child" since the event tends to appear around Christmas off the coast of Peru. Residents along the coast noticed that the temperature of eastern Pacific Ocean waters changed some years around December and January. Although located in a tropical environment, deep, cold ocean currents caused offshore waters to become relatively cool and abundant with fish. Warmer ocean currents in February and March usually temper the cool water. However, in some years this warming event starts earlier (in December), is much stronger, and lasts one or two years. Rainfall is torrential during these cycles, and some arid, desert lands become

Fig 3.8 National Oceanic and Atmospheric Administration (NOAA) 3-meter buoy (Brazilian R/V Antares, photo courtesy Paulo Arlino)

gardens. Many scientists originally thought the El Niño phenomenon only occurred off the coast of Peru, but now it is known as a worldwide, short-term climatic event.

During a non-El Niño year, tropical **trade winds** (steady wind systems present along the Equator) move toward the west across the tropical Pacific. These winds pile up warm ocean water in the western Pacific and cause the ocean surface to be about 0.5 meters (1.6 feet) higher near Indonesia in the west than off the coast of Panama in Central America to the east. The ocean temperature gradation is also great – about 14 °C (8 °F) higher near Indonesia than off the coast of Peru. This is partially due to the upwelling of deep, cool ocean currents that are rich in nutrients and diverse marine ecosystems, creating major fisheries. Rising air above the warmest ocean water, located in the western Pacific, generates precipitation. By contrast, it is relatively dry above the cooler waters to the east. Temperatures in the western Pacific are the warmest ocean waters on Earth, generally around 29 °C (84 °F). Some parts of the western Pacific can be as warm as 31.5 °C (89 °F). By contrast, ocean temperatures in the eastern Pacific average around 23 °C (73 °F).

Life generally lives in sunlit zones. In regions of the ocean above the thermocline, nutrients are quickly taken up by various forms of life, such as plankton. Waste and other debris, from living creatures in this zone, falls into the colder region of the thermocline, and dissolves. This makes deeper, colder ocean waters rich in nutrients. Upwelling brings these rich, deep waters closer to the surface, often followed by large populations of fish. If this upwelling occurs near land, abundant flocks of birds will have a feast.

As warm surface water collects in the western Pacific, the thermocline (the boundary separating warmer surface waters from deeper, colder water) is pushed downward toward the bottom of the ocean. The thermocline is usually found at a relatively shallow depth of around 40 meters (130 feet) in the eastern Pacific, but much deeper, around 100–200 m (330–660 feet), in the western Pacific. When the depth of the thermocline increases, fewer nutrients are delivered from the normal upwelling of cold, deep waters. This disrupts local fish populations and results in fewer fish available for commercial fisheries in the region. These changes are why the Peruvian fishermen, discussed earlier, were some of the first to notice this naturally occurring cycle.

During an El Niño year, trade winds are much less prevalent in the central and western Pacific, particularly west of the International Date Line, which runs north and south roughly from east of New Zealand to the Bering Sea off the coast of Alaska. Calmer trade winds result in the warmer, piled-up ocean water in the west moving back toward the cooler waters of the eastern Pacific Ocean. This distorts three factors – the usual pattern of the jet stream (the strong, concentrated band of fast-moving air in the Earth's upper troposphere); it increases evaporation in the warmer regions of the ocean; and increases tropical convection (the movement of warm air upward and cold air downward), causing a mixing of the atmosphere.

Moisture pumped into the tropical atmosphere during an El Niño year can be carried toward the northeast by the jet stream. This contributes to heavier rainfall across the southern US and altered weather patterns around the world. Rainfall follows the warmer water to the eastern Pacific, and sometimes results in flooding in the South American countries of Ecuador, Peru, and Chile, but can produce drought in Australia, Indonesia, and the Philippines. During the 1997–8 El Niño event, surface water temperatures off the coast of Peru increased by 5 °C (9 °F) [24].

El Niño is normally accompanied by a change in circulation of the atmosphere, called the Southern Oscillation. When they occur simultaneously, it is called **ENSO** (El Niño – Southern Oscillation). Asian monsoons are also associated with ENSO. Rainfall totals can be up to 970 centimeters (382 inches) during the monsoon season of late May to mid-September in Bangladesh, along the northeastern coast of India, and other regions in Southeast Asia. June and July monthly precipitation

Fig 3.9 This NASA image shows colder than normal water (white with large arrow) anomalies in the central equatorial Pacific associated with La Niña. The stronger than normal trade winds bring cold water up to the surface of the ocean
(Image courtesy of NASA)

totals often exceed 250 centimeters (98 inches). El Niño irregularly occurs in 2- to 7-year intervals, and typically lasts 12–18 months. However, the time between events can vary from 1 to 10 years.

La Niña means "The Little Girl" in Spanish, and is called a "cold event" by scientists. It was not until the 1960s that researchers realized this was not just an occurrence off the coast of Peru but was associated with changes throughout the Pacific Ocean. Sometimes El Niño is followed by La Niña, but in other years, it may be followed by conditions that are more normal. La Niña is characterized by unusually cooler ocean temperatures near the Equator in the Pacific. (Warmer temperatures in this same region can signal the arrival of El Niño conditions.) El Niño and La Niña are sometimes referred to as the opposite phases of the ENSO cycle. El Niño is considered the warm phase of ENSO and La Niña is the cooler phase (see Figure 3.9).

Global climate impacts of El Niño tend to be opposite from the effects of La Niña and are most prevalent during the winter months. For example, in the continental US, winter temperatures during El Niño tend to be higher in the North Central States (North Dakota, South Dakota, and Minnesota) and over much of Canada, but particularly around Manitoba and western Ontario. Southern Canada is generally drier than normal during an El Niño winter [25].

On the other hand, El Niño can lead to cooler than normal conditions in the southeastern US (Georgia, Florida, Alabama), the southwest US (Arizona, Nevada, New Mexico), and northern Mexico. During a La Niña year, winter temperatures are warmer than normal in the southeastern US, but cooler in the northwest US (Washington, Oregon, Idaho), and southern British Columbia. It's important to remember that even though it is considered an oscillation, La Niña is not necessarily followed by El Niño. There can be several El Niños in a row, such as occurred in the early 1990s.

In Depth **Southern Oscillation**

Sir Gilbert Walker (1868–1958) documented and named the Southern Oscillation (SO) in the 1930s. He was a British physicist and statistician who taught mathematics at Cambridge University. In 1904, he was assigned to India as the Director of Observatories. While there, he studied the characteristics of the Indian Ocean monsoon. (The monsoons failed to arrive in 1899, which resulted in widespread famine in India.) Sir Gilbert studied and documented vast amounts of weather data to describe the great seesaw oscillation that occurred in surface barometric pressure between the Indian and Pacific Oceans.

The clearest indicator of SO is the barometric surface air pressure at two sites: Darwin, Australia, and the island of Tahiti in the south Pacific. High barometric pressure at one site is usually concurrent with a low pressure reading at the other location, and vice versa, with the pattern reversing every few years. This represents a shift in the mass of air oscillating back and forth across the International Date Line in that region – hence "Southern Oscillation."

3.4.2 Floods

Floods have affected human settlements for thousands of years, particularly as permanent homes were constructed along riverbanks in Africa, Asia, and the Middle East. Floodwaters provided valuable topsoil and nutrients to cultivated fields, particularly along the Nile River Valley in Egypt, the Tigris, and Euphrates Rivers in the Middle East, and the Huang He River in China. In contrast, flash floods and other large, severe storm events have devastated villages and entire regions for centuries.

In the past 30 years (1989-2018), floods in the US have killed an average of 86 people each year. In the past 5 years, that number has increased to an average of more than 100 deaths annually. These numbers are due to increasingly frequent extreme rain and hence flood events. The difficulty of determining exactly how much effect human-induced global climate change has had on extreme weather events is, to say the least, complicated [26]. Weather happens, and climate tells us what to expect based on history. That history is what has changed drastically since the industrial revolution in the late 1800s.

Floods are one of the deadliest natural disasters. The 1993 flood along the Mississippi River was the costliest river-related flood in US history, causing US$20 billion in damages [27]. In 2004, monsoon rains and associated floods killed over 1,500 people in Bangladesh, Nepal, and India, with half of Bangladesh under water at one point during the disaster. In 2005, the flooding created by Hurricane Katrina in the US caused more than US$200 billion in losses and was the costliest natural disaster in the country's history (see Figure 3.10).

Floods can be classified by type and cause. Inland or **areal flooding** usually occurs when moderate rainfall continues long enough to saturate soils and fill drainage systems or intense short-term rains occur (intensity is measured as rain per unit time). **Riverine flooding** occurs when precipitation events, such as moderate long-term rains and high-intensity rains (usually associated with tropical storms), increased snowmelt, or ice jams cause a river to overflow its bank. In flat areas, river flooding can result in inland flooding as the waters spread and cannot infiltrate the soil or drain.

Coastal flooding is probably the most common and newsworthy event perhaps because crashing waves are more exciting than muddy water covering flat lands. Coastal flooding is caused by storm surges, water abnormally higher than the regular astronomical tides (tides influenced only by gravitational pull of the sun and moon without atmospheric interactions). Atmospheric forces are generated by severe storm winds, waves, and low atmospheric pressure. If these occur at normal high tide, waves can reach 20 feet or more, causing loss of life and property second only to hurricane damage.

The highest loss of life occurs with **flash foods**. These are the fast-moving waters taking trees, destroying ditch channels, filling dry riverbeds, and sometimes resulting in deaths. Perhaps surprisingly, most flash flood deaths are vehicle-related. Too many people think they will be able to cross the flooded road, stream, or underpass only to be washed away. Urban areas are prone to flash

Fig 3.10 This aerial photo of flooding just around New Orleans, Louisiana, US, helps us to imagine the extent of the damage from one storm event. Notice that the water quality below the highway is murky from sediments, but above the highway the water is dark and clear. What are some of the reasons water quality is different? (Photograph courtesy of NOAA)

floods because there are too many impervious surfaces for water to drain naturally. Flash floods can occur within hours or even minutes (usually six hours or less) of an intense rain event, a catastrophic dam or levee failure, or the breaking of a log or ice jam. Flash flooding in wildfire burn areas is particularly harmful because there is no vegetation to slow the water and allow soil infiltration. These floods can carry heavy debris and erode soil, creating destructive debris flows [28].

3.4.3 Drought

Drought can be defined by rainfall amounts, agricultural productivity, vegetation and habitat conditions, streamflow, reservoir storage levels, soil moisture, and economic impacts. Flash drought adds unusually high incoming solar radiation, extra high temperatures, and strong winds. Megadrought adds longevity. Drought can be categorized in six ways.

Meteorological drought: A departure of precipitation from normal amounts. Since climatic conditions vary around the world, a two-month dry period in one region (Valencia, Spain) might be considered normal, while it would be considered a drought in Seattle, Washington.

Agricultural drought: The condition where available soil moisture is inadequate to provide the needs for a particular crop or livestock at a particular time.

Hydrological drought: Occurs when surface and groundwater supplies are below normal. It can be measured as streamflow, reservoir levels, or groundwater elevations.

Socio-economic drought: Refers to the situation where water shortages affect the needs of people.

Flash drought: Occurs with unusually high incoming solar radiation, high temperatures, and strong winds causing rapidly increased evapotranspiration [29].

Megadrought: Generally used to describe the length of a drought, and not its acute intensity. In scientific literature the term is used to describe decades-long droughts or multi-decadal droughts.

Snow drought: Defined as period of abnormally low snowpack for the time of year, reflecting either below-normal cold-season precipitation (dry snow drought) or a lack of snow accumulation despite near-normal precipitation (warm snow drought), caused by warm temperatures and precipitation falling as rain rather than snow, or unusually early snowmelt [30].

Worldwide, drought is the most damaging of all natural disasters. Since 1967, drought has been responsible for millions of deaths, and billions of dollars in damages. The US National Weather Service defines a drought as "a period of unusually persistent dry weather that continues long enough to cause serious problems such as human suffering, crop damage and/or water supply shortages" [31]. Severe drought kills crops, dries up lakes and wetlands, and allows wind to carry away topsoil. Dust storms can destroy cropland and carry away valuable plant nutrients. In 2003, severe drought in Ethiopia in eastern Africa dried up local water supplies and forced some to walk up to 10 kilometers (6 miles) for drinking water. Bathing and other sanitation needs were unmet due to a lack of water [32]. In 2007, the city of Atlanta, Georgia (US) experienced a loss of drinking water from depleted reservoirs. If the drought had not lifted, the city could have run out of drinking water [33].

Africa suffered its worst drought of the twentieth century in 1991–2 when 6.7 million square kilometers (2.6 million square miles) were affected, along with 24 million people in eastern Africa. An earlier drought in 1984 and 1985 led to severe famine in the same region, and the loss of over 700,000 lives the next year to starvation [34]. The countries of Southern Africa have faced many periods of drought. By mid-March in 2020, 45 million people were classified as food insecure and another 430,000 faced severe acute food insecurity (acute malnutrition) and needed urgent humanitarian action from an ongoing 2018–20 drought. Forty-five percent of Zimbabwe's rural population, 4.34 million people, faced acute food insecurity or worse (starvation, death).

On April 10, 2001, a massive dust storm developed on the Mongolia–China border and made its way across the Pacific Ocean to Canada and the US. The dust cloud was carried by the jet stream and made its way as far east as the Great Lakes. In the winter of 2006, a similar event dropped tons of Mongolian Desert soil on the peaks of the Rocky Mountains in Colorado (US). The decreased **albedo** (ability to reflect light) of the snow, caused by the dust, increased solar radiation to rapidly melt the snowpack. This event seriously depleted snow reserves for lower elevation irrigators and municipal water providers. By May, 20 percent of the contiguous US, including Colorado, was experiencing extreme droughts [35].

Figure 3.11 is a NASA-NOAA Suomi NPP satellite view of the 2020 Saharan Desert dust and aerosol cloud that formed when strong updrafts were picked up by westerly winds and blown across the Atlantic to as far as the Caribbean. This cloud was unusually dense, and there were concerns over increased air pollution harming people with respiratory problems and/or COVID-19 since the pandemic was still active. These dust clouds are not unusual. Meteorologists call them a Saharan Air Layer (SAL). On the positive side they can have some benefits such as the potential to deposit fertile soils on the land they cover and adding to beaches along shores [36].

There are three main contributors to drought: (a) land and ocean temperatures, (b) circulation patterns of the atmosphere, and (c) soil moisture content. Each of these components is linked to

Fig 3.11 NASA-NOAA's Suomi NPP satellite observed a huge Saharan dust plume streaming over the North Atlantic Ocean, beginning on June 13, 2020. Satellite data showed the dust had spread over 2,000 miles (Original photo courtesy of NASA Worldview)

the others, which can become a vicious cycle [37]. Soil moisture is an important component of drought. High moisture levels in the soil allow evaporation to occur as the heat of the day increases. Warm, moist air rises in the atmosphere and, under the proper conditions, will condense and fall back to Earth as precipitation. Dry soil moisture conditions result in minimal evaporation from the land surface, resulting in less atmospheric moisture, fewer clouds, and less rain. As a drought continues, the pattern repeats so that less and less moisture is available in the ground, extending the drought.

There is no one right way to predict drought, but a variety of tools have been developed and have proven useful. In 1965, the **Palmer Drought Severity Index** was developed by US National Weather Service meteorologist Wayne Palmer. It uses temperature and rainfall data to determine dryness in a region. The index uses 0 as normal and negative numbers to categorize drought; –2, for example, is an indicator of moderate drought, –3 represents severe drought, and –4 is extreme drought. During the summer of 2002, many regions of the western US and Canada were in the –4 range.

In the 1980s, scientists began using remote satellite imagery to map global vegetation. An algorithm (a procedure or formula for solving a problem) called a **vegetation index** was developed by using daily vegetation readings (using 8-, 16-, and 30-day image composites of information) to create detailed maps of vegetation density. This allowed scientists to identify regions where plants were thriving, and other regions where vegetation stress was severe due to a lack of water.

Satellite and surface-based scientific instruments are used to study physical processes in detail. Global climate models have shown that an increase in surface temperatures can lead to an increase in evaporation, leading to more extreme weather activity such as drought. In general, if soil conditions are dry, additional solar radiation will continue to warm the land surface. This will make conditions hotter and the soil drier, continuing a chain of events that can lead toward or prolong a drought. Atmospheric circulation patterns must change to break this cycle. Researchers use remote sensing data from satellites to confirm the direct relationship between ocean surface temperatures and areas of high plant growth (more precipitation) and areas of drought. These two parameters – ocean surface temperatures and plant growth – are directly connected to the Earth's climate system [38].

The ability to predict droughts has also been greatly improved using satellite data. NASA scientists at the Goddard Space Flight Center generate weekly maps of soil and groundwater conditions. According to NASA scientists, these maps are based on terrestrial water storage observations derived from GRACE and GRACE-FO (Twin Gravity Recovery and Climate Experiment-Follow On) satellite data and integrated with other observations, using a sophisticated numerical model of land surface water and energy processes. The drought indicators describe current wet or dry conditions, expressed as a percentile showing the probability of occurrence for that location and time of year. These data are provided in map image or binary form for use globally [39]. The GRACE-FO project is an international collaboration between NASA and the German Research Center for Geosciences (GFZ). Partnerships make these tools possible and provide much needed data to help analyze trends in the Earth's mass and water storage [40].

In Depth   **GRACE TELLUS* (Gravity Recovery and Climate Experiment) [41] GRACE-FO [42]**

The one property of water that does not change, whether it is in its liquid, solid, or gas phase, is its mass. No matter where the water is, from mountain tops to ocean depths, this mass has a gravitational pull.

In March of 2002, a collaboration of the US and German space agencies (NASA and DLR) launched a twin satellite system, GRACE (Gravity Recovery and Climate Experiment) designed to make detailed measurements of anomalies in the Earth's gravitational field. The idea that measuring mass could provide insights into the Earth's processes was made possible in the 1990s with innovations in technology, particularly a global positioning system (GPS). These advances resulted in the development of a new field of space-borne remote sensing that became a reality with GRACE (2002–17) and continues with GRACE-FO (Gravity Recovery and Climate Experiment-Follow On) launched May 2018. Analysis of changes in mass provide detailed information of Earth's hydrology, cryosphere, and oceans that has been so successful, especially in the field of climate science, that they are described as "revolutionary, changing scientist's views of how water moves and is stored around the planet."

Basically, the twin satellites, orbiting one behind the other about 137 miles apart (220 kilometers), experience small accelerations and decelerations caused by changing mass below the spacecraft. This alters the distance between the two satellites by only a few microns (think of a distance less than a human hair). The satellites constantly beam microwave pulses at each other and time the arrival of return signals, keeping a record of the changing distances. A GPS keeps track of where the satellites are relative to Earth's surface, and onboard accelerometers record forces, such as atmospheric drag and solar radiation, that can also influence the satellites.

Analysis of these tiny gravitational shifts has increased our understanding of changes in both ocean surfaces and deep currents which are important components of global climate change. The ability to measure surface water runoff and groundwater storage on land has not only contributed to monitoring and predicting droughts but also of floods. Interaction and exchanges between glaciers, ice sheets, and oceans have confirmed that Greenland and Antarctica are losing gigatons of ice yearly, and more accurately show sea level changes. Even the effects of ice sheet melting and groundwater depletion on the Earth's rotation have been documented.

This is not simple science. It requires the continued collaboration of scientists internationally, both to support and operate the satellites but also to analyze the data and produce the visual and digital information used, to explain what is happening with water on Earth on a real time basis. These data are usable not only for academic understanding through research by scientists but also for governments and lay people to better prepare for and manage what we call natural disasters. This is a fascinating field of study that provides opportunities for many different student interests.(*In case you were wondering, Tellus is Latin for Earth, but can also be a play on words "What does GRACE Tell US?")

The National Drought Mitigation Center at the University of Nebraska-Lincoln, US, houses a variety of drought measuring tools, from simple maps showing drought severity to experimental, satellite-based maps used in climate monitoring. This site provides help to planners, from ranchers to cities and governments, to decide which tools best fit their needs to monitor and prepare for drought [43].

The North American Drought Monitor (NADM), a collaboration between Canada, United States, and Mexico, provides up-to-date state of drought information in these countries. The agencies involved are NOAA's National Center for Environmental Information, NOAA's Climate Prediction Center, the US Department of Agriculture, and the National Drought Mitigation Center. Major participants in Canada and Mexico include Agriculture and Agrifood Canada, the Meteorological Service of Canada, and the National Meteorological Service of Mexico (SMN – Servicio Meteorologico Nacional) [44].

Wintertime drought can be profoundly serious for communities that rely on snowmelt from mountain snows. Low snowfall accumulation in mountainous regions can result in extremely low springtime runoff. Rivers that normally flow at high springtime levels can be reduced to small fractions of flows. This provides less water for storage reservoirs, and inadequate water for summertime needs. Many areas have enough water storage reservoirs to make it through one dry winter, but back-to-back wintertime droughts can leave reservoirs exceptionally low, forests dry, and fire danger seriously high. Summertime rains can help alleviate the problem, but too often such dry cycles continue for many months and even years.

The impact of drought on regional economies, society, and the environment can be severe. From an economic standpoint, losses due to drought routinely range in the billions of dollars annually in the US and were as high as US$39 billion during the three-year drought of 1987–9 [45]. In North America, the droughts of the 1930s and 1950s had the most severe economic and social impacts. Those droughts lasted five to seven years; with the Dust Bowl years of the 1930s resulting in great clouds of dust that blocked out the midday sun. Millions of people left the Great Plains during this time.

The 1930s are the center of John Steinbeck's novel, *The Grapes of Wrath* [46]. The drought affected 20 million hectares (50 million acres) of the Great Plains of central Canada and the US (see Figure 3.12). To really feel the experiences of the people living through the drought years (see Figure 3.13), and to appreciate how strongly our use of the land can change it and us forever, please read *The Worst Hard Time: The Untold Story of Those Who Survived the Great American Dust Bowl* by Timothy Egan. The *Austin* (Texas) *Statesman Journal* said, "In an era that promises ever-greater natural disasters, *The Worst Hard Time* is 'arguably the best nonfiction book yet' on the greatest environmental disaster ever to be visited upon our land, and a powerful cautionary tale about the dangers of trifling with nature."

Drought also hit hard in the northeastern US in the 1960s, such that New York City reservoirs were down to 25 percent of normal capacity. In 1998, Canada suffered severe forest fire damage induced by drought, and northern Mexico experienced the beginning of three years of record low

Fig 3.12 The term "dust bowl" was not an exaggeration. At one time, these clouds were soil, part of carefully plowed fields. Drought and wind turned the soil into dust, scraping across the land, moving to distant places, and on the way devastated towns, houses, fields, and people
(Photograph courtesy of USDA NRCS)

Fig 3.13 *Migrant Mother* by Dorothea Lange (1895–1965) is one of the most iconic images of the Dust Bowl; from the worry on her face to the weary children with dust ground into their clothing and skin, this picture captures the humanity of tragedy. Dorothea Lange was an American documentary photographer and photojournalist. Her work with the Farm Security Administration humanized the Dust Bowl and the Great Depression
(Photograph courtesy of USDA Farm Security Administration)

precipitation (rainfall was 93 percent below normal in the spring of 1999). In 2020, wildfires devastated portions of Colorado and other western US states and Canadian provinces (see Figure 3.14).

To obtain additional information on predicting, monitoring, and mitigating drought, go to the US National Drought Mitigation Center (www.drought.unl.edu), the European Drought Centre (www.geo.uio.no), or for other countries, the NOAA National Centers for Environmental Information (NCEI) covers the globe (https://gis.ncdc.noaa.gov/maps/ncei/drought/global). For more US information, go to the US National Weather Service website (www.nws.noaa.gov/).

Fig 3.14 NASA image of smoke from multiple large wildfires in Québec, Canada. Moderate Resolution Imaging Spectroradiometer (Aqua MODIS) image from July 7, 2002 (Image courtesy of NASA)

3.4.4 Paleoclimatology

Paleoclimatology is the study of past climate. The word comes from the Greek *paleo*, which means ancient, and the word "climate." Paleoclimatology allows for the study of climate before humans began keeping records of such events. Scientists use natural environmental records – such as tree rings, ocean coral reefs, sediments of lakes and oceans, and glacial and ice cap cores – to create a record of ancient climatic conditions.

These natural data provide scientists with evidence of long-term averages and changes that have occurred in the past. For example, instrument records show that our Earth has warmed by 0.5 °C (0.9 °F) since 1860 to the present. However, this record is not long enough (from a climatic standpoint) for scientists to determine if current global warming is caused primarily by human activities, or through natural variations in climate, or both. Paleoclimatology records can help extend this period of record to hundreds and even thousands of years to help evaluate temperature changes of the past 140 years or more [47]. Alarmingly, the 1990s were the warmest decade for all global instrumented records and included the six warmest years in recorded history.

Tree ring records can extend back some 300 years in most areas, and thousands of years in other regions. The study of tree rings, called **dendrochronology**, provides a detailed record of small variations in temperature and precipitation. However, they only record the period of growing conditions on the land across the lifetime of the tree. In the White Mountains of California, for example, the Methuselah Trail provides a glimpse at some of the oldest known trees in the world (see Figure 3.15). These ancient and twisted bristlecone pines (*Pinus longaeva*) grow very slowly and preserve climatic conditions within their annual growth rings. The bristlecone climate record goes back over 4,000 years and is contained within the living and dead growth of the trees.

Researchers in Canada are using paleoecological (the interaction between ancient organisms and their environment) techniques to obtain fossil records of drought from lakes in the southern regions of Alberta, Saskatchewan, and Manitoba in south-central Canada. These techniques compare lake water chemistry and biological community composition with temperature, precipitation, and evaporation changes. Water-soluble chemicals become more concentrated as the climate

Fig 3.15 The Methuselah Grove Trail: Methuselah is a Great Basin bristlecone pine (*Pinus longaeva*) estimated to have germinated in 2832 BCE in the White Mountains of California, US. It was 4,789 years old when sampled in 1957 by Schulman and Harlan. The oldest known non-clonal organism still alive, it is about 4,840 years old. The exact location of this ancient treasure is unknown and there are no official public pictures. This is a picture of the trail walk in an area with old bristlecone pines near Lone Pine, California, US (Photograph by Oke, CC BY-SA 3.0, https://commons.wikimedia.org/w/index.php?curid=1142377)

becomes warmer and drier, leading to distinct changes in aquatic biota. In addition, wetter, cooler weather dilutes water-soluble minerals, and assists in the recovery of less salt-tolerant species. Therefore, examining lake sediments for fossilized biota can provide data to reconstruct past climate regimes. Hopefully, this information can be used to predict drought frequency, duration, and intensity in future years.

In regions where older trees are not available, lake sediments can provide valuable information on lake level fluctuations. These changes can be identified through geologic beach materials. These were naturally deposited, either at higher or lower shoreline locations, depending upon water levels in the lake. Droughts can also affect lake water quality by concentrating salts in less water. This increase in salinity levels can change the type of lake-dwelling organisms able to survive in such water.

Scientists can also analyze pollen grains in lake sediments to determine ancient climatic conditions. An abundance of a certain type of pollen in sediments can signify drier or wetter conditions for ancient vegetation surrounding a lake. For example, a change from an abundance of grass pollen to more sage pollen can indicate a shift from wet to dry climatic conditions.

Another method of studying ancient climates is with sand dunes. These can be analyzed for layers of soil interspersed between sand materials. For soil to develop, the climate needed to include extended wet periods. The presence of soil indicates a longer period of higher precipitation levels.

Researchers also use chemical signals locked in minerals and ice bubbles, as well as in pollen and other biological indicators, to determine historic temperature and precipitation patterns. Coral and ice cores can also serve as "proxy" sources to provide information on natural climate variability. Ice cores can provide an annual record of temperature, precipitation, atmospheric composition, volcanic activity, and wind patterns. Denver, Colorado, is home to the US National Ice Core Laboratory (NICL) where ice cores, retrieved from polar regions of the world, are stored, and studied. The storage facility is held at a constant –35 °C (–31 °F) to preserve core samples.

3.5 The Hydrologic Cycle and the Natural Environment

The hydrologic cycle supplies the water essential to maintain life of animals and vegetation in our freshwater natural environment. It is a dynamic and complex system with three main paths of movement on the land surface – evaporation, infiltration into the soil and groundwater, and as surface water runoff. These paths of water movement greatly affect animals and vegetation. Together, all of this information together provides scientists with a climatic "picture" of wet and dry periods, in a particular region, over a vast period of time.

3.5.1 Evaporation and Transpiration

Evaporation rates are driven by the water vapor pressure gradient (differences) between the evaporating surface and the atmosphere. Ecosystems in humid climates, with heavy plant populations, tend to have low evapotranspiration (ET) rates. This seems counterintuitive, yet anyone living in or visiting the southern states of the US, or India during the monsoon season, has experienced the feeling that sweat will not evaporate from their skin. Evaporative cooling just does not work well in a high humidity situation. Desert ecosystems, conversely, are low humidity and low vapor pressure areas where evaporative cooling is highly effective.

The Northern Hemisphere has vast plant populations that influence climatic events through evapotranspiration. The North America boreal forest is the largest remaining intact forest in the world [48]. It is more extensive than the Amazon rainforest and stretches from Newfoundland across Canada into western Alaska. Stands of aspen, jack pine, tamarack, spruce, and birch mingle with rivers, lakes, and wetlands. Healthy herds of caribou roam the land just as they have done since time immemorial. The forest is home to many large mammals including grizzly bears, moose, wolves, wild sheep, lynx, and wolverines. Over 300 species of migratory birds arrive seasonally to raise their young. Black ducks, whooping cranes, and trumpeter swans all nest in these woods. In short, this is an amazing natural resource with a huge job to do – in terms of not only wildlife, water quality, forest plants, and humans, but also long-term climate change because of the forest's effects on the hydrologic cycle.

In Depth   **NASA Improves Global Climate Models**

When many of us hear the term NASA (the National Aeronautics and Space Administration) we think of Moon landings, Mars rovers, exciting studies of distant planets. What we may not know as much about is NASA's fifty years of monitoring our own Earth. NASA's Aqua Satellite is just one example. It monitors many aspects of the hydrologic cycle to provide better data for global climate models. Information is gathered on atmospheric temperatures and water vapor content, global precipitation, cloud height, size, and water content, size of cloud droplets and ice particles, sea ice concentrations, snow coverage on land, soil moisture content, and interactions between the Earth's land, ocean, air, ice, and biological systems. Visit NASA's International Earth Observing System (EOS) website to explore these amazing and continuing to grow missions (https://eospso.nasa.gov/) [49].

Seasonally, vegetation influences CO_2 levels in our atmosphere. Trees use CO_2 to make food energy, then respire some of it back into the atmosphere at night. The Northern Hemisphere also has large industrialized areas producing greenhouse gases. Trees (and all plants that photosynthesize) help buffer anthropogenic (human) inputs. Loss of the boreal forest in the North and the rainforest in the South potentially could cause irreparable climatic change. The boreal forest is facing industrial development for hydroelectric power, diamond mining, timbering for pulpwood, and oil and gas development. These bring infrastructure needs including roads, homes, airstrips, processing facilities, and pipelines. Eight thousand years ago, intact forests were plentiful. Today, the Russian boreal forest, the North American boreal forest, and the Brazilian rainforest are Earth's only three large, intact, green areas. Balance is being lost.

Larry Innes, interim director of the Canadian Boreal Initiative (a conservation group trying to maintain the forests), says that Canada is central to saving the remaining acres of forest and providing an example for other areas. "So, the question is how to protect and conserve these last forests, and Canada is the key. Canada is well-governed, has a strong conservation ethic, and the necessary conservation legal framework." The same can be said of Alaska, United States.

3.5.2 Infiltration

Vegetative cover, such as forest canopy, shrubs along river corridors, and groundcover on sloped surfaces, can help regulate (slow down) infiltration into the soil and groundwater. Forest canopy and leaf cover, for example, can reduce the impact of raindrops on the land surface, and provide more time for infiltration. Likewise, vegetation on sloped landscapes can trap moisture, and provide more time for infiltration into porous soils. Plant roots help retain the soil, reduce soil erosion, and minimize surface water runoff, especially on sloped landscapes. Roots also improve soil quality creating a structure conducive to water infiltration during precipitation events. Inadequate vegetation cover can allow precipitation to run off sloped land surfaces causing erosion, in turn, leading to even less vegetative cover and the transport of more soil sediments into downstream watercourses.

3.5.3 Surface Water Runoff

In many cases, surface water runoff ends up in rivers which are conduits for transporting surface water downstream toward oceans or seas. Before reaching a river, however, surface flows may be intercepted by riparian vegetation or in wetlands, which serve as important components of a dynamic system that slows the flow of water. These surface water networks collect excess precipitation and effectively decrease the speed of a potential flood event. Wetland degradation and removal of riparian buffers has created serious unexpected damage to ecosystems by stopping this natural **floodwater retention**.

Surface water runoff may flow into wetlands – natural water-collecting **lentic** (or still water) ecosystems that serve a vital role during surface water flood events. Wetlands capture excess water, acting like sponges, and gradually release it back to adjacent lakes, rivers, or groundwater systems. Wetlands serve as natural filters to remove sediments and reduce some pollutants and other impurities. Historically, wetlands have been undervalued, and many were drained or filled with soil to improve farming opportunities. Today, wetlands are more highly valued for their ability to provide valuable ecosystem benefits.

Earlier, we mentioned the importance of snowmelt in the hydrologic cycle. Mountains and **uplands** (other high elevation regions) are the natural "water towers" of the world. These physical barriers intercept atmospheric moisture and force it upward where it condenses into clouds and falls as rain or snow. Almost all major rivers of the world, including the Nile, Rhine, Amazon, and

Colorado Rivers, have their headwaters in mountainous regions. Mountains and uplands provide adjacent lowlands with surface water runoff for irrigation, urban and industrial uses, recreation, and for ecosystem needs. Mountain precipitation accounts for 20–50 percent of the total discharge in humid areas, receiving more than 50 centimeters (20 inches) of annual precipitation, and as much as 50–90 percent of available water supplies in the world's semi-arid and arid zones. Mountain rivers are home to a wide range of biological species. In addition, varying elevations create a variety of microclimates and ecosystems, leading to a rich diversity of plant and animal life in these regions.

3.6 The Hydrologic Cycle and the Human Environment

Predictability is an important aspect of the hydrologic cycle. Typically, humans utilize available water supplies in their local area, and then solve water shortages (either due to water quantity or water quality problems) by transporting more water from other sources or locations. We have examined the myriad of ways humans influence global climate change in this chapter. A large part of how humans alter the hydrologic cycle is through two primary actions – urban development and cultivation of land for crops (land use changes). What effect does this have on the hydrologic cycle, and can past practices be sustainable in the future?

3.6.1 Urban Development

Urbanization results in the sealing of land surfaces for streets, parking lots, sidewalks, and rooftops. This increases **stormwater runoff** and reduces groundwater recharge. Impervious surfaces, such as streets, allow stormwater runoff to increase speed, which reduces potential groundwater infiltration. This often negatively affects groundwater levels and **baseflow** (long-term water availability not associated with immediate storm events) of streams. The high proportion of sealed surfaces within urban areas is associated with heat storage and transfer, which ultimately leads to anomalously high temperature gradients, called the "urban heat island effect." This is one expression of how intensive urbanization can alter the hydrology and climate of an area at a local scale. Built-up environments may cover up to 80 percent or more of the land surface area in an urban setting (see Figure 3.16).

Rivers are often **channelized** (straightened) in urban areas. The purpose of a channelized stream is to collect and remove stormwater runoff quickly and efficiently in an urban area to prevent flooding. This causes water levels in urban streams to rise and fall very quickly in response to rainfall events. In addition, this regulation of surface water runoff can prevent groundwater recharge and replenishment of adjacent wetlands and ponds and can direct surface water – often polluted with urban waste – more rapidly to rivers and lakes. Even more critical can be the loss of stream structure – sandbars, shallow and deeper segments, varying river currents, and meanders (large, wide curves in a river) that provide important aquatic and terrestrial habitat. In addition, river channelization alters river water temperature and humidity levels in the lower atmosphere in the immediate vicinity. This channelization (as well as the large expanse of regional urban impervious material) could reduce or increase convection (the spontaneous rise of air) since bare ground or pavement radiates heat more rapidly to the overlying air. Dry soils typically have a lower albedo (absorb less of the Sun's energy) and provide less evaporation to the atmosphere.

Fig 3.16 Rooftops and paved roads create large areas of impervious surfaces increasing surface water runoff but decreasing infiltration and groundwater recharge. Stormwater runoff from urban areas is often contaminated by fertilizers, pesticides, herbicides, oil from driveways, and various forms of trash
(Photograph courtesy of USGS)

3.6.2 Cultivation of Land for Crops

Land surface characteristics, vegetation, and deforestation for crop cultivation can greatly affect the hydrologic cycle in a region. First, runoff and infiltration into the soil and groundwater are affected by the type of land surface. Second, the amount of vegetative cover is important. The degree of absorption and reflection of sunlight from the land surface will affect evaporation rates, humidity levels, and cloud formation. Therefore, land surface characteristics and vegetative cover affect precipitation patterns in a region.

Mountain watersheds are an intricate part of the hydrologic cycle. Unfortunately, conflict often exists when humans utilize mountain watershed supplies for irrigation. In 1995, for example, 14 international disputes occurred regarding the distribution of scarce water supplies, and regional conflicts are common. An example is occurring in East Africa where Mount Kenya's glaciers are melting and placing water users at severe risk. Mount Kenya is Africa's second highest mountain at 5,199 meters (17,057 feet) and is located just south of the Equator. Disputes have occurred between farmers, who divert water for irrigation in the mountain highlands, and some 2 million people living in the lowlands. In recent years, more water has been diverted for irrigation, leaving less water available downstream. These lower elevation residents rely on the Mount Kenya water tower for cattle ranching, pastures, and tourism in wildlife parks [50].

The result in Kenya is that communities are being forced to look upstream to find adequate water supplies. Farmers are fighting farmers. Those who use pastures for cattle are fighting upstream cultivators, and many are fighting about wildlife and tourism needs. The Kenyan government has ordered irrigation to stop, but this negatively affects the large horticultural economy that exports products to European markets. The Kenyan government is planning their program to improve and better manage Kenya's water resources [51].

Another example is the Cordillera de Vilcanota (Quelccaya Ice Cap) – a traditional source of water for residents of Lima, Peru (see Figure 3.17). The Quelccaya Ice Cap (pronounced kell-KIE-yah) is the largest single glacier in the Peruvian Andes Mountains. Even though it has a summit elevation of 5,670 meters (18,602 feet), the glacier has melted at an increased rate over the past six decades. Since 1970, it has shrunk by 40 percent and the rate of retreat has increased to about 30 feet per year. This is

Fig 3.17 Qori Kalis Glacier flowing out from the Quelccaya Ice Cap in Peru, 2006. The impacts of glacier retreat are felt in ecosystems, economies via loss of tourism and agriculture, and by local people whose cultures, traditions, and livelihoods are threatened. Dr. Lonnie Thompson and his research group have been studying the Qori Kalis glacier since 1974, providing valuable data to better predict what is happening with this glacier. The team has sampled ice cores across the globe.

Their mission statement verbatim is "*Our principal objective is the acquisition of a global array of ice cores providing high resolution climatic and environmental histories that will contribute to our understanding of the complex interactions within the Earth's coupled climate system. Ice core histories from Africa, Antarctica, Bolivia, China, Greenland, Peru, Russia, and the United States make it possible to study processes linking the polar regions to the lower latitudes where human activities are most intense. These ice core records contribute prominently to the Earth's paleoclimate record, the ultimate yardstick against which the significance of present and projected anthropogenic effects will be assessed.*"
(Photograph by Dr. Lonnie Thompson, Byrd Polar and Climate Research Center, Ice Core Paleoclimatology Group, Ohio State University. Read about this exciting group's research at http://research.bpcrc.osu.edu/Icecore/)
(https://creativecommons.org/licenses/by-sa/3.0/)

due largely to increasing air temperatures over the Peruvian Andes and changing precipitation patterns. Another water resource issue, water quality, has also come into play as the retreating ice cap has exposed areas of heavy metals, including toxic lead and cadmium. The metals are now reaching groundwater, making the water undrinkable. Livestock and crops are also being killed. This alarming situation puts the future safety of the water supply of 10 million residents at risk [52].

Summary Points

- The Earth's hydrologic system is a dynamic process powered by energy from the Sun. It is called the **hydrologic cycle** because all water on Earth participates in a continuously moving cycle of use and renewal. It takes place simultaneously in the Earth's atmosphere, waterbodies from wetlands to oceans, plants, soils, and geologic materials.

- **Weather** – the state of the atmosphere at a given time and location. It describes a major part of the hydrologic cycle.

- **Climate** – the state of the weather that can be expected months, years, and even centuries into the future. It helps scientists study what is happening within the cycle and how it may influence life on Earth.

- Weather patterns such as the **El Niño** and **La Niña** are the result of oscillations in the ocean–atmospheric system which result in colder or hotter temperature, and higher or lower rainfall amounts than normal.

- **Global climate change** can be documented through the study of tree rings, ocean coral reefs, sediment cores taken from lakes and oceans, and from ice cores taken from glaciers and ice caps. Satellite missions have added data that greatly improve the ability to analyze and explain the interactions of Earth's water, land, and atmosphere.

- **Paleoclimatology** – the study of these past climates, which helps meteorologists understand natural climate change and can be used to find how humans have altered these changes.

- **River** and **stream floods** are naturally occurring events which replenish floodplain soils and provide spawning and feeding habitats for wildlife. Floods can become disasters when they conflict with human use of the floodplain.

- **Droughts**, extended periods of below normal precipitation, occur worldwide, and are major factors in the health and survival of the Earth's resources – from soils to humans. Prolonged drought can cause human suffering, economic losses, and environmental degradation.

- One early measure of drought severity is the **Palmer Drought Severity Index** based on temperature and rainfall amounts.

- The **North America boreal forest** is the largest remaining intact forest in the world. Large tracks of forest can alter global climates by sequestering carbon lowering greenhouse gases in the atmosphere. Today, the North American and Russian boreal forests and the Brazilian rainforests are the only large, intact, green areas left. Preserving these forests will make a difference in balancing the gases causing global warming.

- Humans tend to alter the hydrologic cycle through many actions. Among these are urban development, cultivation of land for crops, raising of livestock, burning of fossil fuels, industrial processes that emit aerosols and greenhouse gases. Extensive areas of development alter water movement, infiltration into soils, and water quality. Extensive areas of cultivation also alter water movement, and this commonly results in increased erosion and water quality degradation.

Questions for Analysis

1. Why is Earth considered a "water planet"?
2. What are the functions of evaporation and transpiration in the hydrologic cycle?
3. What are the seven types of drought, and how do they differ?
4. What does the study of paleoclimatology add to our knowledge of global climate change?

5. Can parts of the hydrologic cycle be lost and still maintain the natural environment as we know it?
6. What effect does urbanization have on the hydrologic cycle?
7. How can urbanization continue without negative global consequences?

Further Reading

Baver, L. D., W.H. Gardner, and W. R. Gardner, 1972, *Soil Physics*, New York: John Wiley and Sons.

Egan, Timothy, 2005, *The Worst Hard Time: The Untold Story of Those Who Survived the Great American Dust Bowl*, Houghton Mifflin; Illustrated ed. (December 14, 2005) ISBN-10: 9780618346974

Pielou, E. C., 1998, *Fresh Water*, Chicago, Ill.: University of Chicago Press.

Steinbeck, John, 1939, *The Grapes of Wrath*; reprinted 2006, New York: Penguin.

Stream Corridor Restoration: Principles, Processes, and Practices, October 1998, by the Federal Interagency Stream Restoration Working Group (FISRWG) (15 federal agencies). www.nrcs.usda.gov/wps/portal/nrcs/detailfull/null/?cid=stelprdb1043244

Taiz, Lincoln and Eduardo Zeiger, *Companion to Plant Physiology*, 4th ed, http://4e/plantphys.net/

Weil, Raymond and Nyle Brady, 2017, *The Nature and Properties of Soils*, 15th ed, Pearson Education.

References

1 E. C. Pielou, 1998, *Fresh Water*, Chicago, Ill.: University of Chicago Press, p. 246.
2 Patrick Taylor, 2011, "What's causing the poles to warm faster than the rest of Earth?" NASA Langley Research Center in Hampton, VA, USA.
3 Federal Interagency Stream Restoration Working Group (FISRWG), 1998, *Stream Corridor Restoration: Principles, Processes, and Practices*, Springfield, Va.: (15 federal agencies), https://www.nrcs.usda.gov/wps/portal/nrcs/detailfull/null/?cid=stelprdb1043244
4 *Stream Corridor Restoration: Principles, Processes, and Practices*, 10/98, by the Federal Interagency Stream Restoration Working Group (FISRWG) (15 federal agencies). https://www.nrcs.usda.gov/wps/portal/nrcs/detailfull/null/?cid=stelprdb1043244 (accessed 2020, images converted to black and white).
5 National Aeronautics and Space Administration (NASA), 2007, *The Water Cycle*, January, http://earthobservatory.nasa.gov/library/Water/printall.php
6 H. H. Dixon and J. Joly, 1894, "On the ascent of sap," *Annals of Botany* 8, 468–470.
7 E. C. Pielou, 1998, *Fresh Water*, Chicago, Ill.: University of Chicago Press.
8 Ibid.
9 US Geological Survey, 2008, "The water cycle: water storage in the atmosphere," October, http://ga.water.usgs.gov/edu/watercycleatmosphere.html
10 NASA Earth Observatory, 2020, "Extreme rain douses fires, causes floods in Australia," https://earthobservatory.nasa.gov/images/146284/extreme-rain-douses-fires-causes-floods-in-australia
11 Emma Charlton, 2020, "Locusts are putting 5 million people at risk of starvation – and that's without COVID-19," World Economic Forum, June 26, https://www.weforum.org/agenda/2020/06/locusts-africa-hunger-famine-covid-19/
12 Abrahm Lustgarten, 2020, "How climate change is contributing to skyrocketing rates of infectious disease," *ProPublica*, https://www.propublica.org/article/climate-infectious-diseases

13 David Wallace-Wells, 2020, "Global warming is melting our sense of time," *New York Magazine, Intelligencer*, https://nymag.com/intelligencer/2020/06/global-warming-is-melting-our-sense-of-time.html

14 Evelyn C. Pielou, 1991, *After the Ice Age, the Return of Life to Glaciated North America*, Chicago and London: University of Chicago Press, p. 15.

15 NOAA National Centers for Environmental Information, 2020, *State of the Climate: Global Climate Report for Annual 2019*, January, https://www.ncdc.noaa.gov/sotc/global/201913 (accessed 2020).

16 Adam Sobel, quoted in Marshall Shepherd, 2016, "Water vapor vs carbon dioxide: which 'wins' in climate warming?" *Forbes*, https://www.forbes.com/sites/marshallshepherd/2016/06/20/water-vapor-vs-carbon-dioxide-which-wins-in-climate-warming/#556e7f893238 (accessed 2020).

17 American Chemical Society (ACS) Climate Science Working Group, "It's water vapor, not the CO_2," *ACS Climate Science Toolkit*, https://www.acs.org/content/acs/en/climatescience/climatesciencenarratives/its-water-vapor-not-the-co2.html (accessed 2020).

18 NASA, "NASA's AIM spots first Arctic noctilucent clouds of the season," https://www.nasa.gov/feature/goddard/2020/nasa-aim-spots-first-arctic-noctilucent-clouds-of-the-season-polar-mesospheric/

19 E. G. Nisbet, M. R. Manning, E. J. Dlugokencky et al., 2019, "Very strong atmospheric methane growth in the 4 years 2014–2017: Implications for the Paris Agreement," *Global Biogeochemical Cycles* 33, 318–342, https://doi.org/10.1029/2018GB006009 (accessed 2020).

20 National Climate and Data Center, 2020, "Greenhouse gases," https://www.ncdc.noaa.gov/monitoring-references/faq/greenhouse-gases.php#cfc

21 Benjamin Abbott, Kevin Bishop, Jay Zarnetske et al., 2019, "A water cycle for the Anthropocene," *Hydrological Processes* 33. 10.1002/hyp.13544, https://www.researchgate.net/publication/334117602_A_water_cycle_for_the_Anthropocene/citation/download (PDF downloaded 2020).

22 The Water Footprint Network (WFN), 2020, "What is a water footprint?", *Water Footprint Assessment Manual: Setting the Global Standard*, https://waterfootprint.org/en/water-footprint/what-is-water-footprint/ (PDF downloaded 2020).

23 National Oceanic and Atmospheric Administration (NOAA), "What is El Niño?" https://www.pmel.noaa.gov/elnino/what-is-el-nino

24 Bob Henson and Kevin E. Trenberth, 1998, *Children of the Tropics: El Niño and La Niña*. From the Learning about Science Easily and Readily Series (LASERS).

25 C. E. J. Jackson, C. E. O. Jackson, P. L. Jackson, and C. D. Ahrens, 2012, *Meteorology Today: An Introduction to Weather, Climate, and the Environment*, Canada: Nelson Education.

26 S. C. Herring, N. Christidis, A. Hoell, M. P. Hoerling, and P. A. Stott, eds., 2020, "Explaining extreme events of 2018 from a climate perspective," *Bulletin of the American Meteorological Society* 101, 1, S1–S128, doi:10.1175/BAMS-ExplainingExtremeEvents2018.1 http://journals.ametsoc.org/doi/pdf/10.1175/BAMS-D-19-0172.1 (accessed 2020).

27 US Geological Survey, 2007, "Flood hazards – a national threat," January, http://pubs.usgs.gov/fs/2006/3026/2006–3026.pdf (accessed 2020).

28 NOAA National Severe Storms Laboratory (NSSL), 2020, *Severe Weather 101*, https://www.nssl.noaa.gov/education/svrwx101/floods/types/ (accessed 2020).

29 L. Gwen Chen, Jon Gottschalck, Adam Hartman et al., 2019, "Flash drought characteristics based on U.S. drought monitor," *Atmosphere* 10, 498, https://doi.org/10.3390/atmos10090498 (accessed 2020).

30 American Meteorological Society, "Glossary of meteorology," https://glossary.ametsoc.org/wiki/Welcome (accessed 2020).

31 US National Weather Service, 2007, "What is meant by the term Drought?" January http://www.wrh.noaa.gov/fgz/science/drought.php?wfo=fgz (accessed 2020).

32 CNN, 2003, "Silently, starvation stalks millions in Africa," January 25, http://www.cnn.com/2003/WORLD/africa/01/25/africa.famine/index.html (accessed 2020).

33 ABC News, "Atlanta suffers as southeast drought continues," February 10, https://abcnews.go.com/GMA/story?id=3730145&page=1 (accessed 2020).

34 NASA, 2007, "Drought and vegetation monitoring," January http://earthobservatory.nasa.gov/Drought (accessed archived article 2020).

35 NOAA, 2006, "US drought highlights," www.ncdc.noaa.gov/sotc/drought/200605 (accessed 2020).

36 Rob Gutro, 2020, "NASA observes large saharan dust plume over atlantic ocean," https://www.nasa.gov/feature/goddard/2020/nasa-observes-large-saharan-dust-plume-over-atlantic-ocean (accessed 2020).

37 David Herring, 2000, "Dry times in North America," NASA, September 25, http://earthobservatory.nasa.gov/Study/NAmerDrought/NAmer_drought_2.html (accessed 2020).

38 Kevin E. Trenberth and Christian J. Guillemot, 1996, "Physical processes involved in the 1988 drought and 1993 floods in North America," *Journal of Climate* 9, 1288–1298.

39 NASA, 2020, "Groundwater and soil moisture conditions from GRACE-FO data assimilation for the contiguous U.S. and global land," https://nasagrace.unl.edu/

40 NASA Jet Propulsion Laboratory, 2020, "GRACE Tellus, Gravity Rcovery and Climate Experiment," https://grace.jpl.nasa.gov/mission/grace/

41 NASA Jet Propulsion Laboratory, 2020, "GRACE Mission, 15 years of watching water on Earth," https://grace.jpl.nasa.gov/news/89/grace-mission-15-years-of-watching-water-on-earth/

42 NASA Jet Propulsion Laboratory, 2020, "GRACE-FO," https://gracefo.jpl.nasa.gov/mission/overview/

43 National Drought Mitigation Center University of Nebraska-Lincoln, https://drought.unl.edu/droughtmonitoring/MonitoringHome.aspx (accessed 2020).

44 NOAA, National Center for Environmental Information, 2020, "North American drought monitor," https://www.ncdc.noaa.gov/temp-and-precip/drought/nadm/

45 Stephanie Pappas, 2012, "The worst droughts in U.S. history," July 25, https://www.livescience.com/21844-worst-droughts-in-u-s-history.html (accessed 2020).

46 John Steinbeck, 1939, *The Grapes of Wrath*; repr. 2006, New York: Penguin.

47 NOAA, 2006, "NOAA paleoclimatology global warming: the story," December, http://www.ncdc.noaa.gov/paleo/globalwarming/home.html

48 Scott Weidensaul, 2007, "The last stand," *Nature Conservancy Magazine* 57, 2.

49 NASA, 2007, "Drought and vegetation monitoring," January, http://earthobservatory.nasa.gov/Drought/

50 ISAT380E, *Basic Background of Kenya*, https://sites.google.com/site/isat380ekenya/home/basic-background-of-kenya (accessed 2020).

51 ISAT380E, "Water crisis in Kenya vision 2030," https://sites.google.com/site/isat380ekenya/home/current-systems/vision-2030 (accessed 2020).

52 Nicholas Casey, 2017, "In Peru's deserts, melting glaciers are a godsend (until they're gone)," *New York Times Magazine*, https://www.nytimes.com/2017/11/26/world/americas/peru-climate-change.html?smid=pl-share (accessed 2020).

4 Water Quality

A friend of mine, a child of the Delta (an area along the Mississippi River in Arkansas), tells the story of a first-grade art lesson. The teacher told the class to draw a picture of a stream. My friend took crayons in hand and rendered his first-grade version of the stream near his home. The water was brown. His teacher scoffed at his effort and holding it up for the class to see said, "Everyone knows water is blue." He remembers to this day his frustration, because in all his seven years, he had never seen blue water. [1]

Karrie L. Pennington, 'Surface water quality in the delta of Mississippi'

Chapter Outline

4.1 Introduction

When we discuss water quality in today's world, the term "quality" is relative. First, let us define water quality as the chemical, physical, and biological properties of water that are influenced by geology, climate, local environment, and people. This is one of the most important things to realize and remember, water quality is not only about water, but also about water as part of an ecosystem and human use systems. The very nature of water depends on what it is reacting with. Whether it is in an ocean, lake, wetland, soil, sewage treatment plant, or an industrial plant, water reacts to its environment and develops properties that reflect that environment. There are other definitions for water quality, but this one will suffice. Quality is the relative term because "the water quality necessary is related to its intended use".

It is common to designate an intended use for water, and then decide which chemical, physical, or biological properties are suitable for defining that use. Values for these properties are usually compared to a threshold or guideline value determined by a reliable source, such as the World Health Organization (WHO) drinking water guidelines (available on their website) [2]. A drinking water standard is, understandably, the most stringent. For instance, water we drink must have better quality than water used to irrigate crops or lawns. Water for irrigation could have biological contaminants that are not suitable for drinking but are not harmful to a crop. Golf courses in Tucson, Arizona, for example, are irrigated with "gray water" – collected from baths, showers, and laundry water, and used separately from drinking water. Lake and stream water often contain microbes and other organisms not healthy to drink, but not necessarily harmful to fish or wildlife.

Most water is not drinkable by humans without treatment, but most water is not toxic, either. Instead, most freshwater found in nature is somewhere in between. Worldwide, we define water's quality relative to its intended use; therefore, water quality is relative.

An important part of designating water use is by its location on Earth. Is it groundwater or surface water? If surface water, is it located in a lake, reservoir, stream, river, ocean, or somewhere else? Is it located in a mountainous area, valley, floodplain, near a large human population, or isolated from most human activity? Water use is influenced by the number of people who need to use it and in what manner. Will a river be primarily used by fish and wildlife with occasional human contact, or will that river provide drinking water for a major population center?

The US Environmental Protection Agency (USEPA), for example, works to enforce the Clean Water Act by designating states and authorized Native American tribes to determine the use of natural waters. These include water for public water supply, protection of shellfish, fish, wildlife, recreation, agriculture, industry, and navigation. Factors to consider are the physical, biological, and chemical properties of the waterbody, local geographic setting, scenic values, and economic considerations. Criteria must also be adopted to protect water quality for the intended use. Water quality criteria fall under the general headings of human health, aquatic life, biological, nutrients, microbial pathogens, and sediments.

In Depth   The Ganges River: A River of Many Uses

The Ganges River in northern India runs about 1,600 miles from the Gangotri Glacier in the Himalayan Mountains to the Bay of Bengal. Its river basin (watershed) is home to 400 million people. Considered to be a Holy River by Hindus, Buddhists, and Jain (see Figure 4.1), pilgrims come by the millions to immerse themselves in its water, to obtain forgiveness and immortality. When Hindustani people die, many have their bodies brought to the shores of the river to be cremated and their ashes sprinkled in the river. Many believe this will guarantee eternal life. Unfortunately, not even a holy river can support the needs of more than a billion people. The Ganges is polluted with raw sewage, industrial waste, trash, plastics from offerings to the river, and partially cremated human bodies. Even though it cannot meet the water quality standards for these uses, it is used for bathing, daily household chores like clothes and dish washing, irrigation, for drinking water and wallowing by cattle, hydropower, and as a human drinking water source [3]. How can better water quality be regained in this river? How can it be protected? The Central Pollution Control Board (CPCB) of India [4] has this unenviable task, and under the 1986 Environmental Protection Act designated appropriate uses, set criteria, and developed a plan for a water quality monitoring system.

By 2021, 35 years have not been enough to restore the Ganges River. It has been disheartening to see how little they have been able to progress toward their goals. Politicians in 2021 were still promising to clean up the Ganges [5]. The CPCB annual report for 2017–18 lists major activities as pollution inventory, water quality monitoring, and the strengthening of environmental regulation. Their inventory reports that of the 1,343 grossly polluting industries, 1,109 were checked, only 380 met a complying designation, 508 were noncomplying, and 221 were either nonoperational or closed. Treatment plants did not fare well either. Common wastewater treatment (non-sewage) had only 1 adequate rating of 8 checked. Sewage (human waste) treatment plants had adequate ratings in 28 of 68 inspected. The CPCB also had serious concerns about the analytical laboratories doing water quality sampling and analysis, and planned to address their concerns [6].

Fig 4.1 The Ganges River serves many purposes and is heavily and detrimentally influenced by those uses. Varanasi, one of the oldest continually inhabited cities in the world, is situated on the bank of the River Ganges, Uttar Pradesh, India (Photograph by J. M. Suarez available at http://commons.wikimedia.org/wiki/File:Ganges_river_at_Varanasi_2008.jpeg)

Efforts were, as one would expect, more successful in some sites on the river than in others due, according to one report [7], primarily to poor use of technical expertise, poor overall environmental planning, lack of support from religious groups, and most unfortunately, corruption. Environmentalists working to restore the river lamented that their government, which calls itself an economic superpower, shows indifference and neglect when it comes to restoring the country's rivers [8]. This is not a problem limited to India, environmentalists in many countries, including the United States, find themselves expressing the same concerns about critical environmental issues.

The most interesting and perhaps the best demonstration of what is possible in river restoration is the recovery of the Ganges during the 2020 COVID-19 pandemic. During that time, 1.3 billion people were told to stay home, industries were shut down, tourists stayed away, religious celebrations were halted, and the river received far fewer human inputs. The overall water quality improved, especially in industrial areas. Water quality standards for use designations, fisheries health, and bathing, were met in much of the river, and this had not happened in decades. Other factors such as high snowfall melting in the summer heat, reduced irrigation water use, and above average rainfall helped to dilute and transport pollutants, but stopping the input of major pollutants was critical. These authors, knowing that the country could not be locked down indefinitely, concluded with a plea to see this amazing temporary river improvement as an "absolute way towards restoring river ecosystems and the environment" [9].

One other note of caution comes from scientists observing the gradual shrinking of the Earth's glaciers. Glacier shrinking is a clear sign that our climate is warming [10]. The Gangotri Glacier, which supplies most of the water for the Ganges River in dry months, has been receding at an alarming rate of 22 meters (72 feet) per year [11]. Continued intensified human use of the river, combined with diminishing flows, can only bring more human and ecosystem damage.

4.2 The Chemistry of Water

4.2.1 The Molecule

The chemistry of water is important to understand when learning about water quality. Water is an amazing substance, period. Remember from Chapter 1 that water exists in all three states – liquid, solid, and gas – at normal temperatures (0 °C to 100 °C, or 32 °F to 212 °F). It is the only common, pure substance on Earth with this property, and it all starts with its molecular chemistry.

Water has a small **molecular weight** (the sum of the atomic weights of the atoms that form the molecule). We know that water is made of two atoms of hydrogen (atomic weight 1), and one of oxygen (atomic weight 16). This totals to a molecular weight of 18: $(2 \times 1) + 16 = 18$. Let's compare that to another simple molecule, H_2S (hydrogen sulfide) – the rotten egg smelling gas from high school chemistry. Again, we have two hydrogen atoms and now one sulfur atom (atomic weight 34) which replaces the oxygen atom. The molecular weight of H_2S is $(2 \times 1) + 34 = 36$. This molecular weight is twice that of water, yet water is a liquid and hydrogen sulfide is a gas at the same temperatures. Molecules this small are usually gases at ambient (surrounding) temperatures. So, why is water different, i.e., not a gas?

Molecular structure (how the molecules are arranged, why, and the electronegativity of the central atom) is the key. **Electronegativity**, in chemistry, is the ability of an atom to attract electrons to itself. Linus Pauling was the first to develop a commonly used measure of electronegativity. He found that fluorine was the most electronegative element and assigned it an arbitrary value of 4 (today the value is 3.98) with all other elements being less than 4.

Basically, the difference between the electronegativity of two elements determines what type of bond it will form, and the shape of the resulting molecule. When atoms with an electronegativity difference of less than 0.5 are joined, the bond that is formed is a **nonpolar covalent bond**, in which the electrons are shared by both atoms. When atoms with an electronegativity difference between 2 and 4 are joined, the bond that is formed is an **ionic bond**, in which the more electronegative element has a negative charge, and the less electronegative element has a positive charge. When the electronegativity difference is between 0 and 2, the more electronegative element attracts the one shared more strongly, but not strongly enough to remove the electrons completely to form an ionic compound. The electrons are shared unequally, that is, there is an unsymmetrical distribution of electrons between the bonded atoms. These bonds are called **polar covalent bonds.** The more electronegative atom has a **partial negative charge** because the electrons spend more time closer to that atom, while the less electronegative atom has a **partial positive charge**. This is because the electrons are partly (but not completely) pulled away from that atom.

Hydrogen sulfide (H_2S) has a bent molecular structure, much like water. The sulfur has an electronegativity of 2.5, and hydrogen has an electronegativity of 2.1, a difference of 0.4, making them close enough together that there is no polarity, and the easily broken hydrogen bond requires only normal ambient temperature heat to make it a gas at normal room temperatures.

Table 4.1 Some important properties of water discussed in this chapter

Property	Value	Value/1000
Density (d) liquid	1000 kg/m^3	1 g/cm^3
Density (d) solid	917 kg/m^3	0.917 g/cm^3
Specific heat capacity (Cp) (liquid)	4186 J/(kg·K)	4.186 J/g·°C
Specific heat capacity (Cp) (solid)	2060 J/(kg·K)	2.06 J/g·°C
Specific heat capacity (Cp) (vapor)	1890 J/(kg·K)	1.89 J/g·°C
pH neutral	7.0	
Surface tension at 20 °C	0.0728 N/m	

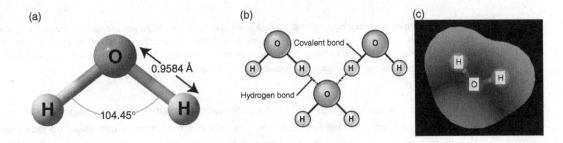

Fig 4.2 (a) Water molecule, note that the 0.9584 Angstroms is the bond length between oxygen and hydrogen and the 104.45 degrees of the angle. (b) Water molecules showing the weak hydrogen bonding (dashed lines) to each other. Covalent bonds, illustrated by bold solid lines, are strong. (c) A 3D model of a water molecule. Electrons near the positive H end are loosely held and can spread out, while negative O holds the electrons closer together, creating the bent polar molecular structure

Chemists call molecules with two poles a **polar** or **dipole molecule**, because it has two poles, one negative (due to oxygen), and the other positive (due to each hydrogen atom). Water is a polar molecule. The oxygen electronegativity, 3.5, is stronger than hydrogen's of 2.1, a difference of 1.4, which is greater than 0.5. This makes the overall molecule polar. Polarity causes the water molecule to bend at a specific angle, 104.45 degrees, giving water its V-shaped structure. The positively charged pole of one water molecule is attracted to the negatively charged pole on another water molecule in solution.

The bonds formed are **hydrogen bonds**. These bonds are not as strong as the covalent (atoms which share a pair of electrons) bonds holding the oxygen and hydrogen together, and are easily and extremely quickly formed, broken, and reformed. Hydrogen bonding causes water molecules to act as units of several molecules rather than as independent, small molecules. Therefore, water "acts" bigger than it is. This accounts for many of water's unusual physical properties, including its changing nature over a wide temperature range (0 °C to 100 °C) (see Figure 4.2).

4.2.2 Physical–Chemical Properties of Water

Universal Solvent

What else do we know about water that relates to quality? (see Table 4.1) Water is called the **universal solvent** because so many substances – from salts to powdered milk – dissolve in it. These substances

(a)   (b)

Fig 4.3 (a) Morning dew on a hydrophobic banana plant leaf. Notice how the water droplets appear to be reaching for each other. This is an example of cohesion, the attraction of water molecules to each other. (b) These water droplets cling to the surface of the pine needles. This is an example of adhesion, the attraction of water to solid surfaces (Photograph (a) by Karrie Pennington, photograph (b) courtesy of USGS)

are hydrophilic, water loving, or tend to mix with water. Of course, many substances do not dissolve in water, and thankfully, that list includes many metals and the organics that make up skin. Substances that do not dissolve in water are called **hydrophobic**; for instance, fats are hydrophobic. Many plant leaves have hydrophobic waxy cuticles that provide a protective layer and act as a water conservation mechanism. Would a magnified drop of water on this surface be perfectly round? Perhaps another property of water to be discussed later will help answer that question (see Figure 4.3).

Water is seldom pure in nature, and is usually a mix of dissolved and suspended materials. Even rainwater contains dissolved materials, dust particles, aerosols, and other items it collects from the air, or from soil and geologic materials. Water's solvent property is also due to its polar nature. Water molecules isolate and surround individual ions in solution by drawing them to either the negative or positive poles. It's as if the ions wear a "water coat" that keeps them apart; therefore, dissolved.

Density

Density is simply weight per unit volume and is usually expressed in kg per cubic meter or grams per cubic centimeter (kg/m^3, g/cm^3). Perhaps you had a teacher who mixed oil and water, allowed them to separate, and then asked "Why does oil float on water? You brilliantly announced that oil is less dense or maybe lighter than water. Indeed, water has a density of 1,000 kg/m^3 and most vegetable oils have a density in the 800 to 920 kg/m^3 range. This property can be especially important in water quality – consider oil spills, for example.

Oil spills can occur when storms damage storage tanks in coastal areas, from sinking cargo ships or tankers, failures on drilling rigs, or from someone carelessly or illegally disposing of oil waste near a storm drain or creek. Such spills can be widespread or relatively small, but all have adverse consequences to water quality, wildlife, and the environment.

In 1989, the sinking of the *Exxon Valdez* was an ecological disaster [12]. The ship lost about 40 million liters (11 million gallons) of crude oil when it ran aground in Prince William Sound, Alaska. Sea otters, whales, some 10 million migratory birds, and a major fishing industry were threatened. No one was prepared to deal with such a large disaster in such a remote location accessible only by boat and helicopter. Tremendous efforts were made to contain and clean up the spill, and then

Fig 4.4 Deepwater Horizon,
A British Petroleum drilling rig
explosion and fire resulted in the
seafloor well gushing oil into the
coastal waters of the Gulf Coast of
the United States for 86 days. This
created the biggest oil spill in
US history
(Photograph by US Coast Guard)

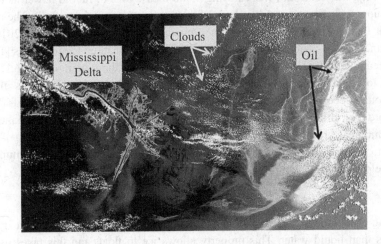

Fig 4.5 On April 20, 2010, an explosion on the Deepwater Horizon oil platform in the Gulf of Mexico killed 11 workers,
sunk the platform, and left its seafloor well gushing. This initiated a massive offshore oil spill – the largest in US history.
This view shows a massive oil slick floating southeast of the Mississippi Delta. The spill continued until July 15, 2010,
when the well was capped. It released 206 million gallons of oil, causing immense surface oil slicks and ecological
damage to the US coastline [13]
(This black and white version of a false-color Landsat 7 image courtesy of NASA)

later led the US Congress to pass the Oil Pollution Act of 1990. The Act strengthened US laws
requiring oil tankers to be adequately equipped, and to have improved communication systems.

Contingency plans and funds were also set up to deal more efficiently with future environmental disasters. These plans have been used in smaller spills but were sorely tested in April 2010 by
the Deepwater Horizon oil spill in the Gulf of Mexico (see Figure 4.4). Coastal ecological damage
from this event was still under restoration in 2021, and will take years to establish an acceptable
recovery to the marshes, the oyster industry, and wildlife [14].

Globally, satellite imagery is used to assess and police oil spills (see Figure 4.5). NASA's
Explore Earth site and the ESA's (European Space Agency) Earth Observation Program have

provided much more than images, and include data about the characteristics of actual oil slicks, technique effectiveness, and too much more to cover here. Explore their websites – since these agencies do vital and diverse work, and they may inspire you in your future decisions and direction of study [15] [16].

Think About It: Oil Spills

The 2020 oil spill on the shores of the small island nation of Mauritius, located east of Madagascar in the Indian Ocean, was devastating to its economy (the loss of tourism that is a major source of income), its food security (which is highly ocean dependent), and the health of its people and their environment. Picture a beautiful island known for its coral reefs and home to several **endemic species**, plants and animals that live nowhere else – a pink pigeon, recently saved from extinction, and a blue-tailed day gecko, a pollinator of a flower so rare that only 250 plants remain. Visualize its white sand beaches, lovely lagoons, and all the marine wildlife and imagine how this thick, oily muck damages your picture.

A cargo ship, the *Wakashio*, was carrying 4,000 metric tons of fuel to power its engines (in comparison, super tankers can carry hundreds of thousands of metric tons of oil). The oil spill released more oil than the combined total from every tanker spill documented in 2019. The Mauritian government urged residents to stay home and leave the clean up to authorities [17], but residents organized themselves and did everything from assembling home-made oil booms (floating barriers to contain and absorb the toxic spill), to physically cleaning up as much of the spill as they could. "People have realized that they need to take things into their hands. We are here to protect our fauna and flora," said Ashok Subron, an environmental activist at Mahebourg, one of the worst-hit areas [18].

Consider, what would you do if a disaster like this happened to your community? Is citizen action a help or a hindrance in this type of situation? Explain your thoughts.

Returning back to density and water structure, notice in Table 4.1 that the density of solid water (ice) is less than liquid water. This property allows ice to float, and this takes us back to molecular structure. Solid water has a defined crystalline structure, rather than the loose association of hydrogen ions in liquid water. This crystalline structure weighs less per unit volume because it has spaces (see Figure 4.6), hence it floats. The ability of ice to float is another property that has tremendous implications in our environment.

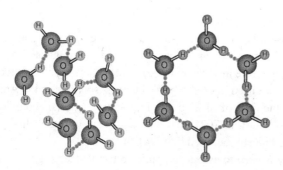

Fig 4.6 Liquid water with no actual interstitial (in-between) spaces because the hydrogen bonds are constantly forming and breaking. When frozen, water has a definite crystalline structure with spaces that make it light enough to float

Think About It: Ice

What if ice did not float? Could life as we know it exist on Earth? Why or why not?

Specific Heat Capacity

Specific heat capacity is the measure of how much heat a substance can store. It is measured as the amount of heat energy needed to raise the temperature of a given amount of substance 1 °C or 1 °K (1 degree Kelvin). Water stores a lot of heat before it turns into a gas. Environmentally, the oceans are giant heat-absorbing bodies that help regulate the Earth's energy balance. The Sun radiates energy onto our planet, and the atmosphere absorbs some of it. However, oceans absorb a great deal of that heat energy, and move it around in ocean currents. Oceans also release heat energy more slowly than the atmosphere. (Heat energy can travel thousands of kilometers (miles) in ocean currents before it is re-released into the atmosphere.) The net effect is that oceans moderate heat energy on the Earth. This plays a critical role in controlling global climate change, particularly the rate of warming. Table 4.2 compares the specific heat of several common materials. Notice that water is given in all three physical states.

Think About It: Water and Thermal Energy

How would the Earth's surface look, and how would weather patterns change, if water did not have a high specific heat capacity? Why do you think liquid water absorbs the most heat?

Table 4.2 Specific heat capacity of some common materials

Substance	Phase	Cp (J/(g·K))	Cp (J/(mol·K))
Air (sea level, dry, 0 °C)	Gas	1.0	29
Aluminum	Solid	0.9	24
Diamond	Solid	0.5	6.1
Nitrogen	Gas	1.04	29.1
Gasoline	Liquid	2.2	228
Hydrogen	Gas	14.3	28.8
Oxygen	Gas	0.9	29.4
Water	Gas (100 °C)	2.080	37.47
	Liquid (25 °C)	4.1813	75.327
	Solid (0 °C)	2.114	38.09

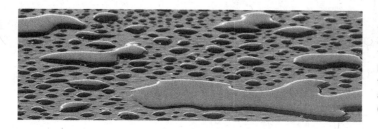

Fig 4.7 Water droplets cling together via cohesion, the attraction of water molecules to each other. You can clearly see how droplets have moved together. They cling to the surface due to adhesion, the attraction of water to surfaces (Photograph courtesy of USGS)

Cohesion and Adhesion

Water molecules are attracted to each other, and this effect is called **cohesion** (see Figure 4.7). This attraction allows the molecules to form a film that is held together by surface tension. Adhesion is the sticking of water molecules to other surfaces. Drop some water onto a non-porous surface and watch the droplets mound up. Have you ever watched a water strider skip across the surface of a pond? Have you noticed that water in a straw has a concave top? Ever tried to float a paper clip on a water surface or throw a flower onto water – they float. These are all examples of surface tension, the result of cohesion and adhesion.

Biologically, surface tension plays a role in transporting water through plants, remember capillarity from Chapter 1? It is especially important in soil water movement through soil pores and in soil water chemistry. Soil clays have ions as parts to their crystalline structures, and water is essential to ion reactions in the soil solution, as well as to biological reactions. It plays an essential role in nutrient movement and a plant root's ability to take up the nutrients from the soil. In terms of water quality, the soil acts as a filter holding back nutrients and pollutants that could enter groundwater. The effectiveness of soil as a filter depends largely on soil texture (the relative percentages of sand, silt, and clay) and the type of clay. Water moves quickly through the large soil pores in sandy soils, and allows nitrates (NO_3) to reach groundwater. Nitrates can cause hemoglobinemia in infants, and this is discussed further in Chapter 11.

pH

One of the most important properties of a solution is its **pH** because it determines the types of chemical reactions that will occur in solution. pH is a unitless number simply written from 0 to 14. pH is a measure of the acid (pH < 7.0) or basic (pH > 7.0) nature of an **aqueous** (contains water) solution. Hydrogen $(H)^+$ ions make a solution acid and hydroxyl $(OH)^-$ make it basic; therefore, when measuring pH, we are measuring the relative amounts of free hydrogen and hydroxyl ions.

Pure water is neutral (neither acid nor base) with a pH of 7.0. However, we know that water in nature is not pure, but contains dissolved substances. Rain contains both carbon dioxide (CO_2) and sulfur dioxide, which make it an acid with a pH reading of about 5.6. This is not to be confused with serious acid precipitation, with pH 4.1–4.4, caused by emissions of sulfur oxides and nitrogen oxides from industrial, power generation, and combustion engines (planes, trains, automobiles, buses, motorcycles, etc.). These oxides form sulfuric or nitric acid. The more acid in precipitation, the more harm it can do to everything – from lakes and trees to buildings and monuments.

In Depth   **pH and Beer**

Brewing is one of our oldest chemical industries. Danish biochemist Søren Sørenson was the director of Carlsberg Laboratories in Copenhagen, which was supported by the Carlsberg Brewing Company. Around 1909, Sørenson developed a system of equations to describe pH, and the pH scale. He needed a convenient way to express acidity after he developed methods to measure pH (Table 4.3). Sørenson worked in collaboration with his wife, Margrethe Høyrup Sørenson, who was also an accomplished scientist, on this and many other projects [19].

Table 4.3 pH values of some common liquids to help understand the ranges of acid or basic substances

Substance	pH value
Battery acid	<1
Gastric juice	2
Lemon juice	2.2
Vinegar	3
Orange juice	3–4
Rain	5.6
Milk	6.5
Distilled water	7
Blood	7.34–7.45
Sea water	8
Egg whites	8.4
Normal soap	9–10
Ammonia	11.4
Bleach	12.5
Household lye	13.6

Specific Conductance

Specific conductance measures the ability of a solution to conduct electricity. It is measured, or corrected, to a temperature of 25 °C (77 °F) and expressed as microsiemens per centimeter (μS/cm). Conductance is highly temperature dependent. Specific conductance provides a standard temperature so that values can be compared, both temporally (time) and spatially (location). Pure water will not conduct electricity, but a solution of ions does. Remember, there is no such thing as pure water in nature.

Salts dissolved in water separate into their component ions. The more ions in solution, the better it conducts electricity. For instance, sodium chloride, also known as table salt (NaCl), ionizes to Na^+ ions and Cl^- ions. Notice that ions are charged, giving them the ability to transmit electricity. Specific conductance provides an indirect measure of the salt content (**total dissolved salts,** or **TDS**) of a solution. Total dissolved solids are often approximated by assuming that the salt content is ~64–70 percent of conductance. For example, if specific conductance is 100 μS/cm, then TDS is 70 percent of 100, or 70 mg/L. The calculation of TDS does not tell which ions are in solution; it only approximates the concentration.

The types of ions present are particularly important for human consumption in drinking water, and also influence soil stability. For instance, humans cannot tolerate excess Cl^- even though it is essential to fluid and pH regulation in normal amounts. Excess levels cause kidneys to have trouble getting rid of acids (a condition called hyperchloremia). Soil has a similar problem with excess amounts of Na^+ in the soil solution. Sodium disperse the smectitic (expanding) soil clays. Dispersion decreases water movement in soils and soil stability for structural (houses, levees, etc.) support. You have probably seen homes with cracked concrete foundations, brick and mortar separation, or even interior splitting of drywall all from building on expansive clay soils without proper precautions. People, animals, soils, and most plants, cannot tolerate large amounts of salt in water. Salt content can make a big difference in the suitability of water for drinking and growing plants.

Dissolved Oxygen

Oxygen that diffuses from the air and into water is called **dissolved oxygen (DO)**. It is essential to living organisms such as fish, zooplankton, many macroinvertebrates, and microorganisms. DO concentrations in natural waters vary with water temperature, time of day, salinity, and seasonally. Theoretically, DO can be between zero and 18 milligrams per liter (mg/L).

Most aquatic systems require a minimum DO in the 5–6 mg/L range to support living organisms. Dissolved oxygen levels vary depending on the type of waterbody. By contrast, DO requirements vary with the types of organisms present. Organic materials – from natural vegetative decay or wastewater disposal inputs – are consumed by microbes. These microbes use the DO to break down organic matter into simpler compounds for energy (food).

The concentration of biodegradable organic materials in water is measured indirectly as **biological oxygen demand (BOD).** This is the amount of DO needed by aerobic microbes to degrade organic matter, and is used as an indicator of pollution [20]. Pristine rivers may have a BOD of 1 mg/L, but even well-treated tertiary treatment (wastewater) can be at 20 mg/L. Poorly treated wastewater can be as high as 200–600 mg/L. BOD is routinely used to test for organic enrichment from wastewater sources. Table 4.4 provides a frame of reference. In the US, the Clean Water Act prohibits releasing wastewater into a body of water if there is inadequate water to properly dilute the expected organic load. This becomes a public health and fisheries problem when water from streams and rivers is pumped for irrigation in low rainfall seasons, and streamflow is reduced.

Chemical oxygen demand (COD), expressed in milligrams of oxygen consumed per liter of solution (mg/L), measures organic materials indirectly. Potassium dichromate, in acid conditions, oxidizes organic matter and is itself reduced in the process. The amount of reduction is an estimate of organic materials, and is usually 1.3 to 1.5 times higher than BOD.

Table 4.4 Biological oxygen demand (BOD) values corresponding to different water quality designations

BOD (mg/L)	Water quality
1–2	Very good
3–5	Moderate
6–9	Fairly polluted
10+	Very polluted

Temperature

Cold water holds more DO than warm water because colder water molecules are less active (less kinetic energy). This inhibits DO loss. An important point to remember is that water temperature strongly influences its ability to hold oxygen. For instance, at 133 Pa (760 mmHg) and 0 °C (32 °F), water can hold 14.6 mg/L of DO. At 30 °C (86 °F), the concentration of DO drops to 7.5 mg/L – nearly half. This is a dramatic change. Notice that atmospheric pressure is also provided above. These same temperatures, at a higher atmospheric pressure, e.g., 139 Pa (795 mmHg), would hold 15.3 and 7.9 mg/L DO, respectively [21].

Water temperature varies much the same way as atmospheric temperature. However, water's high specific heat capacity (discussed previously) provides some buffering and prevents rapid temperature change. Water temperatures have diurnal (daily) changes that are colder at night and warmer during the day. Temperatures also show seasonal variations. As you would expect, most living organisms have temperature ranges in which they thrive. Wide variations can be harmful, if not deadly, partially due to the interaction of temperature and DO.

Turbidity

Turbidity is a measurement of the clarity of water. Streams and rivers can be very turbid, like the "Mighty Muddy Mississippi River," or clear like a mountain stream. Turbidity can be caused by phytoplankton or by sediments suspended in water. Phytoplankton usually produce a green color while sediment creates shades of brown. The area of northwestern Mississippi and eastern Arkansas, known as the Delta, has sediment-laden waters (see Figure 4.8). Locals sometimes call this "water too thick to drink, but too thin to walk on." The Chinese have a saying that roughly translates – "The Yellow River will run clear when hell freezes over."

Perhaps the most significant biological impact of sediments is the reduction of sunlight in aquatic systems. Plants are an especially important part of the ecology of a waterbody and need sunlight for photosynthesis. Light-limited waters seldom have an active phytoplankton population, and high sediment-induced turbidity, at the same time. The ecology of a waterbody can be greatly altered by sediment. In addition to sediment, phytoplankton can contribute to turbidity in water. Generally, water that is cloudy brown in color is high in sediment, while green, turbid water contains phytoplankton and other growths of aquatic life (Figure 4.9).

4.3 Water Quality Failure

When water is no longer suitable for its designated use, there is a water quality failure. The failure can be natural or anthropogenic (caused or produced by humans) and is often a combination of both – a synergistic destruction. Pollution is anthropogenic since humans add the undesirable substances to water. Natural events that cause water quality failure, such as landslides, floods, erosion, and volcanoes, can harm water quality, but are not considered pollution. Mercury, on the other hand, fits both definitions. It enters aquatic systems naturally from the atmosphere but also through industrial pollution. Therefore, high levels of mercury can cause water quality failure both through natural processes and by human activity.

Pollution can be direct with obvious consequences: trash thrown out a car window is ugly and can contaminate a stream; toxic chemicals dumped from a factory cause a fish kill. Sometimes

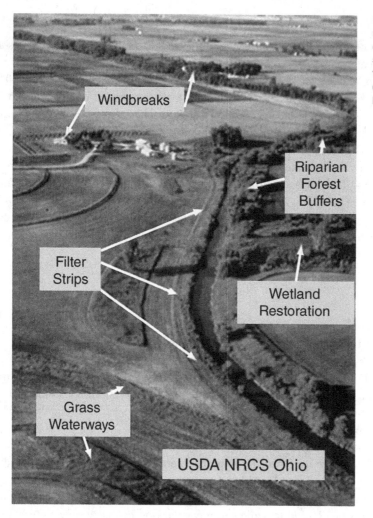

Fig 4.8 Examples of multiple best management practices (BMPs) that offer protection from non-point source pollution to the stream in this agricultural area of Ohio, US (Photograph courtesy of USDA NRCS)

Fig 4.9 Pictured on the left side you can see the Huang He (Yellow River), called the most sediment-laden river in the world, running full of sediment into its Delta in the Bo Hai Sea. The clouds of sediment spread along the coast and into the sea (Photograph courtesy of NASA)

the effects are not an obvious cause and effect relationship. When this happens, it is called a **pollution-induced failure**. For instance, low DO levels result from the decay of an overgrowth of vegetation (the result of fertilizers entering a waterbody from agriculture, yards, or golf courses). Pollution occurred with the fertilizers, but the effect of low DO was not noticed until the plants died and decayed, resulting in a fish kill.

The paths to water quality failure are, unfortunately, many and varied. Remember the discussion of the hydrologic cycle? The massive amounts of water, cycling on, in, and above the Earth, transport everything from soil particles to chemicals. The hydrologic cycle is a transportation mechanism that touches virtually every surface of the Earth. Government agencies consider transport mechanisms (how something entered the water) when regulating pollution sources to prevent water quality failure.

4.3.1 Point Source and Non-point Source Pollution

Where do pollutants originate? We divide them into point and non-point sources. **Point sources** of pollution are anything that can be identified as coming from a specific site – usually a pipe or an accident scene which released a harmful substance, such as an oil tanker spill, or an overturned 18-wheeler that spilled chemicals along a highway and then gets washed into a stream.

Non-point sources of pollution originate over larger geographic areas, such as fields, urban areas, or entire watersheds. Rain and snowmelt carry these diffuse (widespread) pollutants over and through the soil, and into waterbodies. Atmospheric deposition can also be considered non-point source pollution because it is difficult to pinpoint where a specific airborne constituent originated.

Point source and non-point source pollutants were defined by governments so that pollution sources could be identified and regulated. Point source locations, and responsibility, are relatively simple to identify. Non-point sources of pollution, on the other hand, are much more difficult to attribute to a single source. Non-point source pollutants exist over a broad area, can include several types of contaminants, and usually occur only after precipitation events. Specific identification and pollution control methods have been developed to cover both point and non-point sources of pollution.

Water samples from a point source can be analyzed, and the impacts of contaminants can be attributed to an owner. The USEPA issues National Pollutant Discharge Elimination System (NPDES) permits to individuals and companies that add point source pollutants to waters of the US. In 1972, the Clean Water Act authorized these permits to help improve the nation's water quality. Improvements have been enormous, with many waterways recovering from years of neglect.

For example, the summer of 2003 was a crisis time in Rhode Island's Narragansett Bay [22]. Fish and shellfish died by the millions because excess nitrogen, from wastewater treatment plants, caused a phytoplankton bloom that depleted DO in the bay as it decomposed. Remember the example of a pollution-induced response? The Narragansett Bay bloom was the direct result of excess nitrogen. Dissolved oxygen levels declined as an indirect effect of vegetation decomposition caused by microorganisms. Wind also contributed to the problem by rapidly mixing deoxygenated water with incoming surface water. This trapped the fish and increased the total kill – a destructive synergy of wind and point source pollutants.

Citizen action and legislation ultimately reduced wastewater treatment plant pollutants (a point source) into the bay. The Rhode Island Legislature passed a law calling for a 50 percent decrease in nitrogen inputs, from wastewater treatment facilities, by December 2008. The citizens of Rhode Island approved a US$10.5 million Clean Water and Open Space bond to help improve the efficiency of wastewater treatment facilities to comply with the new law.

Note that two significant actions were taken to solve this problem – a law was passed and funding was approved to achieve the law's goal. This is important to note. It is not enough simply to say that a practice is wrong and now it is also illegal; the means to achieve the goal must also be present. The people of Rhode Island put both legislation and money into action to solve a major water quality problem in their state.

Narragansett Bay has undergone significant ecological changes with pollutant reduction efforts and climate change. Decreased wastewater inputs have cut the nitrogen load by just over 50 percent and the waters are much clearer, beaches are open, and that is good. Some fisherman fear that crabs, lobsters, and shellfish suffered long-term impacts because there is less nitrogen for phytoplankton growth, and phytoplankton is the major food source for harvest. However, scientists at a recent conference explained that things are just not that simple, and we are seeing that such issues are complex. Rainfall in Rhode Island has increased nearly an inch every year since 1865, the waters of the Narragansett Bay are warming, and there is less wind to stir the mix to bring nutrients to the surface. Chlorophyll production, an indicator of aquatic plant growth, has decreased since the 1970s and is not related to wastewater treatment plant inputs. These things all influence the primary productivity of the Bay [23].

We've just discussed how some pollution solutions can be straightforward while others are extremely complicated, and problems continue. People continue to break the law and dump waste into lakes and rivers. Fines may be issued when culprits are identified, but these are not always effective deterrents. Worse yet, money cannot always fix what has been damaged. Stronger laws and larger penalties may help, but without enforcement, water quality improvements are limited.

4.3.2 Solutions, Best Management Practices

Section 319 of the US Clean Water Act provides the primary mechanisms and financing for dealing with non-point source pollution in the US.

This federal legislation provides money for management practices that address a particular non-point source pollution problem. These practices are collectively called **best management practices (BMPs)**. BMPs have been developed for all types of industries that produce non-point pollutants – such as forestry, agriculture, urban development, mining, and road building. These industries commonly disturb the Earth, remove native vegetation, and alter stream courses. Precipitation events can lead to accelerated erosion and movement of materials at these locations, and are often treated using "319 dollars" to fund BMPs. An important point to remember is that most of these BMPs are not fully funded. Instead, they are cost-shared, meaning that the local landowner must pay for some of the costs. A landowner who understands how and why a practice is needed, and shares in the costs and installation, helps to ensure the maintenance of BMPs and their long-term effectiveness.

Agricultural, silvicultural (cultivation of trees), and industrial BMPs all include restoration of riparian buffers (such as grasses, shrubs, and other vegetation along streams). This is a critical component because a buffer between a river and a disturbance activity slows the delivery of any sediment, organic materials, or other pollutants produced by that activity. Riparian buffers also allow time for nutrient recycling, pollutant processing, and sediment settling. Riparian areas often include adjacent wetlands, and provide another opportunity for processing (cleaning) surface water runoff before it enters a river or stream (see Figure 4.8).

Buffers around lakes and ponds are an additional and effective BMP. The US Department of Agriculture Natural Resources Conservation Service (NRCS) buffer program is called "Common

Sense Conservation," because the effectiveness is obvious and immediate. There are many kinds of buffers for many different situations. These include the following:

- riparian buffers
- filter strips (areas of close-growing plants)
- grassed waterways
- windbreaks (areas of trees to provide wind protection)
- living snow fences (shrubs and other vegetation to capture blowing snow)
- contour grass strip
- crosswind trap strips
- shallow water areas for wildlife
- field borders
- alley cropping (crops grown between rows of trees and/or shrubs)
- herbaceous (plants that die back in winter) wind barriers
- vegetative barriers

These buffers have many uses in addition to water quality protection. Some buffers provide food and habitat for waterfowl, quail, rabbits, birds, and other animals. Some are physical barriers against erosion and wind damage. Others protect us from snowdrifts on roads. Buffers are usually used in conjunction with other BMPs, such as nutrient management, sediment fences, terracing, and contour plowing. These maximize the effectiveness of all the BMPs used. Rather than **destructive synergy**, we now see **constructive synergy**. Buffers are common-sense solutions, because they do a great deal of good for relatively small investments of land and dollars [24].

There are many types of BMPs for wetlands, rivers, agriculture, industry, forestry management and timbering, urban planning, and construction of all types. In fact, if something needs to be done that influences land, water, or air, someone has probably developed a way to minimize environmental damage using BMPs. It seems logical to legislate the use of BMPs; however, the politics of personal choice too often supersede common sense.

4.3.3 Pollution and the Environment

Pollutants are usually classified by type: nutrients, sediments, chemical toxic substances, microbiological pathogens, oxygen-depleting organics, and heat. How do we recognize these pollutants, in a broad and historical context, as environmental concerns?

Water is essential to any civilization – an unnecessary statement coming from beings that are 60 to 70 percent water. There was a line in one of the (older author alert) *Star Trek: The Next Generation* episodes where a beautiful crystalline creature describes humans as "ugly bags of mostly water." Earth is not just a water planet; it contains billions of water beings who can live weeks without food, but only days without water. When humans were in the hunter–gatherer phase (discussed in Chapter 2), they had minimal impact on the environment primarily because their numbers were low, they were dispersed over a large land area, and they lacked the tools or knowledge for shaping nature. Humans began affecting their environment when they started congregating in one spot, planted crops, increased population densities, developed water delivery systems, and developed tools to support themselves.

Water quality became an issue with population growth. The Industrial Revolution of the late-1800s created plagues of water-related diseases. Human and animal wastes, and industrial pollution, created an urban environment of unsightly, often smelly waters. People assumed that lakes, rivers, and streams were self-cleaning (if the issue was considered at all). You put stuff in; it decomposed, dropped out of sight, or was washed downstream – perhaps even to the sea.

Water-related diseases were rampant, and urban centers became environmental disasters. Fortunately, conditions in some locations improved in the twentieth century as polluted water was transported away from drinking water sources by improved wastewater treatment methods. This was a tremendous achievement for the more-industrialized world.

World War II sparked the chemical industry. Toxic chemicals evolved for making explosives, gases to kill, and other weapons of war. This set up a post-war manufacturing industry in the 1950s and 1960s for fertilizers, pesticides, insecticides, herbicides, and fungicides. In addition, the chemical industry made possible antibiotics, vaccines, insect repellents, body lotions, and even perfumes. The range of these chemical mixtures extends to the deadly, the useful, the lifesaving, and perhaps, the frivolous. Agricultural chemicals have improved millions of lives through more abundant food supplies. Chemicals have helped eliminate and/or cure insect-borne and other water-related diseases, saving millions of lives. Mass-produced antibiotics and vaccines have eliminated many feared childhood diseases in treated populations.

Some of the good from chemicals, however, comes with a high price to the environment, especially to water quality. Safe drinking water is readily available to most people in the more-industrialized world. We can swim, fish, and play in many rivers and streams, although not all. More-industrialized countries have come a long way toward improved water quality. For the most part, we are motivated to protect water from point source and non-point source pollution.

Think About It: Laws and Human Behavior

We have enacted federal, state, provincial, and local laws. We have involved concerned citizens, created friends of rivers organizations, environmental groups, and other resources to improve water quality. However, there still are scoundrels who ignore laws, and do only what is financially expedient, regardless of environmental or human consequences. There are wide-ranging differences of opinion about the best uses of a water resource, or how best to protect a water source from pollution.

Labeling a country as civilized, advanced, or more-developed, unfortunately, does not mean that it is immune to water quality problems. It is, too often, quite the opposite. What are some of the major sources of pollution common to more-developed countries? Where you live, are the sources of pollution primarily individual, industrial, or agricultural?

Nutrients

Nutrients are one of the six types of pollutants mentioned above. A nutrient is either an element or compound (a mix of two or more elements) that is consumed by an organism to grow, repair itself, and create energy. Humans primarily consume the nutrients of carbon, hydrogen, nitrogen, oxygen, phosphorus, and sulfur. Nitrogen and phosphorus are the primary nutrients of plants, although they require 16 essential elements. Excess nitrogen and phosphorus, however, cause accelerated eutrophication by stimulating the plant growth and death cycle. Liebig's Law of the Minimum (discussed in Chapter 1) states that growth of an organism is not restricted by the total resources available, but by the scarcest resource. Research has shown that phosphorus is usually the limiting nutrient in freshwater aquatic systems, while nitrogen is the limiting nutrient in estuaries and oceans.

Phosphorus is the major plant nutrient that attaches to soil, principally through binding with calcium, magnesium, iron, and aluminum-type minerals. Phosphorus is readily available to plants in the dissolved form and increases aquatic plant growth. This plant growth is a good thing until it becomes excessive. Then the entire cycle of excess plant growth – decaying vegetation, decreased DO, and dead aquatic organisms – begins. Just as too much food causes human problems with obesity and accompanying poor health, too much plant food is also a problem.

In Depth   **Phosphorus**

The only significant conversion that phosphorus goes through is between organic and inorganic forms. The most significant form of phosphorus is orthophosphate ($H_2PO_4^{-1}$, HPO_4^{-2}, PO_4^{-3}). Looking at these formulas you can perhaps recognize that soil solution pH will determine the form of orthophosphate in that solution. The two most plant-available forms are $H_2PO_4^{-1}$, HPO_4^{-2} which occur about equally at pH 7.2. Bacteria are primarily responsible for the transformations between organic and inorganic forms.

Nitrogen (N) can also be a problem for water quality, if in excess. It is usually attached to the organic particulates in water, rather than soil sediments. High levels of nitrogen can become a water quality problem because nitrogen enhances plant and algae growth. Plant growth and decay both require DO, which reduces the amount of DO remaining for plants and fish. Algae can also block light from reaching stream bottoms and prevent photosynthesis in beneficial aquatic plants. Ammonia (NH_3) poses a serious threat to water quality because it consumes oxygen dissolved in the water, causing fish to suffocate. Ammonia is more likely to be present in waters already low in DO levels because the oxygen necessary for the normal conversion of ammonia to nitrate or nitrite is absent.

When nutrients are added to an otherwise balanced aquatic system, algal blooms can occur. Fortunately, phosphorus levels in wastewater have decreased due to reduced phosphates in laundry detergents. Point sources, such as wastewater treatment plants, are major nutrient contributors to waterways in heavily populated areas. Once again, advances in treatment have helped alleviate nutrient enrichment problems somewhat. These are discussed in Chapter 11. However, non-point pollution from agriculture, lawns, and golf courses also contribute nutrients to aquatic systems.

Sediments

Sediment can be a natural cause of water quality failure. However, it can also hold attached chemicals that dissolve in water and become a human-induced pollutant. The importance of sediment, both natural and anthropogenic, cannot be overstated. Sediment is considered a physical pollutant because it can "physically" block light penetration into water and prevent phytoplankton and submerged vegetation from becoming established. Since sediment blocks light, it eliminates DO that would be produced by phytoplankton and submerged plants, and reduces the amount of nutrients that would normally be used by these organisms. If high sediment levels are present, visibility is difficult for predator fish. Sediment can also smother fish eggs, reducing the number of young fish, called "fry."

Chemical Toxic Substances

Most people are familiar with the word "pesticide," a type of biocide. The USEPA defines a **pesticide** as "any substance or mixture of substances intended for preventing, destroying, repelling, or lessening the damage of any pest." A **biocide** has the literal meaning of "life killer." There are many "-cides" available at local garden or home supply stores to kill various pests. There are fungicides (for fungi), herbicides (to kill weeds), insecticides (for insects), and algaecides (for algae). There are even non-discriminate killers, called "fumigants" or "sterilants," that can kill, well, indiscriminately.

Many of us (or our parents) have fought with ants in the kitchen, crabgrass in the lawn, or mildew on the roses, and have used a biocide. Unfortunately, pesticides used on farms and suburban lawns can wash into waterways as non-point source pollution. Urban dwellers tend to be aggressive against unwanted vermin, and sometimes overuse a product. Too often, we believe that, if a little is good, then more can't hurt. However, when it comes to fertilizers and biocides, "more" can hurt our environment a great deal. We might have the nicest, green, weed- and pest-free lawn, but a dead creek flows nearby.

Early chemists did not formulate pesticides with environmental protection in mind. Pesticide persistence (how long they remain active in the environment) or their mobility (ease of movement) were not primary interests of these scientists. However, increased environmental concerns have led to the development of pesticides that are less persistent and less mobile in water. These newer products help protect waterways but are not the total solution. Application of these pesticides, and in what doses, makes a big difference in whether a lawn, crop, or golf course is treated or over-treated.

The scientific literature on biocides is immense and broad because the subject covers so many types of materials. It includes the already mentioned "-cides" plus other substances that influence the environment, including industrial petroleum processing and storage sources. These oil-based substances are **hydrocarbons** – long chains of carbon and hydrogen with varying levels of toxicity to life.

Hydrocarbons in water produce an ugly oil slick appearance. In large amounts, such as an oil spill, hydrocarbons are very harmful to any wildlife caught in it. They are not particularly toxic, i.e., hydrocarbons do not kill as a poison would kill but can create severe oil-coating problems on feathers and skin. The long-term cumulative effects are not known.

Another chemical toxic group of substance is the highly toxic **dioxin**, generally defined as **phenols** with chloride groups. Dioxin is a collective term for a group of environmental contaminants that includes certain dioxin, furan, and dioxin-like PCB (polychlorinated biphenyl) compounds. These are by-products of combustion largely from waste incineration, manufacturing processes of **chlorophenol chemicals**, and paper and pulp processes that use chlorine bleach. The range of toxicity, and other human health and environmental effects, are also broad. Dioxins are both hydrophobic and **lipophilic** (fat loving), therefore, they rapidly move up the food chain as it is consumed by higher animals including humans. There is no safe level of dioxins for humans [25] [26].

More recently, scientists have begun studies of emerging chemicals of concern particularly the broad category of **pharmaceuticals**. These are chemicals from antibiotics to vitamins that humans use daily. A 2020 US Geological Survey study of 111 pharmaceutical compounds in 444 streams [27] found human-use pharmaceuticals, primarily metformin, a type-II diabetes medicine; lidocaine, acetaminophen, and tramadol, all painkillers; carbamazepine, an anti-seizure medication; and fexofenadine, an antihistamine and nicotine, in small streams even after the water has been through a wastewater treatment plant.

How we use and dispose of human-use chemicals is something only humans can control. Because pharmaceuticals are designed to affect biological activity, their presence in streams is a potential concern for aquatic ecosystems. More than 60 percent of sites had at least one pharmaceutical compound, at a concentration in at least one sample, that might be of potential concern for fish. These compounds are also found in drinking water, and there is no definitive knowledge about their cumulative risk to humans, making this a much-needed area for future research [28].

Chemical concentration can create human health concerns – from cautious to panic mode. The most basic rule of toxicology is credited to a sixteenth-century Swiss chemist named Paracelsus. He is often quoted as simply stating, "The dose makes the poison." His more complete statement was "All substances are poisons; there is none which is not a poison. The right dose differentiates a poison from a remedy" [29]. This statement is more instructive. Athletes have died from an overdose of water, a problem called **hyponatremia**, caused by abnormally low blood sodium levels. Children have died from too many aspirin [30]. Both items – water and aspirin – are good in the proper amounts and are so common that we would be hard pressed to think of them as poisons. However, dose did make the poison in these cases.

Conversely, if someone tells us that there is an organophosphate chemical in our drinking water, we may panic. Even if the concentration is so low that a measuring instrument can barely detect it, people may become alarmed. In some cases, we have a reason for concern because we do not have data telling us what minimum harmful levels of a contaminate, or a cumulative level of compounds (i.e., pharmaceuticals), can be. The only "best" solution is to try to keep contaminates out of ecosystems as much as possible. Perhaps the biggest difference in our reactions to chemicals is familiarity, followed by knowledge of what levels are harmful or safe to eat or drink. The more we learn, both scientists and citizens, the better able we are to make knowledgeable decisions concerning risk.

Think About It: How Much Should Citizens Be Told?

Long-term cumulative effects of toxic chemicals may or may not be of consequence. How concerned should people be about these effects? What steps should governments play in regulating these long-term effects? How could they be monitored and regulated? Should chemical companies provide their data on possible toxic effects of long-term exposure to their product? Should families living in areas where they are exposed to chemicals, such as agricultural areas or near a chemical plant, be screened for possible effects?

There are two properties of potential biocides, discussed earlier, that determine their reactions in the environment – stability and mobility. The potential danger of a biocide is determined by how fast the chemical breaks down in soil or water. It's important to remember that the products of a chemical reaction can also be toxic, so the degradation process of chemicals is also especially important.

Some chemical toxic substances are rapidly fixed by (attached to) soil particles, clays, and organic materials, and do not easily become dissolved in water. Other toxic substances are not bound (attached to soil particles, clays, or organic materials), and move easily with water through the soil and into groundwater. The chemistry of chemical movement is complex, and is often studied in soil science, weed science, watershed science, and traditional chemistry classes. For more information on chemical movement, see the *Further Reading* section at the end of this chapter.

Heavy Metals and Trace Elements

Heavy metal refers to any metallic chemical that has a relatively high density and is toxic or poisonous at low concentrations. Examples of heavy metals include mercury (Hg), cadmium (Cd), arsenic (As), chromium (Cr), thallium (Tl), and lead (Pb). These metals are naturally found in the Earth's crust and cannot be degraded or destroyed. High doses can occur from inhalation of air near emission sources, from drinking water (e.g., lead pipes), and intake from the food chain. Heavy metals are dangerous because they tend to bioaccumulate (build up) in biological organisms over time, including the human body.

Trace Elements

Trace elements are naturally occurring chemicals needed by organisms, in small quantities, for proper growth and health. However, elevated concentrations of trace elements can be very harmful, and can be derived from paint, pesticides, copper pipes, vehicles, industrial machinery, air pollution, and surface water runoff. Trace element readings are taken from both water and sediment samples because they tend to bind to soil particles in sediment. Water quality standards, for trace elements in water, are generally referred to in terms of their toxicity to fish, as either **acute** (rapid) or **chronic** (takes time).

For example, arsenic is a poison, and we know that lead affects human brain development. Lead also damages the nervous systems of fish and waterfowl, and makes them more susceptible to other problems. We also know that mercury is a neurotoxin (a toxin that acts specifically on nerve cells). Copper is an essential element necessary to life but is toxic in excess. Too much copper harms kidneys, livers, and nervous systems in both fish and humans. Zinc has great value as an essential element and is only toxic in extreme excess. All these metals are relative health risks if too much is ingested – refer back to dosage. Ladies in Victorian times took a bit of arsenic to keep their complexions white. Some must have suffered side effects and shortened their lives, but ignorance and vanity prevailed.

Summing up: chemicals are vital to life itself, but only when used appropriately and in a controlled manner. The excessive use of almost anything can be harmful.

Microbiological Pathogens

Pathogens are microbes capable of causing disease. They can be protozoa, viruses, or bacteria. In freshwater, the most common sources of microbial pathogens are fecal matter; this includes improperly or incompletely treated wastewater, animal feedlots, wildlife, livestock, and leaky septic tanks. Waterborne pathogenic organisms of significance to humans include viruses (e.g., hepatitis A, poliomyelitis); bacteria (e.g., cholera, typhoid, coliform organisms); protozoa (e.g., *Cryptosporidium, Amoeba, Giardia*); and worms (e.g., *Schistosoma*, Guinea worm). These are usually introduced into the environment by fecal contamination from animals and humans. This makes the need for basic sanitation even more important. Worldwide, preventable diarrheal diseases kill 1.6 million people a year and are a leading killer of children under five [31].

Most of us live relatively comfortable lives, so some things are difficult and uncomfortable to imagine. For example, consider basic sanitation and water accessibility. Think about this: one-half of all hospital beds in less-developed countries are used by people with waterborne diseases. Mike Magee, MD, produced the following information from statistics on world health, regarding a typical African village of 1,000 residents [32].

- Six hundred residents will have no sanitary facilities – no bathrooms of any kind, even latrines.
- On any given day, 20 residents will have diarrhea, and 15 of these will be under the age of five.
- A family of six will spend three hours a day hauling water from a distant location.
- Many children (usually girls) will be too busy, carrying water or caring for sick family members, to go to school.
- Conditions will be filthy, and disease will spread rapidly.

It is not difficult to see how pathogens are hard to control under these conditions.

Think About It: Water Inequity

Imagine cooking, dishwashing, laundry – all the things we do every day. Now imagine needing to carry all the water necessary to do these sanitation duties. Suppose you must choose between drinking or cleaning because your water supply is limited. Your water source is not a clean faucet, but instead is a stream, pond, or shallow well, subject to contamination. The choice becomes much easier to imagine. Is it reasonable for people to live this way when others have so much? Are there solutions to these inequities?

Humans are not the only victims of waterborne disease. Fish are susceptible to pathogens, especially if they are injured or overcrowded – just like humans. Bacterial pathogens tend to be opportunists, and look for sites of injury to invade. Most fish diseases produce ulcerative sores, can spread to other fish, and end in death. These disease problems have become important in the aquaculture industry and are being studied by researchers to develop control and prevention methods. Pathogens can also infect other aquatic inhabitants. The problem is the same with any population. A healthy community, one with clean surroundings, adequate nutrients, etc., can conquer disease. An unhealthy community decreases the ability of members to resist illness. How do we decide the way forward to protecting both humans and their environment?

Oxygen-Depleting Organics

Oxygen-depleting organics are organic materials from natural vegetative decay, or from wastewater treatment plant discharge. Vegetation is often washed into waterbodies after crop harvesting and subsequent disking (soil preparation). Rain will wash these pulverized plant residues along with sediment into lakes and streams. On the other hand, if crop residues like corn stalks or wheat straw are left on the ground after harvest (a type of BMP called residue management), less plant residue will wash off fields. This land management strategy protects the ground from erosion by lessening the impact of raindrops, and by slowing surface water runoff. This can result in increased water stored in the soil. Residue management is a BMP, whereas cultivation of crop residues is not a preferred soil management practice.

Crop organic materials are consumed by microbes in water. The microbes use DO to break down organic matter into simpler compounds used for energy (food). The concentration of bio-degradable organic materials in water is measured indirectly as BOD. This is the amount of DO needed by aerobic microbes to degrade organic matter, such as crop residue [33].

Heat

Thermal pollution from power generation is a type of physical pollution. We have discussed the fact that hot water does not hold as much oxygen as cold water. Heat-induced, low-DO levels are a stressor for most aquatic life since aquatic animals evolve in water at temperatures, and DO levels, characteristic of their location. Their enzyme systems, hormones, and other systems were developed to work optimally at those ambient (surrounding or normal) temperatures.

Water temperature is buffered by its heat capacity, so temperature is not quickly changed by weather conditions. Thermal shock is a rapid temperature variation, often caused by manufacturing discharges, and can be very harmful to aquatic inhabitants. Even a 1 to 2 °C temperature change within 24 hours can cause problems with disease susceptibility, reproduction, and even migration.

Warm water on top of colder water (remember density), can act as a thermal barrier for some aquatic animals. A thermal barrier can stop normal movement. Exactly how much temperature influences an organism depends on its susceptibility, the duration of exposure, and the rate at which the temperature change occurs.

4.4 Clean Water as a Human Right

"Water is fundamental to a life of human dignity and is prerequisite to the realization of all other human rights" [34]. In 2002, this statement was made in a report by the World Health Organization (WHO). Today, we continue to see countries with huge population growth, but inadequate infra-structure for human waste removal and treatment, insufficient medicines to cure water-related diseases, unavailable chemicals to remove host organisms, and no formal mechanism to regulate pollution.

It is hard to realize that, in today's civilized world, we have over 844 million people without basic access to clean water, and 2.3 billion without adequate sanitation – the leading causes of water-related disease [35]. The number of people who suffer from water-related diseases every year is difficult to assess, because many countries lack mechanisms for reporting and diagnosing diseases. Water-related diseases, excluding diarrheal diseases, were estimated annually at 250 million cases in the 1990s. Worldwide, diarrheal diseases in 2000 reached 4.37 billion cases [36]. These are terrifying numbers to ponder, especially when they are preventable with technology that exists right now. If only it was that easy.

For example, since 1986 former President of the United States, Jimmy Carter and his wife, Rosalyn, have worked in Africa to improve public health [37]. Their efforts have focused on clean drinking water and improved sanitation. Guinea worm disease, among others, is a major public-health issue in parts of Asia and Africa. When their work started in 1986 there were an estimated 3.5 million cases in 21 countries in Africa and Asia.

Guinea worm disease is a parasitical disease with a complex life cycle. However, if any point in the cycle is stopped, the disease can be eliminated. The cycle begins when humans drink contaminated water. The water often contains fleas that have ingested Guinea worm larvae. The Guinea worm larvae then grow inside a human body and become a worm. This worm grows in size, and actually breaks through the skin of the human host by making a hole beneath the skin with acids. This process burns and hurts the victim terribly. Placing the afflicted flesh into water eases the pain but causes the worm to eject larvae – often into the source of drinking water for the local community. This is where a microscopic water flea ingests the deposited larvae, humans ingest the fleas when they drink the untreated water, and the horrible cycle of contamination starts all over again.

The spread of Guinea worm disease can be stopped by teaching people to filter their water through a cloth, to clean hands, build latrines, and wash their dishes and themselves. There are also chemicals that can kill the water fleas. Efforts of the Carters, and many others, have almost eradicated Guinea worm disease. Such humanitarian projects involve basic technology and education, and save and improve the lives of many. The Carter Center website has links to remarkable stories and explains how and why something that should have been relatively simple to accomplish has taken more than 20 years. The effort to eliminate Guinea worm disease was 99.5 percent completed in 2006, but the goal had to be 100 percent successful to prevent new cycles of contamination. Today, we can celebrate the successful work of the dedicated citizens of the Carter Center and its partners, including the countries themselves, in reducing the incidence of Guinea worm by more than 99.99 percent to only 54 provisional cases in 2019.

4.4.1 The Example of DDT

One of the most dramatic and far-reaching debates in science has been a long battle over the first synthetic pesticide, DDT (dichlorodiphenyl-trichloroethane). DDT is an organochlorine (organic compound with at least one chlorine atom) nerve toxin that kills insects on contact. The exact mechanism is still unknown even though Paul Müller, a Swiss scientist, discovered its insecticidal properties in 1941. DDT is also an environmental hazard, has an exceptionally long half-life (i.e., is persistent), and bioaccumulates in living organisms.

Rachel Carson's *Silent Spring* [38] was a plea to stop, in her words, "the wholesale, indiscriminate use of pesticides, primarily DDT." Her profound words are credited with starting the environmental movement because she clearly identified human's influence on, and our responsibility to, the environment. However, 40 years of research and millions of dollars of other work have failed to prove that DDT caused the widespread environmental and human destruction painted in her book. The increase in the numbers of birds, once harmed by DDT thinning of eggshells, is a very real fact that coincides with the ban of DDT in the US and most other more-developed countries. This is proof enough for many. However, a worldwide ban has been pursued by some, primarily because the use of DDT is considered anathema (detestable) to some people. Why, then, did Paul Müller receive the 1948 Nobel Prize in medicine for his work with DDT?

Paul Müller's work saved thousands of lives during the 1940s. During World War II, soldiers dusted themselves, and their clothing, with a powder of DDT to avoid typhus – a louse-borne disease. It was also used to fight the malaria-carrying *Anopheles* mosquito in the South Pacific, and later back home against agricultural pests. It worked so well during the war that many Pacific Islands became malaria free. To understand fully the magnitude of this achievement, we need to note that malaria was a worldwide problem prior to the use of DDT. The disease existed in the US, Europe as far north as Holland, India, and tropical countries. The annual death toll was in the millions, but the less dramatic effects – on worker productivity, loss of time, the health of pregnant women, women and children with other diseases, and basic health in general – were just as devastating to national stability and productivity.

In 1898, Ronald Ross, a military doctor in India, discovered that the *Anopheles* mosquito was the carrier of malaria [39]. The British government then promoted Dr. Ross and charged him with mosquito control in all the British colonies. His efforts in Sierra Leone in West Africa were extensive, but he did not have sufficient tools for the medical battle. No one had enough tools until 1941 when Paul Müller discovered the insecticidal properties of DDT.

Since then, the more-developed world has eliminated malaria, but many tropical countries are still plagued by the disease. Despite being both preventable and curable, in 2018 malaria killed 405,000 people (67 percent 272,000 were children under 5) while infecting 228 million [40]. Tragically, death and sickness caused by malaria has devastating effects on people in poorer countries ill-suited to cope with the disease.

Think About It: Too Much

DDT was a victim of its success and human excess. The negative aspects of DDT occurred because it was overused in huge amounts, and in situations where it was not the appropriate choice. The idea that if a little works, more should work better won the day. DDT was sprayed indiscriminately into water and the air where it could travel into different ecosystems. Soldiers were said to use it like talcum powder. Could DDT be used today if that use was carefully controlled and limited only to specific purposes? Would there be any benefit that could not be gained in another way?

Overuse of any pesticide can lead to insect resistance. Recent tests have shown that mosquitoes do not stop in rooms sprayed with DDT. However, since they cannot bite if they do not land, many agree it can still be an effective chemical. The judicious use of DDT has the potential to help millions of people; however, its use is so politically charged that it remains banned in many countries. Anecdotally, it is reported that visiting scientists, legislatures, and other guests in malaria-ridden areas, given the choice between a DDT-sprayed tent and an unsprayed tent, always choose the sprayed one [41].

Other insecticides could be effective but are more expensive than DDT. Costs, and other reasons, prevent poorer nations from purchasing insecticides for widespread use. The spread of malaria can be slowed by eliminating mosquito habitat such as draining swamps, clearing forests, and draining or pouring oil on pools of standing water. These techniques are effective to a point but are not without their own environmental effects. Physical barriers, such as mosquito nets over beds, are inexpensive by some standards, but are simply not feasible for millions of potential victims who live in poverty.

No sensible person would promote the kind of wholesale misuse of DDT that sparked this original debate. Once again, however, the potential to improve human life, to end a great deal of misery, and increase the standard of living in poor countries provides cause for continued debate. This is a social and environmental issue. Controversial problems such as the use of DDT are the reason a country needs an educated and informed citizenry to understand issues and take appropriate and effective actions.

Interestingly, the reference provided concerning Ronald Ross (the military doctor in India who discovered the cause of malaria) is the type of history of scientific discovery that inspires and educates by telling a good story. Ronald Ross objected to his portrayal in the referenced book, *Microbe Hunters*, published in 1926, and stopped publication with a lawsuit. He felt that liberties had been taken by the author to make a good story but were not an accurate portrayal of his character. Right or wrong, *Microbe Hunters* tells a good story of adventure, and a true story of discovery in science. Such works have sparked many to consider careers in science.

4.4.2 US Geological Survey National Assessment of Water Quality Program

One of the most comprehensive environmental surveys in the US for water quality has been the Department of the Interior's US Geological Survey (USGS) National Assessment of Water Quality Program (NAWQA) [42]. It included chemicals from fertilizers, pharmaceuticals, heavy metals, pesticides, and many more. This program inventoried hydrologic information in study areas that covered most of the 50 states. It was started in 1991 with a baseline inventory of water conditions in 51 watersheds and aquifers (see Figure 4.10). These study area investigations provided baseline information about surface and groundwater chemistry, aquatic life, stream habitat descriptions, and land use in much of the US. The studies were set up so that information could be correlated and interpreted across state lines, and other boundaries, by using nationally consistent design, sampling, and testing methods. Each study area was to be revisited in five-year cycles to monitor changes.

In 2020, the USGS stated,

> As the USGS Water Resources Mission Area (WMA) leadership looks to the future, we have begun a planning effort to update our water programs to meet 21st century water-resource challenges. As part of these updates, the National Water Quality Assessment (NAWQA) Project's water-resource monitoring, assessment, trends, modeling, and forecasting activities are being strategically integrated into new WMA programs. These new programs include Integrated Water Availability Assessments (IWAAs), the Next Generation Water Observing System (NGWOS), National Hydrologic Monitoring Networks, Hazards, Integrated Water Prediction (IWP), and research to improve our understanding of water-quality processes and water availability impacts of extreme events. The result of this decision is that NAWQA activities for the third decadal cycle will wind down by the end of FY 2021 and there will not be a fourth NAWQA decadal cycle [43].

Reports from NAWQA studies are still excellent sources of both specific scientific data and general summaries. All reports are available for download from the USGS website, along with fact sheets and other information. This includes a pesticide primer and information to help members of the press understand the language of the reports. It also provides suggestions how to accurately

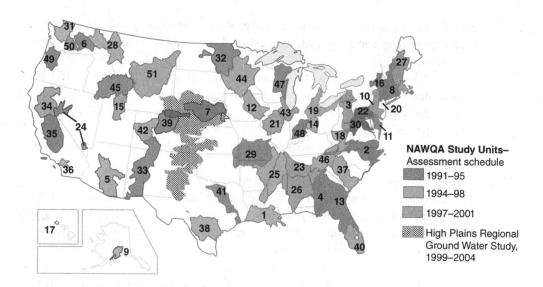

1 -Acadian-Pontchartrain Drainages (LA, MS)	27 - New England Coastal Basins (ME, MA, NH, RI)
2 -Albemarle-Pamlico Drainage Basin (NC, VA)	28 - Northern Rockies Intermontane Basins (ID, MT, WA)
3 - Allegheny and Monongahela River Basins (MD, NY, PA, WV)	29 - Ozark Plateaus (AR, KS, MO, OK)
4 - Apalachicola-Chattahoochee-Flint River Basin (AL, FL, GA)	30 - Potomac River Basin (DC, MD, PA, VA, WV)
5 - Central Arizona Basins (AZ)	31 - Puget Sound Basin (WA)
6 - Central Columbia Plateau (ID, WA)	32 - Red River of the North Basin (MN, ND, SD)
7 - Central Nebraska Basins (NE)	33 - Rio Grande Valley (CO, NM, TX)
8 - Connecticut, Housatonic and Thames River Basins (CT, MA, NH, NY, RI, VT)	34 - Sacramento River Basin (CA)
9 - Cook Inlet Basin (AK)	35 - San Joaquin -Tulare Basins (CA)
10 - Delaware River Basin(PA, NY, NJ)	36 - Santa Ana Basin (CA)
11 - Delmarva Peninsula (DE, MD, VA)	37 - Santee River Basin and Coastal Drainages (SC, NC)
12 - Eastern Iowa Basins (IA, MN)	38 - South Central Texas (TX)
13 - Georgia-Florida Coastal Plain (GA, FL)	39 - South Platte River Basin (CO, NE, WY)
14 - Great and Little Miami River Basins (OH, IN)	40 - Southern Florida (FL)
15 - Great Salt Lake Basins (UT, ID, WY)	41 - Trinity River Basin (TX)
16 - Hudson River Basin (NY, CT, MA, NJ, VT)	42 - Upper Colorado River Basin (CO, UT)
17 - Island of Oahu (HI)	43 - Upper Illinois River Basin (IL, WI, IN)
18 - Kanawha-New River Basins (WV, VA, NC)	44 - Upper Mississippi River Basin (MN, WI)
19 - Lake Erie-Lake Saint Clair Drainages(IN, MI, OH, PA, NY)	45 -Upper Snake River Basin (ID, MT, NV, UT, WY)
20 - Long Island-NewJersey Coastal Drainages (NJ, NY)	46 - Upper Tennessee River Basin (GA, KY, NC, SC, TN, VA)
21 - Lower Illinois River Basin (IL)	47 - Western Lake Michigan Drainages (MI, WI)
22 - Lower Susquehanna River Basin (MD, PA)	48 - White River Basin (IN)
23 - Lower Tennessee River Basin (TN, AL, GA)	49 - Willamette Basin (OR)
24 - Las Vegas Valley Area and the Carson and Truckee River Basin (NV, CA)	50 - Yakima River Basin (WA)
25 - Mississippi Embayment (AR, KY, LA, MS, MO, TN)	51 - Yellowstone River Basin (MT, WY, ND)
26 - Mobile River Basin (MS, AL, GA)	

Fig 4.10 US Geological Survey National Assessment of Water Quality Program (NAWQA) watersheds, original study units
(Map courtesy of USGS)

present information to the public. The implications of their results for pesticides are a good example of what can be learned from the studies. The USGS study results are summarized as follows [44]:

- Pesticides and **degradates** (degradation products of the original pesticide that have been transformed in the environment) are likely to be present at detectable levels throughout most of the year in streams that have substantial agricultural or urban land use in their watersheds.
- Streams are more vulnerable to pesticide contamination than groundwater in most hydrologic settings, as indicated by much more frequent detections in stream water.
- The frequent detection of pesticides and degradates, in shallow wells in agricultural and urban areas, indicates that groundwater may merit special attention in these land use settings. Shallow groundwater is used in some areas for drinking water, and this shallow water can move downward into deeper aquifers. Early attention to potential groundwater contamination is warranted because the movement of groundwater is usually slow, and contamination is difficult to reverse.
- Pesticide occurrence in streams and groundwater does not necessarily cause adverse effects on aquatic ecosystems or humans. The potential for effects can be assessed by comparing measured pesticide concentrations with water quality benchmarks, which are based on the concentrations at which effects may occur.

Notice that the above summary only includes detection of chemicals. There are no chemical concentrations listed, and no interpretations of the impacts on the environments in which they were found. To get that type of data, the reports must be read for specific areas. There are many peer reviewed papers for each aspect of the NAWQA studies that are invaluable in interpretation and use of the collected data.

The USGS is a tremendous resource for environmental data of all types. They have real-time water quality data accessible online for many streams and rivers in America, maps, and reports of investigations and studies. Anyone interested in environmental data, particularly in the US, should become familiar with their work.

4.5 Who...Me?

Imagine a spring stream, fed by snowmelt, cold and blue with white caps, rushing out of the mountains. The stream flows into a mountain valley where it meanders through meadows and wetlands, and finally to an outlet in a larger river. Moose and deer walk through the valley, eating along its waters. Hawks and eagles soar overhead looking for a meal. The sounds of a mountain jay, a moose bellowing in the background, surround you. Add your own touches – an ideal scene, a place of beauty for hiking, fly fishing, and animal watching. The natural inclination for humans is to move in, develop a place to live, construct roads for easier travel, and perhaps harvest some of the trees to help with expenses. The scene changes, as it often does. Will water quality be negatively impacted?

The immediate reaction is to say, "Yes," and lament that this should not be allowed. Throughout history, early settlers cut trees, broke ground, and changed their environment. Populations grew, and the magnitude of change increased quickly. The change they caused was primarily related to surviving and living. Some were making homes and farms. However, others were simply after money. Often, these individuals lived in cities far away from the area where natural resources were obtained, extracted, and, too often, exploited. For example, the owners of mines during the 1800s and later paid little or no attention to water except as a tool for obtaining gold and silver ore. Initial lumber operations had free access to clear-cut ancient trees, build roads and shelters, and harvest vast sections of land. Trees were floated down rivers to sawmills without regard to bank stabilization or aquatic life. Was this all wrong? Does it matter if it was? History is what has already happened in the past. Judging history is an intellectual activity but learning from history can make a difference today.

Humans have long viewed nature as something to be dealt with, to be molded into what is needed, to be controlled. There is no moral judgment here – people need to survive and thrive, and the natural world provides the resources needed. However, times change, population increases, and survival is not necessarily the key issue when dealing with nature today. Now is the time in history when judgment and thought must come into play since most humans have risen above mere survival. Decisions about how to develop the land and water resources can be tempered with the need for conservation – to preserve resources for future generations.

Today, we know about minimum impact development and many other ways to conserve and protect the environment – whether farming, mining, lumbering, or building homes. Ignorance or expediency are not a valid excuse for the disregard or destruction of the environment. Water resources are known to be limited and, in many cases, fragile. As such, water quality issues cannot be ignored at any phase of the development of a city, suburb, resort, town, farm, ranch, industry, or energy development (fracking, coal mining, dams).

It's a fact that we can only protect a limited amount of land from human development. Therefore, causing as little damage as possible is the next best option. There are places that should be protected because they have value as a last remaining wilderness – old-growth forests, stretches of desert, and homes for wildlife. However, even in national parks in America, Canada, or Kenya, it is difficult to protect the environment from human use. Literally millions of people pass through Yellowstone and Yosemite National Parks in the US every year. Minimizing their impacts, while allowing as many people as possible the pleasure of viewing some of the most beautiful country on Earth, is a tremendous task.

Personal responsibility is something many people do not think about on vacation. Whether you are hiking or driving, it's important to pack out what you pack in. It may seem like a small thing until you multiply your impact by a million people – each one trying to have an experience in "natural" surroundings. You see a faster way down a hill and take off. Others follow and soon there is a trail – not a properly developed trail, but an erosive channel for water to wash through, producing a gully. A gully forms where none should be. Think about your ideal scene discussed earlier before you step off the path, drop a wrapper, whack at tree branches, or prepare a campsite by moving rocks, or stomping plants to make a tent spot. The idea "leave no trace" works well for the environment.

We all find it easier to blame big business, politicians – anyone other than ourselves – for water quality degradation. There is plenty of blame for those just mentioned, particularly because their impacts are larger than any individuals, but it is harder to face the consequences of our own choices. Yes, it's important to hold industries, agriculture, oil drillers, miners, lumber, chemicals, power generators, and developers accountable, and to higher standards than in the past. However, accountability must also apply to those who benefit, even to us.

Summary Points

- **Water quality** can be defined as the chemical, physical, and biological properties of water that are influenced by geology, climate, local environment, and by people.

- Desired water quality is usually defined by its intended designated use. The **designated use** determines the level of water quality required. Designated uses include drinking water or public water supply, protection of shellfish, fish, wildlife, recreation, agriculture, industry, and navigation.

- The important **properties of water** relating to its designated uses include its solvent abilities, density, specific heat capacity, pH, cohesion, adhesion, dissolved oxygen content, turbidity, and temperature.

- **Water's molecular chemistry** makes it unique. It is a polar molecule, i.e., has a positive and a negative pole. Polarity accounts for water's unique properties. It allows individual molecules to bind to each other and to ions in solution so that it can be found as a gas, liquid, or solid at normal temperatures. It is the only naturally occurring substance on Earth with this ability.

- Other unique properties of water are its **high specific heat capacity,** its **variable density based on temperature**, its **universal solvent character**, and its **cohesive** and **adhesive** properties.

- **Water quality failure** occurs when water no longer meets the requirements for its designated use. Both natural and human influenced causes can result in degraded water quality.

- Both **point source** pollution, from a known source such as a pipe from an industry, and **non-point source** pollution, from surface runoff over a large area such as farmed fields or paved areas, have specific best management practices (called **BMPs**) for their control.

- **BMPs** have been developed for many land uses including agriculture, silviculture, mining, urban development, construction, industry, and road building.

- **Pollutants** are usually classified by type: nutrients, sediments, chemical toxic substances, microbiological pathogens, oxygen depleting organics, and heat. Each produces different effects on aquatic systems.

- Excess **nutrients: nitrogen** and **phosphorus** are the most common, and generally cause excessive plant growth and die-off. This ultimately limits the amount of dissolved oxygen in the water. Dead plants are an example of **oxygen-depleting organics**. Effluent wastes are also in this group.

- **Sediments** decrease water clarity and limit plant productivity disrupting the natural food webs that support all aquatic life.

- **Chemical toxic substances**, whether pesticides, pharmaceuticals, or industrial hydrocarbons, can be directly poisonous to aquatic life and potentially harmful to humans.

- **Pathogens** are microbes in water that cause disease. The impact of these diseases on human life can be devastating.

- **Heat**, usually an industrial cooling byproduct, increases the rate of chemical reactions, lowers dissolved oxygen levels, and can harm fish communities especially the life cycles of cold-water fish such as salmonids.

- Clean water is a fundamental **human right. Preventable waterborne diseases** still kill millions each year.

Questions for Analysis

1. Density is an important physical concept. Discuss the environmental importance of density differences between oil and water, and between water and ice. Use examples.
2. Which environmental factors influence dissolved oxygen levels in fresh water? Which of these can be altered by human activity? Give examples.
3. Differentiate between point and non-point source pollution. What are some control methods for each and why are they different?
4. How are pollutants classified? What are the classes of pollutants? How would you rate the importance of these pollutant types? Justify your answer.
5. Discuss the impact of waterborne diseases in less-developed countries. What are some of the obstacles to eliminating waterborne diseases?
6. Provide an example of one person's (your choice) attempts to improve any aspect of the Earth's natural environment. Discuss your thoughts, back them with facts, and provide references.

Further Reading

Carson, Rachel, 1962, *Silent Spring*, Boston, Mass.: Houghton Mifflin.

De Kruif, Paul Henry, 1926, *Microbe Hunters*, New York: Harcourt Brace.

Gleick, Peter H., 2004, *The World's Water: The Biennial Report on Freshwater Resources*, Washington, DC: Island Press.

Pielou, E. C., 1998, *Fresh Water*, Chicago, Ill.: University of Chicago Press.

References

1 Karrie L. Pennington, 2004, "Surface water quality in the Delta of Mississippi," in *Water Quality Assessments in the Mississippi Delta*, eds. Mary T. Nett, Martin A. Locke, and Dean A. Pennington, ACS Symposium Series No. 877, Washington, DC: American Chemical Society, p. 30.

2 World Health Organization (WHO), 2006, *Guidelines for Drinking-Water Quality*, 1 addendum to 3rd ed, electronic version, www.who.int/water_sanitation_health/dwq/gdwq3rev/en/, September 2007.

3 Diane Raines Ward, 2003, "Water wants: A history of India's dams," PBS, September 14, 2003, https://www.pbs.org/wnet/wideangle/uncategorized/the-dammed-water-wants-a-history-of-indias-dams/3098/ (accessed March 2021).

4 Central Pollution Control Board, www.cpcb.nic.in, August 2007.

5 Archana Chaudhary, May 2, 2019, "A journey down the Ganges in the age of Modi," *Bloomberg Businessweek*, www.bloomberg.com/features/2019-ganges-journey/

6 Central Pollution Control Board, India, 2019, *Annual Report 2017–2018*, Ministry of Environment, Forest & Climate Change, www.cpcb.nic.in

7 Amanda Briney, 2020, *Geography of the Ganges River*, ThoughtCo, February 11, thoughtco.com/ganges-river-and-geography-1434474

8 Joshua Hammer, 2007, "A prayer for the Ganges," *Smithsonian Magazine*, www.smithsonianmag.com/travel/a-prayer-for-the-ganges-173708201/ (accessed 2020).

9 S. Lokhandwala and P. Gautam, 2020, "Indirect impact of COVID-19 on environment: A brief study in Indian context," *Environmental Research* 188, 109807. Advance online publication, https://doi.org/10.1016/j.envres.2020.109807

10 NASA Earth Observatory, "Glacial retreat," https://earthobservatory.nasa.gov/images/7679/glacial-retreat

11 Qadri, Altaf, August 2020, *Ganges River Flows with History and Prophecy for India*, Associated Press, https://apnews.com/0a2dfe48fd4c9bbe48981f5b9a191e68

12 USEPA, www.epa.gov/oilspill/exxon.htm, August 2007.

13 NASA GSFC Landsat/LDCM EPO Team, NASA Landsat Image Gallery, https://landsat.visibleearth.nasa.gov/view.php?id=78061

14 Schleifstein, Mark, 2020, "Restoring oyster beds and saving stranded dolphins: See how $28.7 million in BP oil spill money will be spent," www.nola.com/news/environment/article_17debd6c-da7d-11ea-b2d4-43c9a8714188.html

15 NASA Explore Earth, 2020, www.nasa.gov/topics/earth/index.html

16 European Space Agency (ESA), 2020, www.esa.int/

17 BBC News, 2020, "Mauritius oil spill: Fears vessel may 'break in two' as cracks appear," August 10, www.bbc.com/news/world-africa-53722701 (accessed August 2020).

18 Associated Press, 2020, "Mauritius struggles to contain oil spill polluting its seas," www.france24.com/en/20200809-mauritius-struggles-to-contain-oil-spill-polluting-its-seas (accessed August 2020).

19 Chemical Achievers, "The human face of the chemical sciences: Søren Sørensen," www.chemheritage.org/classroom/chemach/electrochem/sorensen.html

20 Lenore S. Clescerl, Arnold E. Greenberg, and Andrew D. Eaton, 1998, *Standard Methods for Examination of Water and Wastewater*, 20th ed, Washington, DC: American Public Health Association.

21 US Geological Survey, *Field Manual*, Table 6.2–6, August 2007, http://water.usgs.gov/owq/FieldManual/Chapter6/6.2.4.pdf

22 Save the Bay Center, www.savebay.org/advocacy_nutrients_reduce.asp, Providence, Rhode Island, August 2007.

23 Rhode Island Sea Grant, 2018, "Narragansett Bay is changing in more ways than one," presented at the 2017 Ronald C. Baird Sea Grant Science Symposium, https://seagrant.gso.uri.edu/narragansett-bay-changes/ accessed 2020.

24 US Department of Agriculture (USDA) Natural Resources Conservation Service, 1999, "Conservation corridor planning at the landscape level," from Managing for Wildlife Habitat, Part 190 of the National Biology Handbook, ftp://ftp-fc.sc.egov.usda.gov/WSI/pdffiles/Conservation_Corridors_Manual-Cover.pdf. August 2007

25 US Food and Drug Administration, 2019, "Dioxins and PCBs," www.fda.gov/food/chemicals/dioxins-pcbs

26 United States Department of Agriculture (USDA) Science Staff, 2015, Office of Public Health Science Food Safety and Inspection Service, "Dioxin Fiscal Year 2013 Survey: dioxins and dioxin-like compounds in the U.S. domestic meat and poultry supply," www.epa.gov/dioxin/dioxin-resources (PDF downloaded 2020).

27 P. M. Bradley, C. A. Journey, D. T. Button et al., 2020, "Multi-region assessment of pharmaceutical exposures and predicted effects in USA wadeable urban-gradient streams," *PloS ONE* 15, 1: e0228214, https://doi.org/10.1371/journal.pone.0228214 (PDF available at https://journals.plos.org/plosone/article?id=10.1371/journal.pone.0228214, downloaded 2020).

28 USGS, 2020, "Pharmaceuticals common in small streams in the U.S.," www.usgs.gov/center-news/pharmaceuticals-common-small-streams-us?qt-news_science_products=1#qt-news_science_products

29 C.D. Klaassen, M.O. Amdur, and J. Doull, eds., 1986, *Casarett and Doull's Toxicology: The Basic Science of Poisons*, 3rd ed, New York: Macmillan, p. 4.

30 P. Gaudreault, A. R. Temple, and F.H. Lovejoy, Jr., 1982, "The relative severity of acute versus chronic salicylate poisoning in children: a clinical comparison," *Pediatrics* 70, 566–569.

31 Our World in Data, 2017, "Global deaths," https://ourworldindata.org/what-does-the-world-die-from

32 Mike Magee, 2005, *Healthy Water: What Every Health Professional Should Know About Water*, New York: Spencer Books.

33 Clescerl et al., Standard Methods.

34 WHO Committee on Economic, Social and Cultural Rights, 2002, *The Right to Water*, 29 session. Geneva, Switzerland: WHO.

35 United Nations, 2018, "Sustainable development goal 6 synthesis report 2018 on water and sanitation," www.unwater.org/publications/highlights-sdg-6-synthesis-report-2018-on-water-and-sanitation/ (PDF download 2020).

36 Peter H. Gleick, 2004, *The World's Water: The Biennial Report on Freshwater Resources*, Washington, DC: Island Press, pp. 7–8.

37 The Carter Center, 2007, "Guinea Worm Eradication Program," www.bing.com/search?form=MOZTSB&pc=MZSL02&q=The+Carter+Center and www.cartercenter.org/health/guinea_worm/index.html

38 Rachel Carson, 1962, *Silent Spring*, Boston, Mass.: Houghton Mifflin (A beautifully illustrated 40th Anniversary Edition was published in 2002 and was still available in paperback in 2020.)

39 Paul Henry De Kruif, 1926, *Microbe Hunters*, New York: Harcourt Brace.

40 WHO, 2019, "The World Malaria Report at a glance 2019," www.who.int/news-room/feature-stories/detail/world-malaria-report-2019 (PDF download of full report available at www.who.int/publications/i/item/9789241565721, accessed 2020).

41 Richard Tren and Roger Bate, 2001, *Malaria and the DDT Story*, London: Institute of Economic Affairs (PDF available at https://papers.ssrn.com/sol3/papers.cfm?abstract_id=677448, downloaded 2020).

42 National Water-Quality Assessment (NAWQA) 2020, www.usgs.gov/mission-areas/water-resources/science/national-water-quality-assessment-nawqa?qt-science_center_objects=0#qt-science_center_objects

43 USGS, 2020, "New water-quality directions," www.usgs.gov/mission-areas/water-resources/science/new-water-quality-directions?qt-science_center_objects=0#qt-science_center_objects

44 Robert J. Gilliom, Jack E. Barbash, Charles G. Crawford et al., 2007, *Pesticides in the Nation's Streams and Ground Water, 1992–2001*, USGS Circular No. 1291, Reston, Va.: US Geological Survey.

5 Watershed Basics

> We know from science that nothing in the universe exists as an isolated or independent entity. Everything takes form from relationships, be it subatomic particles sharing energy or ecosystems sharing food. In the web of life, nothing living lives alone.
>
> Margaret Wheatley, *Turning to One Another* [1]

Chapter Outline

5.1 Introduction

A **watershed** (also called a catch basin or catchment) is the land that drains into a body of water. The receiving bodies of water may be as small as a farm pond or as large as a sea. It could be the mighty Mississippi River or even the mightier Amazon; it could be like the relatively small Walden Pond in Massachusetts or massive Lake Superior on the Canada and US border. The size of a watershed can vary from a few hectares to thousands of square kilometers (acres to thousands of square miles).

The geology, land cover, size and shape, relief, latitude, chemistry, regional climate, local biological communities, and land use are especially important factors in shaping a watershed. The activities of humans and associated modifications of land use, especially during the past century, all influence the health of a watershed and its wetlands, ponds, lakes, reservoirs, streams, and rivers.

An **open watershed** drains into an ocean. Sometimes multiple river mouths exist, creating a multiple open watershed. The Amazon River is a good example of a multiple open watershed (see Figure 5.1). A **closed watershed** empties into an inland body of water, such as a lake or inland sea (see Figure 5.2).

Fig 5.1 Amazon River watershed, South American continent: an open watershed

Fig 5.2 This closed (endorheic) watershed receives water from the surrounding land, but it does not leave the lake except by evaporation. Water flows into the lake from the river in the left side of the picture. This is Üüreg Lake, Mongolia
(Photograph courtesy of NASA/ GSFC/METI/ERSDAC/JAROS and US/Japan ASTER Science Team)

5.2 Watershed Delineation

Watersheds are usually **delineated** using topographical maps to determine ridges or other high points (breakpoints) separating the different watersheds. Topographical maps show surface relief as **contours**. The closer together the contour lines are, the steeper the relief. Contours on a topo map that are far apart represent flatlands. Delineation can be difficult in flatlands where the height of ridges is measured in inches or centimeters.

Using paper topographic maps to study local areas has been a major effort since the 1800s. The USGS National Geospatial Program has developed a collection of historical topographical maps of the United States dating back to 1884. They have produced high-resolution scans of more than 178,000 maps that are now available in electronic form along with their new generation of current topographic maps. Historic maps show us what the land looked like before development and show how the changes progressed over time. Many disciplines from scientists, environmentalists, to historians can use these maps in their work. Topographic maps remain essential for everyday use in environmental science, for instance, for ecosystem restoration. Industries, governments local and regional, agriculture and forestry use them for planning [2].

The history of map-making is fascinating. The first country to create topographical maps of an entire country was France. It took four generations of the Cassini family, presiding over the first attempt to survey and map every meter of their nation, to complete the project. The Cassinis used the science of triangulation to create a nearly 200-sheet topographic map, which French revolutionaries nationalized in the late eighteenth century. This, Professor Jerry Brotton, professor of Renaissance studies at Queen Mary University in London says, "is the birth of what we understand as modern nation-state mapping . . . whereas, before, mapmaking was in private hands. Now, in the Google era, mapmaking is again going into private hands." Professor Brotton believes that there are no perfect maps, only maps that capture our understanding of the world at discrete moments in time [3]. The topics in his book span stone maps to Google Earth images. To learn more, you can read Brotton's book *A History of the World in 12 Maps* [4].

Today, sophisticated geographical information systems (GIS), with spatial analysis software and digital elevation lines, can be used to delineate watersheds quickly and accurately based on topographical features. Accurate digital terrain models (DTM) can be produced by light detection and ranging (LiDAR) technology, aerial photography, and global positioning system (GPS) technology when used simultaneously. These tools can measure vertical elevation changes within 7–8 centimeters (~3 inches) of accuracy [5]. This small scale helps to create maps that are accurate even in extremely flat landscapes.

In Depth   Mapping with Technology

LiDAR

A wide variety of instruments and techniques are used to delineate and assess watersheds. LiDAR stands for light detection and ranging. Here's how it works: A scanning laser unit is mounted to an airplane or helicopter. The aircraft flies in one direction as the laser emits a stream of light pulses in a downward and side-to-side motion. These light pulses are perpendicular to the flight path, and toward the land surface, so that the reflected pulses record the time it takes to reach the surface. The slant angle, from the line directly beneath the plane, is also recorded along with GPS data and altitude from the plane's inertial measurement unit (IMU).

Post-processing converts the collected LiDAR data into usable forms. The slant distance between the aircraft and the ground, for each returned pulse, is calculated as one-half the time it takes for a laser pulse to travel from the aircraft and back, multiplied by the speed of light. Each slant distance is then corrected for atmospheric conditions, and for the roll, pitch, and yaw of the aircraft using the IMU data. This post-processing makes the slant distance as accurate as possible.

GPS data (discussed next) are processed separately and imported into the LiDAR solution. Then, each corrected slant distance is transformed to a ground surface elevation. Additional software is used to generate land contours, quality control of the digital terrain model, and to verify aspects of the data, such as calibration parameters [6]. These data are used to create accurate digital watershed maps or models.

GPS

The Navstar GPS can provide accurate latitude, longitude, and height locations of anything on Earth. The system comprises a constellation of satellites orbiting about 20,000 km (~ 12,400 miles) above the Earth's surface, which transmit ranging signals on two frequencies in the microwave part of the radio spectrum [7]. The system is maintained by a control segment through a system of ground monitor stations and satellite upload facilities. To use the system, a receiver is necessary. These receivers can be civilian or military. GPS receivers receive satellite signals; they don't transmit or bounce signals off the satellites. GPS can support an unlimited number of users because they are passive (receive-only) units.

A satellite in the system generates a unique digital code sequence of 1s and 0s. These are precisely timed by an onboard atomic clock – picked up by a GPS receiver's antenna and matched with the same code sequence generated inside the receiver. The receiver matches up the code sequences to determine how long it takes a signal to travel from the satellite to the receiver. These timing measurements are converted to distances using the speed of light (about 300,000 kilometers per second, or 186,000 miles per second), the same speed as radio waves.

The receiver can determine its latitude, longitude, and height by simultaneously measuring distances to four or more satellites with known locations (included in the signals transmitted by the satellites). The receiver also synchronizes its clock with the GPS time standard. This sequence provides accurate readings of location and time. Using the measurement of distances, to determine a position, is known as trilateration. This should not be confused with triangulation, which involves the measurement of angles. A GPS provides 24-hour global coverage. Importantly, it is not influenced by rain, snow, fog, or sandstorms.

GIS

A GIS is a powerful mapping, analysis, and database tool. It consists of a computer system, software, and data. Software is commercially available and is very good. Data can be begged, borrowed, created, or purchased. Many commonly used data layers, such as rivers, roads, and political boundaries, are available for free. GIS is widely used in watershed delineation, and numerous software programs are available. GIS data are organized into layers of information for a location (such as a watershed) that are gathered and organized into a database (see Figure 5.3). The data can then be used to produce maps, graphs, or charts of the area's information. These are smart maps, because any data entered about a location can be retrieved, analyzed, and viewed along with the map or as part of the map.

Computer graphics, combined with database information, create a powerful analysis tool where spatial relationships between different database layers can be examined. Common database layers include geographic information such as roads, watersheds, rivers, railroads, city locations, and street names. Census data are commonly used for sociological or demographic analysis. The US Environmental Protection Agency (USEPA) has data layers which display all permits for waste disposal into rivers and streams. Cities chart their stormwater drainage systems, bus routes, school locations, and other

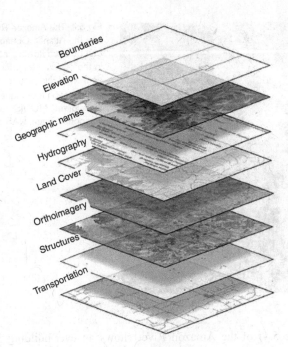

Fig 5.3 GIS data layers are built one on top of the other to complete the information for a specific location. Each layer also has tables of information to provide details about the layer. These are the eight base layers used in the USGS National Map (Credit: Sexton, Pamela Ann. Public domain)

information using GIS maps. Wetland areas have been digitized in many parts of the world for use as data layers. Virtually any data related to a location can become a GIS data layer. When combined, the data layers can show information useful in as many ways and situations as can be imagined.

For example, migratory birds often forage in wetlands for food on their journey. Wetland locations within a watershed (including type, extent, food sources, predator populations, nesting trees, etc.) can be layered with known migratory flight patterns for a particular species to predict the birds' ability to complete their migration. Information about a species could also be added as a layer, and food preferences, nesting habits, and habitat preferences could refine the analysis. The usefulness of these tools in analyzing spatially related parameters is endless. Chapter 1 has an In Depth section on using technology to delineate ecosystems.

5.3 A Comparison of Erosion from Two Major Watersheds

The potential datasets of a watershed are broad and varied. Similarly, the physical size of some of Earth's largest river systems, such as the Amazon and the Mississippi, are truly amazing. The size of these two watersheds results in high levels of sediments in river water from erosion. We'll look at the Amazon first. The Amazon River watershed (see Figure 5.1, shown previously) is 6,915,000 square kilometers (2,670,000 square miles), or about 40 percent of the total land area of South America. The magnitude of the Amazon is difficult to imagine. It is the largest river in the world by volume, with greater total river flow than the next eight largest rivers combined. It also has the largest watershed in the world. The Amazon carries 20 percent of all the freshwater emptied into the Earth's oceans. Think about that for a minute and try to visualize a flow so large that it creates a freshwater plume some 320 kilometers (200 miles) out into the ocean. This river starts with small mountain streams in the Peruvian Andes and grows as it flows through its watershed to the Atlantic Ocean.

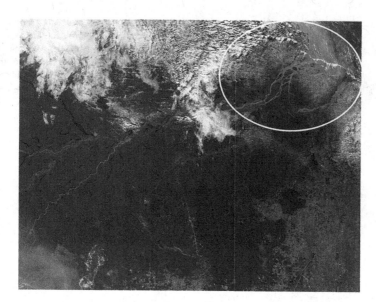

Fig 5.4 The Amazon River flowing into the Atlantic Ocean. Sediments can be seen flowing into the ocean in the oval in the upper right corner section of this picture (Photograph by Jacques Descloitres, MODIS Land Rapid Response Team, NASA/GSFC)

A satellite photo (Figure 5.4) of the Amazon River shows an ever-building land mass forming from the sediments of the watershed flowing into the Atlantic [8]. Removal of the rainforest lays land bare and subject to erosion, so the amount of sediment delivery from surface water runoff can be expected to increase. The satellite image presented is a pictorial example of the impact of land use on water quality. The annual sediment load of the Amazon is 363 million Mg (329 million tons). Erosion would be 13 Mg/ha (11,600 lbs/acre) to supply this much sediment.

A little less than 1 million tons of suspended sediment enters the Atlantic Ocean each day and has formed Majaro Island – a river island about the size of Switzerland just off the coast of Brazil. Sediment can be any size – from boulders to clay – depending on the definition used. Soil scientists generally use the term *particle size* to differentiate sediments. They consider gravels to be >2 mm, sands 0.05–2 mm, silts 0.002–0.05 mm, and clays <0.002 mm. The type and size of sediment found in a river depends upon the soils in the watershed and in the banks of the river. Gravels and sands are heavy and drop quickly out of the water. Silts and clays are lighter and can stay suspended in water for hours to days, depending upon the clay type.

The Amazon watershed contains a huge river plain, part mountain land, part rainforest. It is sparsely populated for most of its course, with few cities on the river's shores. However, over 28 million inhabitants live in the watershed with agriculture and fishing providing most jobs in the region. Slash-and-burn agriculture is extensively utilized and involves cutting down trees and burning them to expose the fertile soil for cultivation. Depletion of the soil requires the farmers to move to a new forest and start over, leaving land exposed to erosion. Erosion is a major source of sediments (see Figure 5.5).

The Mississippi River watershed, in comparison, is the third largest in the world, exceeded in size only by the watersheds of the Amazon River and Congo River in central Africa (see Figure 5.6). The watershed covers more than 3,225,000 square kilometers (1,245,000 square miles), including all or parts of 31 states and 2 Canadian provinces. It drains 41 percent of the 48 contiguous states of the United States. The annual sediment load is 300 million Mg (270 million tons). Erosion to supply this sediment would be 93 Mg/ha (83,000 lbs/acre). This would be roughly equivalent to dumping 20 Mg (18 tons) of sediment into the Gulf of Mexico every 2 seconds [9]. The average

Fig 5.5 This remarkable scene is the Amazon River where two tributaries contribute quite different water qualities. The dolphin is enjoying the tannin-colored clear water while on the edge of the heavy sediment-laden water. The tannin color comes from decaying vegetation in the Rio Negro watershed, which is largely rock with little sediment. The sediment-loaded water of the Rio Solimoes comes out of mountains containing glacier flour (rocks ground fine by glaciers) and other soil sediments. The difference is visually dramatic

Fig 5.6 Mississippi River watershed

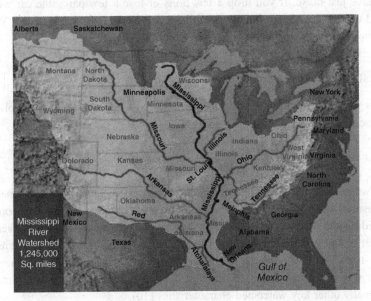

pickup truck can carry 225 kilograms (500 pounds). Imagine trying to unload 72 pickups filled with sediment every 2 seconds.

Think About It: Two-River Comparison

The Amazon watershed has about 13 Mg/hectare of erosion (11,600 lbs/acre), while the Mississippi watershed has about 93 Mg/hectare (83,000 lbs/acre). What type of watershed features, or land use practices, could account for the much larger erosion rate in the Mississippi River watershed?

These two rivers represent some of the largest watersheds of our world. However, understanding how individual events or activities influence a particular waterbody usually involves studying smaller watersheds. The sheer logistics of developing an experimental design to understand an ecosystem the size of the Amazon River watershed, even with the assistance of GPS, GIS, and satellite imagery, would be daunting to anyone.

5.4 Watershed Structure

Watersheds, on the simplest level, are composed of uplands, lowlands, and a water feature. Each of these has **abiotic** (physical) and **biotic** (biological) components that we will explore. It is important to realize that a watershed is not a neatly labeled set of independent characteristics. The abiotic and biotic components come together to form an interacting system that is greater than the mere sum of its parts. This is an ecosystem, discussed previously in Chapter 1.

A watershed ecosystem is like a symphony of notes coming together to make music, or the parts of an automobile joined in just the right way to make it run. If you leave out or change a few notes, soon you have just noise. If you drop a few bolts or lose a few parts, the car will not run. Knowing how and why an ecosystem functions can help prevent the loss of key parts. Without the proper parts or notes, the watershed ecosystem falls apart.

5.4.1 The Physical Abiotic Environment

Three components provide the physical basis for a watershed – the climate, geomorphology, and hydrologic processes.

5.4.2 Climate

The difference between climate and weather is temporal (time-related). Determining climate requires data that are measured over an extended period, including average and extreme conditions in temperature, humidity, and precipitation (with type and amount), winds, and cloud cover. Weather is immediate. What is happening today? Current temperatures and meteorological events make up weather, and long-term weather trends establish averages that become climatic regimes. Climate heavily influences watershed vegetation communities, streamflow magnitude and timing, water temperature, and many other key watershed characteristics [10].

5.4.3 Geomorphology

Geomorphology is the study of the Earth's surface today. It is the field of study that examines the geologic structure of an area and its development. Geologic structures are landforms such as mountains, valleys, flats, and depressions. They evolve through a combination or series of causal events, such as tectonic plate movement, earthquakes, floods, volcanic eruptions, glacial formation and movement, waves, weathering, erosion by wind and water, and more recently human activities. It is sometimes difficult to see how these agents work because many are not present in today's landscape. Most of the existing landforms were created in the last 1.6 million years during the glacial cycles of the Pleistocene and the warming cycles of the Holocene epochs. The effects of these events are what we observe today.

Geomorphology is largely dependent on the type of geologic materials present and the soil that forms from them. Volcanic materials, water-deposited alluvium, or wind-deposited eolian sediments; all form different characteristics through the actions of deposition, removal, and

Fig 5.7 Cleveland Volcano eruption on the Aleutian Islands, Alaska, US (Photographed by Jeff Williams from the International Space Station, May 23, 2006)

Fig 5.8 The Snake River Canyon near Twin Falls, Idaho, US, provides an excellent view of layered parent material. The canyon has basalt layers mixed with sandstones and loose rocks. The basalts in the river continue to erode. This formation is called Pillar Falls
(Photograph by Karrie Pennington)

weathering (Figure 5.7). **Weathering** is the breakdown of materials from wind, water, freezing, thawing – in short, the actions of climate over time. The **parent materials** (geologic) can be hard bedrocks such as granites or basalt, or the more easily weathered sedimentary formations (see Figure 5.8). Soils that form from the weathering of parent material have a wide range of characteristics developed partially from the varying mineralogy of the parent materials. The rock fractures and cracks in the parent material, along with the water transport characteristics of the soil present, determine how water moves through the land portion of the watershed – both above and below the ground.

5.4.4 Hydrogeomorphology

Hydrogeomorphology is concerned with the source and flow of water within the landscape because each shapes the other. Water in a watershed can come from precipitation, as rain or snow, or it can be groundwater or surface water runoff. The source of water is a very important part of watershed chemistry, as well as watershed hydrology (water movement). The hydrogeomorphology method of evaluating and classifying landscapes is used to define specific landform types, their functions, habitats, vegetation types, soil characteristics, and hydrology. Hydrogeomorphology assessment tools are being developed for evaluation and classification of wetlands to help in restoration efforts.

5.4.5 Hydrologic Processes

Hydrologic processes can be the powerful forces of hurricane waves or floods, or as gentle as the flow of a meandering stream. The power of water is amazing – it can cut through rock or flow softly over it. One hydrologic process is catastrophic while the other is persistent, yet both processes are important in the shaping of the watershed. Less obvious, but no less powerful than waterfalls and raging rivers, is the impact of raindrops on exposed soil. The drops loosen the soil and wash it away, into furrows, ditches, and rivers – changing the landscape and the river.

5.5 The Biological (Biotic) Environment

5.5.1 Soils

The biological starting point for any watershed is the soil. Soil is a living system full of microbes, fungi, insects, and earthworms. Soil is home to gophers, snakes, and rodents – from ground squirrels to prairie dogs and plants ranging from fungi to trees. Soil is much more than "inert dirt." The components of our Earth, including its soils, are not isolated units. The pedosphere concept shows the soil as the meeting place of the atmosphere (air), lithosphere (rock), biosphere (plants, animals, microbes), and hydrosphere (water). Just looking on a larger scale, soil is part of global cycles of atmospheric gas exchange, rock weathering, water storage and partitioning, and terrestrial lifeforms, whether on the soil or in it (see Figure 5.9).

Soils have distinct structural forms derived from the weathering of the parent material from which they formed. They are modified by climate, the plants and organisms that live in and on the soil, and by the landscape over time. The interactions of the pedosphere determine the type of soil that develops in a particular location. Distinctive layers of soil, called **horizons**, form from the processes of **additions, deletions, transformation**, and **transportation** of the various mineral and organic soil components (see Figure 5.10). Basic soil horizons are classified as O, A, B, and C. There are many different soils – the US recognizes 20,000 soil series. A watershed usually contains more than one soil and can contain many more depending on its size and how it was formed. The International Soil Reference and Information Center (ISRIC) has soil data available in different formats for much of the globe (www.isric.org/).

Soils can be mineral or organic. A typical mineral soil is 45 percent mineral, 5 percent organic, and 50 percent pore space, which can be filled with water or air. The mineral material is a combination of sand, silt, and clay. The relative percentages of sand, silt, and clay make up the soil's texture. Soil texture, color, and structure – the arrangement of soil particles into larger units (called

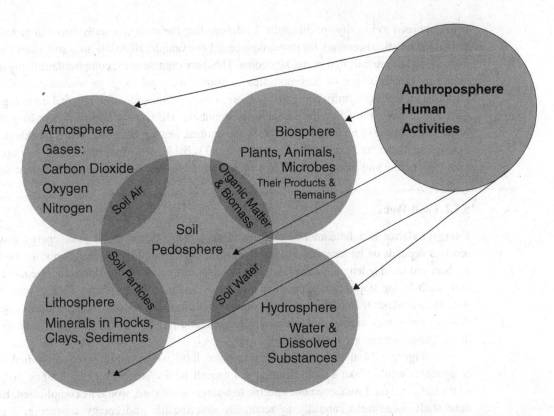

Fig 5.9 Pedosphere interaction with the atmosphere (soil air), biosphere (plant, animal, microbial life, and their products and remains), lithosphere (rocks), and hydrosphere (water). Soil is like the pot where the soup of life comes together. It is the foundation where animals stand or burrow, where plants put down roots and take out nutrients and water, where water flows, dissolving and moving materials in the soil, from minerals to added chemicals down into rock aquifers.

We chose to add the human element, anthroposphere, because humans interact with all five of the other spheres and we are uniquely capable of improving or harming all of these. Making our connections, recognizing, and accepting our responsibilities to the Earth we live on is crucial to its and our survival

(This diagram was modeled after the diagram of the late Ray R. Weil, with respect and admiration for his enduring contributions to our Earth) [11]

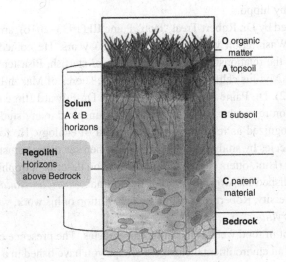

Fig 5.10 This is a representative soil profile with horizons identified. Not all soils have all horizons and not all soils form from the geologic material they are on (consider transported materials by wind (aeolian), water (alluvial), or gravity (colluvial). Decomposed organic matter (OM) and duff – a mix of partially decomposed leaves, pine needles, etc. – make up the layer above the mineral soil.

The A horizon begins the mineral portion and can contain varying amounts of OM and are often dark in color. B horizons contain minerals weathered in place or leached from above. C horizons are more like parent material but with plant roots, microbial activity, and water-movement weathering often apparent. The term "solum" is used for the A and B horizons where there is the greatest biological activity. The term "regolith" refers to the unconsolidated material above bedrock peds), are used to identify specific soils. Understanding the soils in a watershed will provide a great deal of information necessary for its management. For example, Histosols are a soil order. Order is the highest level of organization in soil taxonomy. These are organic soils; composed primarily of organic matter in various stages of decomposition – from leaves and twigs to undistinguishable humus. Histosols are light and fluffy compared to more solid, mineral soils. You can feel these organic soils give as you walk, providing the sensation of bouncing. Histosols are formed over long periods in water that has little or no oxygen present. When drained, aerobic microorganisms quickly decompose the organic matter, and cause the soils to shrink and subside, sometimes drastically. It is important to know this before you plant crops, build a house, a road, or any other structure on a Histosol.

5.5.2 Food Webs

Energy balance is a fundamental concept in ecosystems. How much solar energy a watershed receives depends on its location on Earth – specifically its latitude, which determines the height of the Sun and the day length. In the Northern Hemisphere, south-facing slopes receive more sunlight than north-facing slopes (the opposite is true in the Southern Hemisphere). Energy comes into the watershed ecosystem from the Sun, and plants transform it into food. Feeding patterns organize the flow of energy into, across, and out of a watershed, and provide energy for the organisms that live there. These patterns are **food webs**.

Figure 5.11 depicts a simplified aquatic food web. Adding even simplified terrestrial components would make a more detailed and complicated system. You can imagine that showing all the links in a food web, even one specific to a single watershed, would be complicated. Ecologists must study these webs carefully to accurately describe life and energy sources in a particular ecosystem.

There can be many species in the food web of any ecosystem. A species is a **keystone species** when it has a larger impact on the entire system than its numbers would suggest. The stage of development of the species, and the condition of the watershed, such as during a drought versus a flood, can change whether a species is a keystone species or not. The hippopotamus could be a keystone species because it is terrestrial and aquatic, and it wreaks havoc by uprooting plants and creating mud wallows. This significantly changes the land around it. During the rainy season, the adult hippo could be a keystone species. One adult hippopotamus would have a larger physical impact on an ecosystem than a baby hippo.

This concept was developed by Dr. Robert Treat "Bob" Paine III (1933–2016), an American ecologist with the University of Washington where he taught for 36 years. He coined the term "keystone species" while studying the relationship between a species of starfish, Pisaster ochraceus (sea stars) and a species of mussel (Mytilus californianus) found on the shores of Makah Bay on the Olympic Peninsula (see Figure 5.12). Dr. Paine is described much like Dr. Edward Birge (discussed later) as an out-in-the-field, hands-on researcher, who shepherded and inspired many students.

Dr. Paine's work was recognized as revolutionary in the field of ecology. He found other examples of important keystone species by studying what happened when sea otters vanished due to fur trading, predators, or pollution. (Hint, otters eat sea urchins.) Can you describe a trophic cascade that could have occurred? He published his work in 1966 in the journal *American Naturalist*. The leading ecologist at Princeton University, Robert MacArthur, in recognition of his work, wrote Paine a letter saying, *"This changes Everything"*.

Another important concept in food webs is the **indicator species**. The presence or absence of an indicator species can indicate an environmental change. Perhaps you have fished in a particular

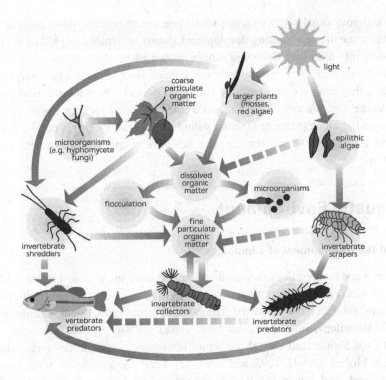

Fig 5.11 This simplified aquatic food web shows relationships between organisms in an ecosystem, in this case a lake. All the animals play essential roles in keeping the food chain healthy. Primary producers get energy from the sun, plants, algae, and phytoplankton, to make sugars from carbon dioxide and water and give off oxygen used by other aquatic organisms. Primary consumers are herbivores like ducks and mayfly nymphs but the most common are zooplankton tiny critters that eat phytoplankton.

Carnivores and omnivores make up secondary consumers eating other animals and or plants. These are young or small fishes and crayfish that eat zooplankton and insects. Tertiary consumers are larger fish, and even larger fish, perhaps otters that eat secondary consumers. Humans would be tertiary consumers, mostly omnivores.

Not to be overlooked, dead and decaying organisms are broken down by decomposers into nutrients that can be used by other lifeforms. Generally, these food webs are stable unless something removes a major player, and the web is broken (Courtesy USDA NRCS)

Fig 5.12 Dr. Bob Paine in his element. His idea was that if you removed the star fish (which he did with a crowbar and then threw them into the sea!), the mussels they eat would over-proliferate and limit other lifeforms. He told a reporter from the Seattle Times that "Experimental manipulation is not only more interesting, it's much more fun." Dr. Paine did find that algae and limpets were crowded out by mussels, and proved that removal of a keystone species could start a "trophic cascade," (chain reaction) that could drastically alter an environment [12] (Photograph courtesy of his daughter, Anne Paine)

pond for years, and your usual catch included bass. You notice that the water in your pond has gotten very muddy since the new housing development started construction uphill of the pond, and the soils are eroding into the water. Soon you cannot catch a bass.

In this location, the bass may be an indicator species. Bass hunt for food by sight, but if the water is too muddy, they can't see to catch food and could die out. Their absence is an indicator of environmental change – in this case, excess erosion. Scientists often look for a species that would naturally be in a specific **niche** (place or function within an ecosystem). If it is not there, they try to find out why. This is one method to determine what has gone wrong in an ecosystem, and what is needed to restore it.

5.6 The Aquatic Environment

5.6.1 Lakes and the Development of Limnology

Just as watersheds vary in size, from thousands of square kilometers to just a few hectares, the receiving bodies of water can also range from small ponds to huge lakes or rivers. Each waterbody is part of an ecosystem and each has a watershed. The study of inland lakes, ponds, streams, rivers, and wetlands is called **limnology**, from the Greek *limne* (lake) and *logos* (study). François Alphonse Forel (1841–1912) of Switzerland coined the term "limnology" in his three-volume monograph on Lake Geneva, published in 1892, 1895, and 1909 [13]. Forel was a professor at the University of Lausanne, Switzerland, and began his lake investigations while lecturing on physiology and anatomy. His work presented concepts and methods of study still used today.

A few years earlier, in 1887, Stephen Alfred Forbes (1844–1930) of the University of Illinois at Urbana, wrote an essay entitled "The lake as a microcosm," presenting lakes as understandable communities. He proposed that field observations of a complex system, such as a lake, could be examined to determine its properties, the predator–prey relationships, food chain structures, and other relationships.

This was a new and controversial concept. Previously, biological systems as large as lakes were considered too complex to describe in any meaningful scientific way. Forbes was able to combine science with creative thinking to formulate a new field of study. He merged field observation with his knowledge of physics, chemistry, and biology – whichever basic science was appropriate – to help explain the field observations. Forbes' work encouraged scientists to study lakes as biological, chemical, and physical systems, and to look for mechanisms to explain how a lake system worked – the science of limnology.

In Depth   The Lake as a Microcosm

Stephen Alfred Forbes, 1887, "The lake as a microcosm"

Note: We have chosen to share the first two paragraphs of this essay, just as the author wrote them, and have included the last paragraph because it gives a clear picture of our role as humans – the ultimate manipulators of watersheds. Even though the English is a bit different from the way we write today, the entire essay is remarkable, the concepts applicable, and the insights invaluable [14]

A lake is to the naturalist a chapter out of the history of a primeval time, for the conditions of life there are primitive, – the forms of life are, as a whole, relatively low and ancient, and the system of organic interactions by which they influence and control each other has remained substantially unchanged from a remote geological period.

The animals of such a body of water are, as a whole, remarkably isolated, – closely related among themselves in all their interests, but so far independent of the land about them that if every terrestrial animal were suddenly annihilated, it would doubtless be long before the general multitude of the inhabitants of the lake would feel the effects of this event in any important way. One finds in a single body of water a far more complete and independent equilibrium of organic life and activity than on any equal body of land. It is an islet of older, lower life in the midst of the higher more recent life of the surrounding region. It forms a little world within itself, – a microcosm within which all the elemental forces are at work and the play of life goes on in full, but on so small a scale as to bring it easily within the mental grasp.

Have these facts and ideas derived from a study of our aquatic microcosm any general application on a higher plane? We have here an example of the triumphant beneficence of the laws of life applied to conditions seemingly the most unfavorable possible for any mutually helpful adjustment.

In this lake, where competitions are fierce and continuous beyond any parallel in the worst periods of human history; where they take hold not on the goods of life, merely, but always upon life itself; where mercy and charity and sympathy and magnanimity and all the virtues are utterly unknown; where robbery and murder and the deadly tyranny of strength over weakness are the unvarying rule; where what we call wrong-doing is always triumphant, and what we call goodness would be immediately fatal to its possessor, – even here, out of these hard conditions, an order has been evolved which is the best conceivable without a total change in the conditions themselves; an equilibrium has been reached and is steadily maintained that actually accomplishes for all the parties involved the greatest good which the circumstances will at all permit.

In a system where life is the universal good, but the destruction of life the well-nigh universal occupation, an order has spontaneously risen which constantly tends to maintain life at the highest limit, – a limit far higher, in fact, with respect to both quality and quantity, than would be possible in the absence of this destructive conflict. Is there not, in this reflection, solid ground for a belief in the final beneficence of the laws of organic nature? If the system of life is such that a harmonious balance of conflicting interests has been reached where every element is either hostile or indifferent to every other, may we not trust much to the outcome where, as in human affairs, the spontaneous adjustments of nature are aided by intelligent effort, by sympathy, and by self-sacrifice?

Limnology became well known as a field of study in the late 1800s in Germany, with the establishment of the first limnological institute. In the US, Edward Birge and Chancey Juday established similar research efforts at the University of Wisconsin in Madison around 1891. Edward Birge recognized that aquatic systems could be better understood if their biology, physics, and chemistry were studied. He recruited scientists in these fields, as well as geologists, to help integrate their knowledge into what we study today as limnology. Several factors influenced today's limnologists and broadened the field of limnology from its traditional focus on natural lakes. These factors include:

- Increasing emphasis on research related to the effects of pollution on aquatic resources, and on ways to restore and manage these resources.
- Increasing diversity in the disciplinary backgrounds of limnologists.

Fig 5.13 Edward A. Birge, Ph.D.: "Limnologist in his element" Notice the patched raft does not stop the investigation
(Photograph with permission from University of Wisconsin-Madison libraries)

- Increasing variety in the types of lakes studied.
- Increasing focus on other types of inland aquatic ecosystems, such as streams, wetlands, and reservoirs.

Edward Birge the scientist was an interesting person, but Edward Birge the man was, as mom would say, a hoot (see Figure 5.13). Birge's student and future colleague, Robert Pennak, relates an anecdote about working with Birge in the field to illustrate this point (cited in Annamarie L. Beckel):

> After a Model A car, we were using had been turned on its side by slippery road conditions, Birge's comment was "... dammit Pennak, put it back on its wheels, the survey must go on!" We went out in all kinds of weather. If it rained that day, it didn't make any difference, you went out anyway. If it was windy, you went out anyway. We used these heavy old oak boats in those days. There were no life preservers. The boats were heavy, but none of us ever worried about having the boat dump over. If it had dumped over, I'm sure we would have drowned. . . . We never gave it a thought and worked along blissfully [15].

Robert Pennak went on to the University of Colorado where he worked in freshwater invertebrate studies. His book *Collegiate Dictionary of Zoology* is a reference classic and has been reprinted over 30 times [16]. His *Freshwater Invertebrates of the United States* is still used as a reference text [17]. Pennak's accomplishments rivaled those of his mentors. Sadly, his death in 2004, at age 92, marked the end of students taught by the team of Edward Birge and Chancey Juday.

There are many reasons to admire Edward Birge, but an important one was his ability to build interdisciplinary teams of scientists to work in the growing field of limnology. His most important recruit was his partner – Chancey Juday. Juday was an accomplished biologist and was able to continue field research and teaching while Birge was busy administrating. They were a grand pair, and both are considered among the founding fathers of limnology.

Birge and Juday realized that the water systems they studied were too complex to be adequately addressed by any one discipline. Birge used his experience to search the University of Wisconsin's faculty for chemists, physicists, geologists, and biologists. More importantly, he used his reputation to recruit them to work as a team. These teams had many different ideas and methods but were successful in their studies because they shared the common goal of understanding limnology. Birge and Juday established an international communications network among scientists, and a strong educational program at the University of Wisconsin. Edward Birge died in 1950 at the age of 98. It might be a good lesson for us to consider – these scientists, who spent their lives in vigorous outdoor activity and never gave up working, lived satisfyingly long lives.

5.6.2 Rivers and Streams

The aquatic environments of rivers and streams are studied separately from that of lakes because the ecology of lakes and rivers are significantly different. The most significant differences are flow of water and the location of energy fixation (where "food" is processed). Flowing aquatic systems are called **lotic**, while still water systems, such as lakes, are called **lentic**.

Water Flow

River flow is unidirectional (downstream) because water flows downhill. However, nature is seldom that simple. There are areas where **backwater flooding** can be a problem. An example of this occurs in the Delta area of Mississippi (see Figure 5.14). This area is a floodplain formed from thousands of years of flooding from the Mississippi River on the west and the Yazoo system on the east. The deep alluvium has a slope, north to south, of about 0.2 meters per kilometer (1 foot per mile), which is very flat. This 1.6 million hectare (4 million acre) expanse of flat ground is amazing to see. Anecdotal evidence indicates that, in flood situations before the levees were constructed, a small

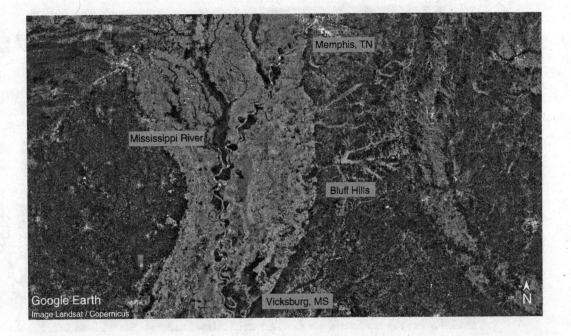

Fig 5.14 Google satellite image of the Mississippi Delta. The Mississippi River forms the western boundary and the bluff hills the eastern. Traditionally it is said that the "Delta" starts in Memphis, TN, and ends in Vicksburg, MS, a distance of about 200 miles. It was home to B.B. King and is often called the "Home of the Blues" for its many blues performers over the years
(Photograph courtesy of Google Earth)

boat could travel from the town of Greenville on the Mississippi River side, to Greenwood on the Yazoo River side, a distance of about 145 kilometers (90 miles). There are no hills to get in the way.

Another geographic anomaly is the river pattern and drainage. All rivers in the Delta run parallel to the much larger Mississippi River, are within its floodplain, and drain into the Mississippi at the same location. This is called a **yazoo system**. Now, take this system and add a massive levee, constructed on the eastern side of the Delta which blocks all inland flooding that normally came from the Mississippi River. In addition, an extension of the levee cradles the southern end of the watershed while the bluff hills span the eastern border. Can you see an interesting picture developing?

So, what is backwater flooding and what creates it? Flows in the Mississippi River can become so high that they actually force floodwater to move back upstream into the Yazoo River, then into the interior rivers – flooding the southern end of the Delta. The levees were built to prevent flooding. Normally, the water would have spread over a large area of river bottomland, but the levee acts like the sides of a huge bathtub – filling up the confined channel of the Mississippi River until it finds a low point where the Yazoo River would normally drain into it. Now, however, floodwaters flow backward, reversing the flow of the Yazoo River; hence, *backwater flooding*.

The US Army Corps of Engineers constructed gates (openings) in Steele Bayou (Figure 5.15) which allow water to flow out of the Delta times of the year. However, when the

Fig 5.15 The US Army Corps of Engineers built the Steele Bayou gates to protect the Delta from backwater flooding. Having no way to drain, water builds up on the land side of the gates. The river flow cannot get in, but water draining the watershed also cannot get out
(Photograph courtesy of Dean Pennington)

Mississippi River gets very high, the gates are closed to prevent backwater flooding. Remember that nature is not so easily thwarted; the area will still flood because the interior rivers can no longer drain, backing water up onto the land. This is also called backwater flooding because the water is backed up into the rivers and spills over as they overfill. Most of the actual Mississippi River water can be held out but flooding still occurs. This situation fits both definitions of backwater flooding – one by water flow and the other by backing up against a constructed dam, or in this case, a floodgate. This is an interesting problem that brings the phrase "Don't fool with Mother Nature," and the corollary "You can't win" to mind. A fascinating history of efforts to contain the Mississippi River can be found in John M. Barry's book *Rising Tide: The Great Mississippi Flood of 1927 and How It Changed America* [18].

Unlike lakes, rivers can be flashy. That is, they can have rapid changes in flow when it rains or with rapid snowmelt. For example, the Rillito River in Tucson, Arizona, is usually a dry gulch. However, a heavy rainstorm can create a raging river that you can watch develop in minutes. This phenomenon can occur in any area but tends to increase with increased development in the river's watershed. An increase in impermeable surfaces such as roads, parking lots, and rooftops, can cause water to move more rapidly to the river – resulting in this flashy behavior.

Another difference between the aquatic environments of lakes and rivers is that water entering a river is in transit immediately. Lake water moves but stays in more or less one place for a while. On the other hand, mountain streams rushing downhill are in constant motion. Conversely, river systems in areas with very little slope more closely resemble a lentic (still water) system during periods of low flow.

Location of Energy Fixation

Our discussion of lakes pointed to the fact that most energy is produced by plants and phytoplankton living in a lake, while the watershed provides nutrients via sediment transport. The primary producers and consumers live in the same environment. Like lakes, rivers get most of their energy from the watershed. However, organisms in rivers are mostly energy consumers, such as fish. The primary producers tend to live in the river substrate (material on the river bottom), and these are benthic organisms or macroinvertebrates. Rivers with a great deal of vegetation have usually received heavy and detrimental amounts of fertilizer from their watershed. This can be complicated in low-flow situations because a lake-like situation develops within a river or stream, without the mitigating influence of flowing water to carry materials to the depositional zone. The entire river system becomes a depositional zone – an area where water velocity slows down, and small particles fall out of suspension to be deposited on the river bottom.

Lotic (flowing water) systems are subject to the influences of their watershed, the atmosphere, and living creatures – including humans. These influences are the same as for lakes except modified by flowing water. Natural changes from mudslides, heavy rains, and fires can also take a toll on a riverine system.

5.6.3 Wetlands

Wetlands come in a wondrous variety of shapes, sizes, and types. They can be found in locations from the frozen north to the southern tropics. There are fens, bogs, salt and freshwater marshes, vernal pools, floodplain depressions, organic and mineral flats, playas, and tundra wetlands. The term "wetlands" conjures up visions of alligators, snakes, mud, mosquitoes, and still and dark water for some. Others see prairie potholes, depressional basins, and playas dotting the heartland, creating small and wet islands in the landscape. There are bogs and fens, peat wetlands rich in organics from vegetation in various stages of decay.

The more arid lands have riverine wetlands and vernal pools. Coastal areas have estuaries and marshes full of animal life - from crabs scurrying across the bottoms to avoid the great herons, to fish of all kinds, alligators, nutria, pelicans, seagulls, and shrimp. Tundras are permafrost areas containing life adapted to ice and long periods of semi-darkness. Tundra wetlands are usually pools or temporary waters from snowmelt.

Wetlands can be bottomland hardwood forests with ancient cypress trees, swamp oaks, shrubs, vines, flowers, and hanging moss. Emergent wetlands vary, but can have tall grasses, cattails, wild rice, rushes, and sedges. Riparian wetlands follow a river or stream and come in several varieties depending on location, climate, and geomorphology. They may be willow thickets or forested with cottonwoods, water oaks, nutall oaks, or rushes and grasses.

Wetlands in the watershed are often transitional areas between uplands and water, such as a riparian wetland or an estuarine marsh. They may appear to be isolated areas in the landscape, for example, a prairie pothole, or a playa basin. The fact is that wetlands are not isolated areas within the watershed but are an integral part of it. Wetlands have a variety of functions in a watershed because there are a variety of wetland types in different locations within the many different watersheds.

Wetland functions can be divided into biological, biogeochemical, physical, and hydrologic. The following is a partial list of wetland functions to be discussed more fully in the chapter on wetlands:

- Provide a diversity of habitats for wildlife, breeding, nesting, and foraging.
- Process organic matter and nutrients, nutrient cycling, and biological uptake.
- Help maintain water quality by filtering sediments and contaminants, and serve as natural buffers for streams, lakes, and rivers.
- Attenuate floodwaters and slow the rate of flow of waters going into streams and rivers.
- Replenish groundwater by providing a place where water can have time to permeate through the soil into an aquifer.
- Provide open space and esthetic value.
- Provide special areas for recreation, birdwatching, hunting, fishing, walking, photography, etc.
- Serve as educational and research areas to study the widely diverse wildlife, plants, and animals.
- Serve as nurseries and feeding areas for young fish and provide nutrients in food webs.
- Protect coastal areas from excessive erosion.

5.7 Watershed Function

It is crucial to understand the natural functions of a watershed to know how changes will influence those functions. For example, wetlands above and near the mouth of the Mississippi River, and all along the Gulf Coast, were lost as development occurred on the coast and especially in the New Orleans area. Two major functions of wetlands are floodwater retention and storm surge abatement. Without the wetlands to slow floodwaters, a city like New Orleans is more vulnerable to storm damage than in the past. This change in the watershed had very specific and far-reaching consequences during Hurricane Katrina in 2005 when floodwater inundated the city.

Think About It: Urban Planning

Few of our large municipalities grew through careful planning. Instead, most developed around and with industry and people. Today, we must be in a defensive position to deal with what we have built. Knowledge is the key to understanding if proposals by your city, state, or federal government are logical and workable. Knowledge is what you need to know how to vote and how to make wise decisions regarding issues that affect water and the environment. What are some issues that you, as a city planner, would try to deal with? How would you plan to reach your goals? Do you need to include any other people, groups, industries, or government entities, and if so, why?

We have examined the components of watersheds. Now, we'll tie them together to discuss the concept of watershed function. The three major functions of watersheds are as follows:

1. Movement and storage of water, nutrients, and energy.
2. Cycling and transformation of nutrients and energy.
3. Providing an opportunity for ecological change and succession.

5.7.1 Movement and Storage

The process of movement and storage in a watershed has five components that can and do occur individually or simultaneously. Materials moving through a watershed can be in various forms that influence their availability for use. For instance, phosphorus in its mineral form may be plentiful, but it is not in a form that plants can use; therefore, it is not available. **Availability** is the first important component of movement and a major part of nutrient cycling. Before any material can move freely, it must first be detached from anything holding it in place. Leaves must fall from the tree, or become detached, before they can become mobile in water. **Detachment** is the second essential component. **Transport**, component three, refers to the actual movement of the materials, regardless of form. **Deposition**, component four, is where movement stops. (Dissolved components may not be deposited and may be available to aquatic lifeforms.) The last component is integration into an organism, usually a plant or macroinvertebrate, i.e., something eats it one way or another.

Water is our primary mover so let's examine the path of water in a watershed. Keep in mind the hydrologic cycle components covered in Chapter 3. Rainfall can be intercepted by plants, where it may be simply slowed down, or caught and used. When used, it becomes part of evapotranspiration and is lost. Water reaching soil will infiltrate until the soil becomes saturated. Having no place to stop, the water becomes surface runoff. This water makes a path for itself by developing the familiar channels that eventually lead to streams, rivers, lakes, or oceans. **Sediment transport**, a major watershed function, cannot be separated from water transport. Sediments are **alluvial** (water-transported), or **eolian** (wind-transported) soil material. Sediments add nutrients and soil material to the watershed. They only become a problem when their rate of accumulation is high due to accelerated erosion, or when they reach a waterbody and introduce too much sediment and/or fertilizer.

5.7.2 Cycling and Transformation

Cycling and **transformation** of materials is largely a microbial function of movement and storage. Microorganisms are part of every ecological system, and exist in and on humans and other animals, in soil, in water, in and on plants, virtually everywhere. They are tiny little engines producing, mediating, and transforming all kinds of matter and energy through chemical reactions, most often oxidation–reduction reactions.

Discussions of nutrient cycling usually include nitrogen, carbon, and phosphorus because of their importance to water quality, soil fertility, and soil quality. A cycle refers to the transport, adsorption, transformations or chemical reactions, and loss and gain of the element studied. It includes everything that happens to the element in the occupied environment. These nutrients are exchanged between the land, water, and living organisms in an open cycle. Nutrients come in and go out.

Carbon is usually a part of the food webs and is an important part of cycling and transformation. Life on Earth is based on carbon, nitrogen, and oxygen. Carbon and energy move from plant forms to animals, or to decay, by microorganisms. Only about 1 percent of the solar energy reaching plants is used to produce food, and only about 10 percent of that food energy is transported between the trophic (various) levels in the food web. It is not an efficient system. Decomposition of plants and animals creates organic detritus, which is further broken down by microorganisms into usable food forms for fish, turtles, other small animals, and macroinvertebrates. Death and consumption, combined with transformation and transport, are what keep minerals available in the food web.

Nitrogen cycling involves several transformations depending upon its form and the presence of microorganisms. Nitrogen is plant available as nitrate and as ammonium. These ionic forms can be taken up by plant roots. Atmospheric nitrogen is a gas (N_2) and cannot be assimilated by plants. However, it can be fixed by microorganisms, called rhizobium, that live in nodules on the roots of legumes such as soybeans and clover. The rhizobia microorganisms cannot fix nitrogen without being associated with a plant. The plant and the microorganism have a **symbiotic** relationship, i.e., one that is beneficial to both organisms.

Nitrification is the process of converting ammonia (NH_3), to nitrite (NO_2^-) then to nitrate. All these reactions are aerobic (require oxygen). Denitrification is the anaerobic (does not require oxygen) process of converting nitrate and nitrite to nitrogen gas where it can be lost back to the atmosphere. Organic nitrogen forms, and dissolved organic nitrogen forms, come from plant and animal degradation. Nitrogen forms are not readily absorbed by soil particles and are considered mobile because they move with water.

Phosphorus is very different from nitrogen, in terms of cycling and transformation, because it is readily attached to many elements found in soils and does not move with water (is not mobile). Plant-available phosphorus is the ion form (PO_4^{-3}). This form also readily binds to soil organic matter, calcium, manganese, iron, and aluminum. Dissolved inorganic phosphorus occurs in water in small amounts. Most phosphorus in aquatic systems either is attached to sediments, organic matter, or is part of algae, phytoplankton, fungi, or other microorganisms.

Both nitrogen and phosphorus can be the limiting elements in aquatic systems; for review see Chapter 4. A balanced system has both nitrogen (N) and phosphorus (P), usually in an N: P ratio of 15:1. If the ratio is less, nitrogen is limiting; if more, then phosphorus is the limiting element. Freshwater systems are more often phosphorus deficient because it is less available than nitrogen. Ocean systems are most often nitrogen deficient, which is why they can be harmed by large inputs of wastewater.

5.7.3 Ecological Change and Succession

Ecological succession is a dynamic process. Today's ecologists consider a mosaic of successional stages within a watershed to be normal. Vegetation varies with landscape position, elevation, and spatial scale. Plants and energy are major factors in watershed function. Different plants provide a variety of habitats for wildlife and can promote biodiversity in the animal realm as well as the plant environment.

5.8 Water Quantity

Watershed functions are greatly affected by water quantity. The volume of water within a watershed can range from a small pond or wetland complex, covering a few hundred square meters (hundreds of square feet), to a large system like the Mississippi River watershed, which extends from Canada to the Gulf of Mexico. Groundwater levels often fluctuate, and river flows can vary greatly. Floods, drought, climate, and human activity all affect the quantity of water in a river, lake, wetland, or aquifer.

For example, dam construction, irrigation, and municipal/industrial uses along the Missouri River and one of its tributaries, the Platte River in the north-central US, have greatly reduced flows along the Nebraska–Iowa border in the past 150 years. In 1804, during their Voyage of Discovery, Meriwether Lewis and William Clark commented on the flows at the confluence of the Platte and Missouri Rivers, just south of present-day Omaha, Nebraska, and Council Bluffs, Iowa, as described by Stephen Ambrose:

> Lewis wrote a five-hundred-word description of the Platte ". . . running a mile wide and an inch deep, just bursting with animal and plant life." What most impressed Lewis was the immense quantity of sand the Platte emptied into the Missouri, and the velocity of the current. He measured, as he always did: "whereas on the Missouri below St. Louis a vessel would float at four miles an hour, and on the Missouri from five and a half to seven, depending, on the Platte that vessel would make at least eight miles an hour. Assuming it never ran aground, which it would at every bend or sandbar" [19].

Nearly 200 years later, author and historian William Least Heat-Moon retraced the route of Lewis and Clark, and described in his book, *River Horse*, a much different scene along the Missouri River near the same location:

> The bluffs were now lower and farther away, often blocked from view by trees, and the tilled bottomlands steadily came closer to the water until they eradicated the green margin to expose low banks of soft brown earth that today require miles of stone revetment to withstand the engineered current. We might as well have been in a wide canal . . . the impression heightened by old wooden markers giving mileage to the Mississippi, figures more than fifty miles greater than the distance these days, a difference made by channeling the Missouri and cutting out meanders. A shorter river, of course, is a less abundant river [20].

Let's briefly consider groundwater. Groundwater aquifers do not necessarily follow watershed boundaries. Deep aquifers, such as the Ogallala in the High Plains of the US, or the Guarani Aquifer of South America, generally lie beneath numerous watersheds and do not necessarily follow land-surface features. For example, the Ogallala Aquifer lies beneath portions of eight states, and the Guarani Aquifer lies beneath four countries – Brazil, Argentina, Uruguay, and Paraguay. Alluvial aquifers, on the other hand, are located beneath and adjacent to many rivers and streams. These tributary aquifers follow the course of rivers and streams since precipitation and river flow are its sources of water. An exception would be water artificially transported by humans from other watersheds.

Groundwater can have significant human and environmental impacts in a watershed. For example, shallow groundwater can provide baseflow (as discussed in Chapter 4) that could improve streamflow and water levels in wetlands and ponds. This emerging groundwater can beneficially improve water habitats for a wide range of invertebrates and other aquatic species. On the other hand, the availability and use of groundwater for domestic, industrial, and agricultural purposes can greatly change land uses in a watershed. Think of large irrigated farms that use groundwater in center pivot systems. Or, consider communities that can grow and thrive because of the availability of groundwater resources. What types of land use changes would occur in these situations due to the availability of groundwater in a watershed?

Guest Essay By Dr. Milada Matouskova

Milada Matouskova graduated from Charles University in Prague, Czech Republic, with a degree in geography and biology (1995) and completed her PhD thesis "Ecohydrological monitoring of streams as a basis for river rehabilitation" there in 2003. She joined the Department of Physical Geography and Geoecology at Charles University in 1997 after her foreign study of geography at the Johannes Gutenberg University in Mainz, Germany. She has taught courses specializing in the restoration of water ecosystems, management, and protection of water ecosystems, and a seminar on hydrology. She has been involved in research projects including hydromorphology, assessment of river habitat condition, and human impact on the river network, flood protection, quality of freshwaters and river sediments, and restoration of freshwater ecosystems.

Dr. Matouskova was awarded the Faculty of Science Award for excellent results in research management and in pedagogical activities in 2006. She is involved in the organization of international seminars and workshops within the Erasmus Program, which promotes upper-level student study in foreign countries to increase their world experience and knowledge.

An Example of Water Ecosystem Restoration and Its Influence on Flood Protection in the Czech Republic

The relationship between river restoration and flood protection can be seen on two levels. On the one hand, there is the potential effect of restoration measures on holding back and slowing down water discharge during floods. On the other, floods can act as an effective restoration factor in nature.

The main aim of restoration is the revitalization of, and care for, the optimum water regime in the landscape. The priority is to restore the retention ability of the landscape, which corresponds with the aims of flood protection. Flood events in Europe in 1993, in 1995 in the Rhine River basin, in 1994 in the Po River basin, in 1997 in the Oder River basin, and in 2002 in the Elbe River basin have brought about a change in the understanding of flood protection (see Table 5.1). Besides traditional technical approaches, eco-hydrological solutions close to nature are being applied. Attention is particularly being paid to the reduction of extreme flood flow. The strategy of providing the necessary space to rivers is generally supported. There should not be any further growth in urbanized areas in floodplains [21]. At the same time, however, it is necessary to change the approach of people to flood dangers because there have always been floods and always will be.

"The strategy for flood protection in the Czech Republic," prepared by the government of the Czech Republic in 2000, emphasized the necessity of the combined approach, using both technical and non-technical measures to increase the protection of people and property from the consequences of flood situations. The strategy clearly states that, apart from traditional technical

Table 5.1 Flood damage on main European rivers

River	Year	Total losses to society (millions $)	Total insured losses (millions $)
Rhine	1993	2,000	800
Po	1994	9,300	300
Rhine	1995	2,000	780
Oder	1997	5,275	785
Elbe	2002	18,500	3,000

flood protection measures, other measures are available, such as water overflow in areas where this is feasible. It is possible to obtain financial funding for such flood protection restoration measures from the European Union structural funds for improving landscape structures to support biodiversity, ecosystem stability, and water retention of the landscape.

The question of the retention ability of the landscape was widely discussed in the Czech Republic after the floods in 1997 and 2002. Floodplains represent natural, "cheap," effective, and permanent retention areas. If there is the possibility of overflow, they can reduce culmination discharges in lower river sections and help slow down the course of flood waves. However, the main aim of the restoration measures is not to reduce discharge only during floods. Their primary importance is the long-term increase in the retention ability of the landscape, i.e., during all types of water situations.

In terms of water retention, the structure and character of the land use are also important. The Czech Republic experienced fundamental changes in its landscape structure after World War II, particularly between the 1960s and 1980s. During these years – under the Communist regime – individual plots of land were put together into large tracts of land (50–200 ha) without any connection to the character of relief. Balks and turf were removed along rivers. The area of natural meadow decreased significantly in floodplain regions. These measures are associated with the large-scale drainage of agricultural land and alterations to rivers, particularly smaller ones, in agricultural land.

In the Czech Republic, the amount of river network modification carried out by humans is, on average, less than 30 percent; in the case of smaller rivers running through agricultural land, this figure reaches almost 40 percent (see Figure 5.16). These interventions undoubtedly affect the runoff regime and reduce the retention ability of the land, as shown, for instance, by the analysis of rainfall runoff trends in the mountainous areas of the Šumava (Black Forest). Flood protection in the second half of the twentieth century consisted mainly of carrying out technical measures such as building water reservoirs, protection dikes in towns and villages, increasing the volume of the cross profile of rivers, and strengthening their banks and riverbeds.

At the end of the twentieth century, the human approach to water ecosystems changed fundamentally, with emphasis being placed on the protection and restoration of natural river habitats. The restoration methods in practice concentrate on the creation of "near natural" riverbeds and the renewal of riparian belts. These restoration alterations to riverbeds usually result in reducing their discharge capacity, i.e., the opposite of what happens with technical flood protection. However, when one calculates flood losses, near-natural riverbeds often experience less damage than sections changed by human intervention. In the case of shallow natural riverbeds, the water overflows the riverbed during floods. The energy of the water flow is distributed to the

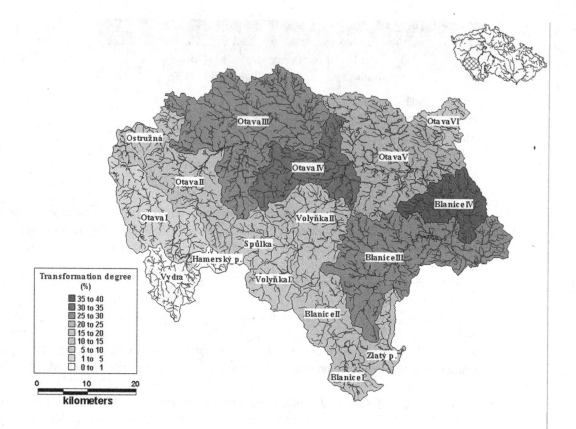

Fig 5.16 River network modification in the Otava River basin

riverbed and the bank zone. In the case of rivers straightened and strengthened by human intervention, the riverbed is the main path where the flow is concentrated. Modified riverbeds in urban areas are sized to take only n-year return period floods, usually 50- to 100-year return periods, which means that bank strengthening is usually not able to withstand catastrophic floods.

The fundamental prerequisite for successful, "complete" restoration is to provide the water ecosystem with sufficient space. In practice, this point is one of the largest problems, and not only in urban areas, which generally lack space. This is just one of the many reasons why restoration approaches are different in urban areas and non-urban areas. Not even the sections of rivers running through urbanized areas can be completely excluded from restoration because a river must be viewed as a continuous object from the spring to the mouth of the river. Therefore, partial restoration is recommended in such cases.

The above-mentioned flood events in the Czech Republic show that it is not possible to rely on only technical measures because many of them are not able to withstand more than 100-year return period floods. Of course, non-technical, passive flood protection measures carried out in non-urban areas cannot protect built-up floodplains from floods, but they can reduce the need for technical changes to riverbeds in urban areas. The factor of slowing down the so-called fast runoffs is also important as it can protect urbanized units located on confluences. It is therefore appropriate to look for the so-called retention compensation areas where water could be held for a certain period. The principle of controlled flooding could be applied.

Traditional hydro-technical flood protection measures include building polders, which can be flooded during floods. From the ecological point of view, it is advisable to build multifunctional half-dry polders which hold a relatively small amount of water in the given depression for the whole year and which are filled to their full capacity only during flood flow.

The creation of bypass flood channels is another option for increasing the retention ability of the landscape, which also functions as flood protection for towns and villages. During flood discharges, bypass channels can be used to transfer a certain amount of water away from urbanized areas. From the ecological point of view, it is advantageous if at least a small amount of discharge remains in the bypass channel throughout the whole year. Another alternative to bypass channels is the restoration of old river arms.

Retention ability is undoubtedly improved by the overflow of flood discharge into the floodplain, which can be either controlled or spontaneous. Dikes built away from the river, draining and flooding canals, and ditches are usually built for controlled overflowing. The space between dikes can be used as extensively managed meadows, floodplain forests, and areas for sport and recreation. Spontaneous overflow can be used only in non-urban areas with suitable vegetation.

Research conducted on model river basins, which have been restored and subsequently affected by floods, is considered vital. The river basin of the Borová brook near the town Český Krumlov in south Bohemia can serve as an example. The Borová brook was restored in two stages, in 1997–8 and in 2000. The main aim of the restoration was to change the riverbed character from an ameliorative riverbed into a near-natural one. A new shallow flow profile was created to allow overflow onto surrounding meadows. The Borová brook water basin was subsequently hit by a 100-year flood in August 2001. Only small flood losses were recorded on the restored brook because water overflowed into the flooding area, with an average width of 20 meters, which reduced both the speed of flow and the erosion ability of the water. The culmination discharge was reduced by almost 20 percent, which limited potential flood losses.

However, floods can act as an effective restoration factor. This was clearly shown in the Czech Republic by the flood in 2002 (see Figure 5.17). Thanks to natural fluvial–morphological processes, significant changes occurred, particularly in sections of modified riverbeds, which could be, to a certain extent, considered restoration. The banks of the Vltava River in Prague near Modřany can serve as an example. The flood in 2002 completed disintegration of the old regulation fortification made of stone tiles, which had been built for towing boats from the bank and which no longer fulfilled its function. There is a wide floodplain strip with alluvial growth behind the altered bank. It was, therefore, suitable to remove the non-functional technical fortification of the banks to restore the original near-natural character. However, for reasons unknown, the representatives of the river management decided after the flood in 2002 to restore the original technical fortification of the riverbanks.

The Branná River basin offers a similar example of a flood functioning as a restoration factor. The river springs in the Rychlebské Mountains in the northeast part of the Czech Republic. This typical mountainous river basin was hit by a flood in 1997. During the July flood, the water overflowed the banks of the altered (straightened and strengthened) riverbed onto the nearby meadows below the village of Ostružná. The width of the overflowing water reached between 10 and 30 meters. The 1997 flood caused a sudden change of the course of the riverbed. The "old" modified riverbed was blocked with gravel sediment. A new riverbed was gradually formed 2 to 10 meters from the modified one (see Figures 5.18 and 5.19). The new riverbed has a natural character in this section.

Fig 5.17 The catastrophic flood on the Vltava River in Prague in 2002. The flood was caused by two areas of low pressure and associated frontal systems moving across central Europe in a short period of time. Both areas of low pressure brought extreme rainfall to the territory of the Czech Republic. There was 5 billion cubic meters of rainfall in the Vltava River basin alone in 8 days. The flood claimed 17 lives in the Czech Republic, 225,000 people had to be evacuated, 753 towns and villages were affected, and the damage reached CZK$73.3 billion (Photograph by X. Kender)

Fig 5.18 The Branná River: original, blocked riverbed

The Czech Republic has experienced a fundamental change to its flood protection strategy. Besides technical measures, non-technical approaches, which concentrate mainly on the restoration of the retention ability of the landscape, are being supported. This aim corresponds with the main aims of the restoration of water ecosystems. Although the interconnection of flood protection and restoration is logical at the theoretical level, it is not always applied in practice. One of the biggest problems is that the space that can be returned to rivers is limited. It is necessary to find a compromise between the technical and near-natural solutions. Effective restoration measures can help to reduce the extremity of flood waves. We consider the restoration of the natural fluvial

Fig 5.19 The Branná River – the newly formed riverbed, during the flood in 1997 (photograph taken by Milada Matouskova in fall 2005). The flood in 1997 hit the east part of the Czech Republic, particularly the Morava and Odra River basins, where a total of 2.5 billion cubic meters of rainfall fell. Torrential rains in mountainous areas caused landslides and erosion, and changed riverbeds. Plains, on the other hand, suffered extensive flooding. The flood affected 538 towns and villages, claimed 60 lives, and the losses reached CZK$62 billion

morphological characteristics of riverbeds in non-urban areas in conjunction with the controlled or spontaneous overflowing of floodwater onto floodplains as the optimum solution [22]. However, the comprehensive restoration of river basins should not be ruled out.

5.8.1 Your Watershed

A watershed is a dynamic system, subject to constant change – physical, chemical, and environmental – on a daily basis. Water and gravity combine to cut and erode land features, sometimes very slowly and indiscernible to the naked eye. However, change can also occur in the catastrophic swiftness of flood, hurricane, or tornado.

Watersheds are vulnerable to human intervention. Land use changes, timber harvesting, water diversions from rivers, dam construction, and other significant human actions can drastically alter the quantity and quality of water resources in a watershed. In addition, fish and other wildlife can be severely impacted by such changes. Go to the USEPA "How's my waterway?" application at: www.epa.gov/waterdata/hows-my-waterway (accessed 2020) to find the location and boundaries of your watershed if you live in the US. Is your watershed experiencing rapid population growth, mining, or timber harvesting issues? Are water diversions increasing? What about water quality issues? It's important to be aware of water resource challenges in your watershed because they directly influence your lifestyle and your environment.

Summary Points

- A **watershed (catchment** or **catch basin)** is the land that drains into a body of water (stream, river, pond, lake, or ocean).

- **Watershed characteristics** are determined by its geomorphology, climate, and hydrologic processes along with land use by humans. A watershed is an **ecosystem**; all of its components working together to determine its functionality.

- **Structurally**, watersheds are simply uplands, lowlands, and a water feature. The complexity of this mix varies.

- The **Amazon River** and the **Mississippi River** watersheds are two of the largest in the world. The land use in each is very different resulting in different water quality characteristics.

- **Water quality** in the receiving waterbody is a function of what happens in the watershed, both natural and altered activities, as well as what happens in the waterbody itself.

- The **biology** of the watershed starts within its **soils,** which are alive with **microorganisms, worms, insects, fungi, and burrowing mammals.**

- **Soils** as the pedosphere are the sites of interaction between the atmosphere, hydrosphere, lithosphere, biosphere, and anthroposphere. These are the cornerstone of global reactions, influencing local and regional weather, and ultimately to the hydrologic cycle and global climate change. Their roles in food sustainability, water quality, and quantity are essential. Soils are formed by these global interactions manifested through climate, parent material, organisms, and topography or relief over time. Soils have distinctive structures of horizontal layers which influence properties and use. Understanding soils is key to understanding and managing your environment – from the small scale of your lawn to the entire earth.

- **Limnology** is the study of inland waterbodies, primarily lakes and rivers.

- The major differences between lakes and rivers are water movement and food sources. Lakes are **lentic** (non-flowing) while rivers are **lotic** (flowing). Lakes receive input from the watershed but produce much of their food within the lake. Rivers receive most of their food from the watershed.

- **Wetlands** are transitional features between uplands and water in many landscapes. They are land that is saturated long enough during the growing season to produce redoximorphic features in the soils, and support hydrophytic vegetation.

- **Wetlands** do many things (called **wetland functions**). Some of these include floodwater attenuation, nutrient cycling and transformation, habitat for wildlife, maintaining water quality, replenishing groundwater, and providing recreational opportunities.

- Ultimately, **how well a watershed functions** determines the health of the ecosystem. Watershed function depends largely on land use. Planning to maintain watershed functions should be part of any development in the watershed.

Questions for Analysis

1. You have been hired to design a confined animal unit to house a maximum of 1,000 milk cows. The land is in a watershed with a moderately sized river with year-round flow. What are some potential impacts of the confined animal unit? What are some watershed characteristics you should consider to minimize the environmental impacts of your design? What practices can you incorporate into your plan to protect water quality?

2. Which soil characteristics are most important in determining biological activity?
3. How can healthy wetlands influence flooding within a watershed?
4. Ecosystem health is highly dependent on watershed health. Explain.

Further Reading

Ambrose, Stephen E., 1996, *Undaunted Courage*, New York: Simon and Schuster.

Barry, John M., 1998, *Rising Tide: The Great Mississippi Flood of 1927 and How It Changed America*, New York: Touchstone.

Brotton, Jerry, 2013, *A History of the World in 12 Maps*, New York, New York: Viking Adult.

Croker, R. A., 2001, *Stephen Forbes and the Rise of American Ecology*, Washington, DC: Smithsonian Institution Press.

Forbes, S. A., 1887, "The lake as a microcosm," *Bulletin of the Scientific Association of Peoria, Illinois*, 77–87; repr. in Illinois Nat. Hist. Survey Bulletin 15, 537–550.

(As controversy surrounds many of the people we have long admired for their work, you must decide whether to throw out everything they have done, or to use what you consider valuable. Because this is a personal decision, we still include this reference regarding John Muir:)

Muir, John, 1911, *My First Summer in the Sierra*, September 2007, www.yosemite.ca.us/john_muir_writings/my_first_summer_in_the_sierra/index.html

Pielou, E. C., 1998, *Fresh Water*, Chicago, Ill.: University of Chicago Press.

Weil, Ray R. and Nyle C. Brady, 2016, *The Nature and Properties of Soils*, 15th ed, Upper Saddle River, N. J.: Pearson Education.

Wheatley, Margaret, 2002, *Turning to One Another: Simple Conversations to Restore Hope to the Future*, January, Berrett-Koehler Publishers.

References

1 Margaret Wheatley, 2002, *Turning to One Another: Simple Conversations to Restore Hope to the Future*, January, Berrett-Koehler Publishers.
2 USGS, National Geospatial Program, "Historical topographic maps – preserving the past," www.usgs.gov/core-science-systems/ngp/topo-maps/historical-topographic-map-collection?qt-science_support_page_related_con=0#qt-science_support_page_related_con (accessed 2020).
3 Uri Friedman, 2013, "12 Maps that changed the world, is there such a thing as a perfect map?" *The Atlantic*, www.theatlantic.com/international/archive/2013/12/12-maps-that-changed-the-world/282666/ (accessed 2020).
4 Jerry Brotton, 2013, *A History of the World in 12 Maps*, New York, New York: Viking Adult, SBN-13: 978-0670023394.
5 Robert F. Brinkman and Chris O'Neill, 2000, *The Military Engineer*, no. 605, May–June.
6 Ibid.
7 Richard B. Langley, 2006, Department of Geodesy and Geomatics Engineering, University of New Brunswick, http://gge.unb.ca/Resources/HowDoesGPSWork.html
8 National Aeronautics and Space Administration (NASA), "Visible Earth," May 2007, http://visibleearth.nasa.gov/

9 Ray R. Weil, and Nyle C. Brady, 2016, *The Nature and Properties of Soils*, 15th ed, Upper Saddle River, N. J.: Pearson Education.

10 US Environmental Protection Agency (USEPA), "Watershed ecology," https://cfpub.epa.gov/watertrain/moduleFrame.cfm?module_id=20&parent_object_id=516&object_id=516 (accessed 2020).

11 Ray R. Weil, and Nyle C. Brady, 2016, *The Nature and Properties of Soils*, 15th ed, Upper Saddle River, N. J.: Pearson Education.

12 Sam Roberts, 2016, "Robert Paine, UW ecologist who identified 'keystone species,' dies at 83," *New York Times*, www.seattletimes.com/nation-world/robert-paine-uw-ecologist-who-identified-keystone-species-dies-at-83/ (accessed 2020).

13 François Alphonse Forel, 1892, 1895, 1904, *Lac Génève*, vols. 1–3, Lausanne, Switzerland: University of Lausanne; see also Committee on Inland Aquatic Ecosystems, National Research Council, 1996, *Freshwater Ecosystems: Revitalizing Educational Programs in Limnology*, Washington, DC: National Academies Press.

14 S. A. Forbes, 1887, "The lake as a microcosm," *Bulletin of the Scientific Association of Peoria, Illinois*, 77–87; repr. 1925 in Illinois Nat. Hist. Survey Bulletin 15, reprinted at: www.wku.edu/~smithch/biogeog/FORB1887.htm.

15 Annamarie L. Beckel (ed.), 1987, "Breaking new waters: a century of limnology at the University of Wisconsin," *Transactions of the Wisconsin Academy of Sciences, Arts, and Letters*, Special Issue.

16 Robert W. Pennak, 1964, *Collegiate Dictionary of Zoology*, New York: Ronald Press.

17 Robert W. Pennak, 1953, *Freshwater Invertebrates of the United States*, New York: Ronald Press.

18 John M. Barry, 1998, *Rising Tide: The Great Mississippi Flood of 1927 and How It Changed America*, New York: Touchstone.

19 Stephen E. Ambrose, 1996, *Undaunted Courage*, New York: Simon and Schuster, p. 150.

20 William Least Heat-Moon, 1999, *River Horse*, Boston, Mass.: Houghton Mifflin, p. 243.

21 P. H. Nienhuis and R. S. E. W. Leuven, 2001, "River restoration and flood protection: controversy or synergism?" *Hydrobiologia* 444, 85–89.

22 D. Hulse and G. Stan, 2004, "Integrating resilience into floodplain restoration," *Urban Ecosystems* 7, 295–314.

6 Groundwater

If we lived in a desert and our lives depended on a water supply that came out of a steel tube, we would inevitably watch that tube and talk about it understandingly. No citizen would need to be lectured about his duty toward its care and spurred to help if it were in danger. Teachers of civics in such a community might develop a sense of public responsibility, not only by describing the remote beginnings of the commonwealth, but also how that tube got built, how long it would last, how vital the intake might be if the rainfall on the forested mountains nearby ever changed in seasonal habit or amount. It would be a most unimaginative person, or a stupid one, who could not see the vital relation between the mountains, the forests, that tube and himself.

Isaiah Bowman, American geographer (1878–1950) [1]

Chapter Outline

6.1 Introduction

Groundwater is a fascinating and yet somewhat unfamiliar aspect of water resources. We tend to think of groundwater as a separate resource, not connected to surface water, because we cannot see it and it is difficult to measure. The more we learn about using water, both surface and groundwater, the more we realize that both are often connected by interactions with wetlands, streams, lakes, and rivers. Surface and groundwater are connected, as all water on Earth is connected, through the hydrologic cycle.

Groundwater comprises about 30 percent of the world's freshwater resources – sometimes found at shallow depths in river valleys, but in other locations hundreds of meters (feet) beneath the land surface. Precipitation becomes groundwater in a variety of ways – through infiltration at the land surface, through the bottom of streams, wetlands, ponds, lakes, and reservoirs, and from saltwater intrusion in some coastal regions of the world. Groundwater can return to the land surface naturally under the force of gravity or through well pumping for human use.

Approximately 2 billion people around the world rely on groundwater as their only source of drinking water [2]. In Canada, groundwater provides over 30 percent of the domestic freshwater used by its entire population. On Prince Edward Island in eastern Canada, the figure is 100 percent. In Germany, two-thirds of the population relies on groundwater, while in the US about 43 percent of all irrigation water comes from groundwater [3]. Unfortunately, many countries, including China, India, Mexico, New Zealand, Pakistan, most of the Middle East, the United States, North Africa, and portions of Canada have regions experiencing serious groundwater depletions. Declining aquifer impacts reach far beyond the activities of humans, with groundwater depletion now having serious consequences on the environment. Many water professionals consider groundwater management to be the greatest water resource problem of the twenty-first century.

6.2 The Physical Environment

Most groundwater originates as precipitation that falls on the land surface and then infiltrates below ground. The process of precipitation becoming groundwater is called **recharge** (see Figure 6.1) [4]. Varying amounts of precipitation may seep into the soil below the land surface and still not completely saturate it. This upper layer of soil, that is neither continually nor completely saturated, is the **unsaturated zone** or **vadose zone** (from Latin *vadose*, meaning shallow). Water found in this region is commonly called **soil water**, and most of it can be taken in by plant roots. However, air, and not water, fills most soil pore spaces in the unsaturated zone most of the time.

Below the unsaturated zone, where all the pore spaces, cracks, and other openings between geologic materials are filled with water, is the **saturated zone** or **phreatic water** (from Greek *phrear*, meaning spring or well). The term **groundwater** is used to describe this water-filled

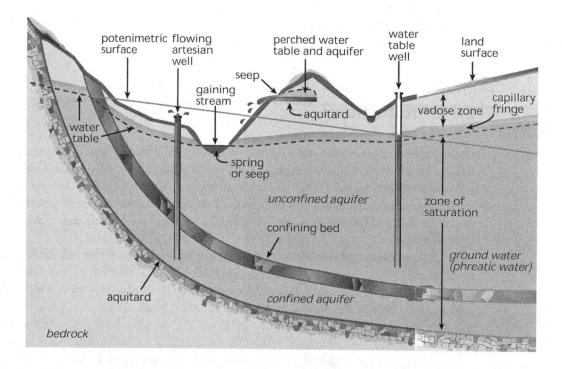

Fig 6.1 Pictorial representation of an aquifer and commonly used groundwater terms (for definitions see text) (Image courtesy of *Stream Corridor Restoration: Principles, Processes, and Practices*, October 1998, by the Federal Interagency Stream Restoration Working Group (FISRWG)) [5]

Table 6.1 Porosity of common geologic materials	
Material	**Porosity (%)**
Soil	55
Gravel and sand	20–50
Clay	50–70
Sandstone	5–30
Limestone	10–30
Fractured igneous rocks	10–40

Fig 6.2 Relative permeability of some common geologic materials from high(gravel), to low (shale). High permeability results in high transmissivity

geologic region. An **aquifer** is a large geologic formation of groundwater capable of yielding enough water to supply the needs of humans.

The top of the saturated zone is called the **groundwater table** (it is also the top of the aquifer and may be simply called the **water table**). An interesting point to remember is that the elevation of the groundwater table does not necessarily follow the elevation contours of the land surface. Water moves from high areas to low areas underground due to the force of gravity, just as it does above ground. However, the path of groundwater movement may be blocked by confining materials, **aquitards** (such as clay or bedrock) that block or impede groundwater movement.

Aquifers are **recharged** by precipitation seeping downward through the unsaturated zone to the groundwater table. Recharge quantities and rates are greatly influenced by the soil and geologic materials beneath the land surface. Aquifers of unconsolidated sand and gravel, found in the alluvial fill material of some river valleys, have interconnected cracks and pore spaces. These openings are large and numerous enough to retain groundwater and allow it to move freely. These materials are said to have a **high porosity** (see Table 6.1). Consolidated material (meaning the geologic substances such as grains of sand, silt, or other materials that are cemented together, such as sandstones) are tighter formations and have **low porosity** (the ability for groundwater to reside between the grains of sand and gravel) and **low permeability** (see Figure 6.2). Groundwater moves very slowly, if at all, through this type of material. Fractured rock aquifers, such as granite and basalt, are geologic formations with cracks that allow for the movement of groundwater (see Figure 6.3).

The rate of groundwater movement in an aquifer is called its **transmissivity**. A formation that allows the relatively rapid movement of groundwater is said to be **highly transmissive** or have a **high transmissivity**. In some cases, groundwater may move several meters or feet in a single day through **highly transmissive** material. In geologic settings with **low transmissivity**, such as clay or shale, groundwater may only move a few centimeters or inches in a century. It's important to note that groundwater recharge characteristics can vary greatly all over the world, and even in the same local area. Human activities, such as the construction of roads, buildings, parking lots, and other impervious surfaces, can greatly alter the recharge rates of an aquifer in a region. Remember that the

Fig 6.3 Groundwater can flow through fractured rocks. Water enters the highly permeable Lava Fields of Craters of the Moon area and travels nearly 200 years to emerge as waterfalls in the Middle Fork of the Snake River here at Thousand Springs, Idaho. Groundwater does not occur at great depths because the weight of the overlying geologic material compresses the pore space so tightly that groundwater cannot penetrate. At best, groundwater is found only within a few kilometers (miles) of the Earth's surface
(Photograph courtesy of Idaho Department of Parks and Recreation and The Nature Conservancy)

direction of groundwater movement does not necessarily follow the direction of surface water flows due to the characteristics of the aquifer layers.

Groundwater generally moves toward a lower elevation beneath the land surface. In some locations, groundwater can eventually move – under the force of gravity – into a river, stream, lake, wetland, or ocean. This process of groundwater moving back to the land surface is called **groundwater discharge**. Once discharged to the land surface, groundwater will ultimately evaporate to the atmosphere to renew the hydrologic cycle. In unique situations, water pressure can force groundwater upward toward the land surface. This can occur when groundwater is contained in a confined aquifer by impervious (watertight) geologic layers, such as a clay layer or bedrock. If enough **hydrostatic** (water) pressure exists in the aquifer to force groundwater up a crack or fissure until it meets that land surface, an **artesian spring** or **flowing artesian well** is created (see Figures 6.4 and 6.5).

Groundwater discharge can provide significant amounts of water to surface water systems. During dry periods, groundwater discharge may be the primary source of water maintaining streamflow, also called **baseflow**. Groundwater discharge can also help maintain water levels in wetlands, ponds, and lakes. During wetter periods, groundwater table elevations may increase due to the recharge of surface water runoff from storms and other precipitation events. These hydrologic events can increase groundwater discharge.

The **residence time** of recharged groundwater is the length of time water resides in an aquifer before discharging to the land surface. Residence time varies greatly. Some groundwater may remain underground for a few days or weeks, or may reside in deeper aquifers for thousands of

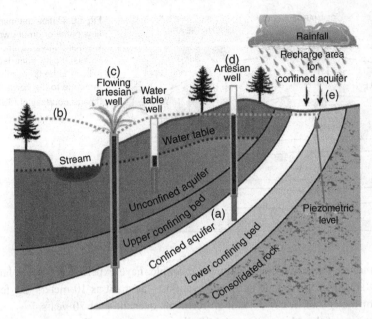

Fig 6.4 Important points to notice in this illustration of artesian wells are (a) the confined aquifer where pressure can build, (b) the piezometric level (potentiometric surface in Figure 6.1) is the level to which water will rise in a well drilled into an artesian aquifer, depicted by a dashed imaginary line), (c) if the water pressure is high enough to raise the water to the land surface, the well will be free-flowing, (d) this well is artesian, but the water pressure did not bring water to the surface, so it is not free-flowing, and (e) because the aquifer (a) is confined, its recharge comes from surface precipitation events. Without pumping, groundwater levels in the well fluctuates with the level of the water table (Image courtesy of USGS, converted to black and white)

Fig 6.5 Artesian well, free-flowing near Rapid City, South Dakota, US, since 1956. Notice the landscape around this well. Does it look like cropland? What could be the primary use of this water supply? (Photograph taken April 7, 2016, courtesy of South Dakota Water Science Center, USGS)

years, sometimes called **fossil water**. By contrast, the average residence time of water in a stream may only be a few days or weeks (see Figure 6.6).

Excessive groundwater pumping can reduce groundwater levels to a point that cannot be fully replaced by recharge. This situation of over-pumping is called **groundwater mining**, or **ground-water overdraft**, and is an increasing problem in many regions of the world. Groundwater mining can be a temporary condition, or can last for decades or longer if recharge rates are inadequate to replace depletions caused by well pumping. In some locations, groundwater levels have declined so far that it is uneconomical to continue pumping, called **economic depletion of groundwater**. In this situation, other water sources may be required, or in the worst situation, previous water use practices must be abandoned.

Excessive groundwater pumping can cause drastic effects on landscapes and humans. **Land subsidence** can occur when large quantities of groundwater are pumped from an aquifer. Certain types of geologic material – such as fine-grained sediments – settle, or compress, when

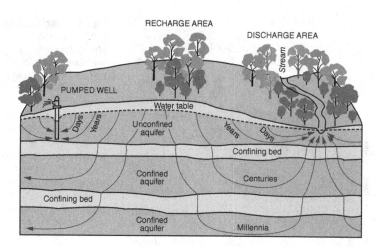

RECHARGE AREA

DISCHARGE AREA

Fig 6.6 Notice the many different flow paths of groundwater around wells and streams, and the varying times, ranging from days to millennia, the flow paths take from recharge to discharge (Image courtesy of FISRWG)

groundwater is removed. Parts of Mexico City, for example, have experienced severe land subsidence. Gradual, but steady drops in land surface elevations – as great as 10 meters (33 feet) – have occurred in parts of the city of nearly 20 million residents over the past 70 years.

Over pumping of the Mexico City Aquifer has caused this subsidence and has created a multitude of problems with the city's water supply and sewerage systems. The main cathedral in Mexico City, the yellow-domed *Basilica de Guadalupe*, leans off center because of the effects of land subsidence. Some regions of the city suffer subsidence of nearly 0.3 meters (1 foot) per year, caused by groundwater pumping from beneath the ancient lakebed of Lake Texcoco, now covered by urban development [6].

In the San Joaquin Valley of central California, extensive groundwater pumping for irrigation has caused the land surface to drop nearly 9 meters (30 feet) since 1925 (see Figure 6.7). This has resulted in damage to buildings, bridges, aqueducts, and highways, and has cost US millions of dollars. The downtown area of Las Vegas, Nevada, has subsided over 2 meters (6 feet) in southern Nevada. Subsidence is not only a serious and expensive issue for facilities on the land surface but can permanently prevent historic levels of recharge from occurring in depleted aquifers. In the US alone, more than 44,000 hectares (170 square miles) in 45 states have experienced subsidence issues.

Saltwater intrusion is another problem related to over pumping of groundwater aquifers. In coastal regions, excessive groundwater pumping caused by population and economic growth sometimes create water demands that cannot be replenished by recharge. In these situations, ocean saltwater can intrude into the freshwater aquifer at a greater rate than normally occurs, contaminating it with saline water. This is happening in many communities of the Atlantic and Gulf States in the US, and in Spain, Italy, Turkey, India, Australia, and other locations around the world, since nearly 70 percent of the world's population lives in coastal areas.

6.3 Interaction of Surface Water and Groundwater

Surface water and groundwater often interact in a wide range of physical, chemical, and biological processes. These can also occur in a variety of climatic and physical settings. For example, the process of surface and groundwater interaction has primarily been studied in large alluvial aquifer and stream situations. However, issues of drinking water contamination, eutrophication of lakes, loss

Fig 6.7 This site is in the San Joaquin Valley southwest of Mendota, California, US. It is the approximate location of maximum subsidence in the United States identified by the research efforts of Dr. Joseph F. Poland (pictured). Signs on the pole show approximate altitude of land surface in 1925, 1955, and 1977, a total subsidence of 9 meters (27 feet) (Photograph courtesy of USGS)

of wetlands, and other changes in aquatic environments are occurring in headwater streams, small rivers, lakes, wetlands, and coastal regions.

Groundwater often intersects with the shoreline of a river, lake, or other waterbody, creating a saturated zone beneath the land surface in this area. Generally, precipitation passes quickly through any thin, unsaturated zone near the land surface, and forms groundwater mounds adjacent to the surface water (see Figure 6.8). This can result in a short-term increase of groundwater discharge to a surface waterbody.

Transpiration of plants near a stream or lake shoreline sometimes has the opposite effect. Plant roots can penetrate the saturated zone of an aquifer and intercept groundwater slowly moving toward a river, lake, or wetland. Transpiration is a highly variable daily and seasonal event and can affect groundwater somewhat like a well being pumped. In some locations, it is possible to measure the diurnal, or 24-hour, flow changes in the direction of groundwater movement during the seasons of active plant growth.

For example, increased plant transpiration during high-growth summer months reduces groundwater movement toward a waterbody. Daily, groundwater movement will increase into the

Fig 6.8 Mississippi River Lock and Dam Number 7 with the I-90 Mississippi River Bridge downstream. The volume of water in this river determines groundwater recharge. The amount of recharge varies with distance from the river and aquifer characteristics
(Photograph courtesy of USGS)

surface water during the night but then decrease during the high-transpiration daylight hours. At the same time, surface water may be drawn from a lake and into the shallow groundwater during the daytime, as plant roots draw upon larger quantities of groundwater, during hot summer daytime hours [7]. These periodic changes in the direction of groundwater movement can also occur over a longer period – seasonally, or between wet and dry cycles. Recharge may occur during wetter periods of precipitation, but groundwater discharge may predominate during dry periods. This variation of groundwater recharge and discharge can be extremely variable, and important, particularly along the hydrologic boundaries of small surface waterbodies.

6.3.1 Interaction of Groundwater and Streams

Streams and rivers interact with groundwater in three basic ways:

(1)   discharge from groundwater – called a **gaining** or **effluent** stream;
(2)   loss of surface water to groundwater – called a **losing** or **influent** stream; and
(3)   a combination of both situations as described above [8].

These groundwater/surface water interactions are dependent upon the time of year, precipitation events, and variability of the geology along a particular stream segment, lake, or wetland. For surface water to move into a groundwater system, the height, or elevation, of water in the stream must be higher that the contours (elevation) of the groundwater table. However, the opposite

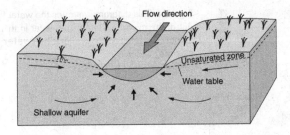

Fig 6.9 A **gaining or effluent** stream. Notice that the arrows in the groundwater system point toward the stream, indicating a gain of water
(Image courtesy of USGS)

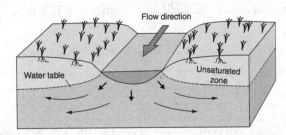

Fig 6.10 A losing or influent stream. In this case, the arrows point away from the stream indicating a loss of water
(Image courtesy of USGS)

conditions must be present for groundwater to seep into a stream, pond, or wetland. Groundwater pumping in these types of systems can have a profound effect on the direction of groundwater movement – particularly near the shoreline of a surface waterbody (see Figures 6.9 and 6.10).

Groundwater interacts with rivers, streams, lakes, and wetlands in a variety of ways, dependent upon the climatic setting and terrain. For example, in mountainous regions, the decay of plant roots and fractures in rocks may allow significant amounts of moisture to move through the subsurface. In some locations, this creates **hillside springs**, or elevated groundwater levels at lower parts of mountain valleys. Snowmelt and precipitation events can exceed the infiltration capacity of limited groundwater aquifers, and create overland flow in various settings. At the base of some mountainsides, groundwater might accumulate and create wetlands, called *fens*.

The high gradient and coarse textures of river bottoms in mountain streams often cause a frequent exchange of surface and groundwater. As surface water flows rapidly over rough streambeds (through pools, over rocks, and around logs), some surface water may flow beneath the steep sections of the stream (called *riffles*) to mix with shallow groundwater creating the **hyporheic zone**. The hyporheic zone is a place of interactions, an **ecotone**. It is where two distinct environments – stream water and groundwater – come together to form a gradient of in-betweens, in the region of sediment and porous space, located beneath and alongside a stream bed (see Figure 6.11) [9].

The hyporheic zone has been identified as critically important to the health of watersheds. It provides an ideal habitat for biofilms of bacteria and fungi to develop on rocks and pieces of wood. These bacteria and fungi cycle nutrients back into the stream that are critical to primary producers at the base of the aquatic food web. Higher up the food web, adult salmon excavate the streambed with their tails to lay eggs where they are safe from the dangers of predators and swift currents. The stream water flushes through the eggs, providing oxygen and controlling the incubation temperature. Larval salmon stay in the shallow hyporheic zone in their early stages of development. Many aquatic insects also inhabit this zone.

There are many other functions of the hyporheic zone besides habitat functions. Pollutants in the water are biodegraded, absorbed, or diluted and water temperatures are regulated as warmer

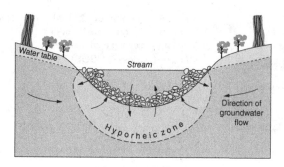

Fig 6.11 This diagram illustrates the water flow into and out of the hyporheic zone. The water in this area is sometimes referred to as interstitial **(water in small spaces)** water
(Image courtesy of USGS)

surface water interacts with colder groundwater. In addition to nutrients, it is a zone for cycling carbon and energy [10].

Think About It: Hyporheic Zone

As discussed, the area where groundwater and surface water interact is called the hyporheic zone (see Figure 6.11). It often has chemical and biological characteristics markedly different from adjacent surface water or groundwater. These streambeds and banks of streams and lakes are unique environments and are altered when groundwater levels decline. Why would the characteristics of the hyporheic zone be different from surface water or groundwater? Do organisms live in groundwater?

Some mountain streams flow across alluvial valleys, created by sand, gravel, and silt accumulations left by centuries of floodwater deposition. Surface water often infiltrates to groundwater in these highly permeable locations, and are readily found in arid regions of the world. This stream seepage is generally the primary source of recharge water for an alluvial aquifer. The extent of recharge in this type of system can be difficult to predict due to changes in streambed characteristics and the difference in **hydraulic head** (elevation and water pressure) between the stream and the aquifer.

Some streams do not have well-developed floodplains, and the amount of surface and groundwater interaction is limited to flooding and evapotranspiration. In these conditions, the amount of interaction varies seasonally. A large river, such as the Mississippi in the US, with wide river terraces and prevalent sources of regional groundwater recharge, has a readily available source of groundwater discharge to a stream system year round. On the other hand, a small river, such as the Olifants River in South Africa, with its relatively shallow alluvial aquifer, is more likely to experience seasonal flow changes.

The direction of movement of water from a stream into an aquifer, and from an aquifer into a stream, can be consistent along the entire length of a river. It can also vary by river reach (segment) or can be altered by a precipitation event. Individual storms can change the flow direction very quickly. Temporary flood flow peaks moving down a river channel can also rapidly recharge adjacent unsaturated groundwater zones (called **bank storage**). These temporary flows also provide water for plants and reduce the transpiration uptake from groundwater. The sources of temporary flood events occur from storm precipitation, snowmelt, or the rapid release of water from an upstream reservoir.

If such a flood does not overflow riverbanks, most of the bank storage will return to the stream in a few days or weeks. If floodwater extends over and beyond the riverbanks, some will recharge the groundwater aquifer. If distances from the river are great, it may extend the length of time needed for the groundwater to discharge back into the stream. This could take weeks, months, or even years. These processes tend to reduce the volume of flood peaks and can reduce the magnitude and intensity of storm events on river flows. Some riverine systems are in a constant state of readjustment from the interactions of bank storage and overbank flooding.

6.3.2 Interaction of Groundwater and Lakes

Lakes have three basic interactions with groundwater:

(1) Some lakes receive groundwater discharge throughout the entire lakebed.
(2) Some lakes have seepage loss to groundwater throughout the entire lakebed.
(3) Some lakes receive groundwater discharge in part of the lakebed, and in other areas have seepage loss to groundwater.

The interaction of surface and groundwater in lakes differs significantly from that of streams and rivers. First, the water levels of natural lakes (not reservoirs, which are created by humans) do not generally change very quickly unless a significant storm event occurs in the region. Bank storage does not generally occur as frequently, and evaporation is generally greater in larger lake systems than in rivers. Lake water is often not replenished as quickly as water in a stream. Therefore, lake sediments may contain greater volumes of organic material, which tend to "seal," or reduce permeability of lake bottoms. This can limit both groundwater discharge to a lake and groundwater recharge from the lake. Such sediments also limit biogeochemical processes between surface and groundwater systems in these locations.

6.3.3 Interaction of Groundwater and Wetlands

Wetlands are like streams and lakes in that they often receive groundwater discharge, provide recharge to groundwater, or can do both during various times of the year. Wetlands can perform these same functions during weather cycles (drought vs. floods, for example). Wetlands located in land depressions can have groundwater interactions similar to lakes and streams. However, some wetlands are located on sloped land, such as fens, that receive discharged groundwater commonly containing a continuous supply of dissolved chemicals. By contrast, a wetland may be in upland regions on drainage divides, sometimes called **bogs**, and receive much of its water supply from recent precipitation.

The period and extent of water in a wetland varies by its location, climate, and recent weather patterns. Groundwater discharge and surface water flooding are common to **riverine** wetlands (wetlands adjacent to rivers). Predictable tidal cycles affect wetlands located near coastal regions. Both flooding and incoming tides create water level changes that make understanding water movement in these wetlands difficult.

The depth, frequency, seasonality, and duration of water level changes in a wetland make up the **hydroperiod**. The hydroperiod of a wetland greatly affects the type of vegetation, nutrient cycling, and types of invertebrate, fish, and bird species present. These wide-ranging hydrological and ecological characteristics provide us with many different wetland types [11].

The interaction of groundwater with the surface water of lakes and wetlands can be quite different. Generally, lakes are shallow around shorelines, and wave action frequently removes fine-grained sediments. This allows surface water in a lake and adjacent groundwater to interact freely. On the other hand, wetland bottoms are usually composed of finely grained and decomposed organic material, which sets up a barrier to easy passage of surface water to groundwater, and vice versa. This will cause the movement of water between the two systems to occur more slowly. However, highly conductive fibrous root mats in wetland soils can create a flow path for water. Therefore, root uptake of soil water creates a significant interchange between surface water and soil water (see Chapter 9 – Wetlands).

6.3.4 Karst Topography and Groundwater

Karst terrain presents unique contrasts to the surface water/groundwater systems previously described. A Karst aquifer is created when water-soluble rocks, primarily limestone and dolomite, dissolve beneath the land surface. These pathways can carry groundwater and are sometimes large enough for humans to walk through (for example, the Carlsbad Caverns in New Mexico, Mammoth Cave in Kentucky, or Postojna Cave in Slovenia). Groundwater recharge is very efficient in a Karst aquifer system because flow paths are generally large. However, some fractures may be quite small, making predicting groundwater movement difficult. In some locations, groundwater moves in underground streams within highly developed Karst systems. In other areas, thin fractures limit the movement of water, and groundwater flow may be extremely slow. Streams can entirely disappear into a Karst aquifer and may just as rapidly reappear at a downstream location. The study of Karst aquifer systems can be exciting for people interested in combining hiking, spelunking, and scuba diving with science.

6.3.5 Water Supply Interactions

Too often, water supplies are not adequate at the locations and times when they are most needed. In response, humans have engineered works of small and great size to distribute water supplies – both surface water and groundwater – to locations of need. Transport of either type of water can have an impact on the other. Diverting water from a stream for irrigation will change the historic return flow patterns of water in the stream. Surface water that would have flowed downstream may now be consumed by a crop, or may percolate below the plant root zone and become groundwater.

This groundwater may travel very slowly, perhaps a meter or a few feet per day, and then return to the stream months or years later. In another example, water may be pumped from a shallow well to provide drinking water to a community. Normally, this groundwater may have discharged to a wetland or lake. Now that the groundwater has been pumped for municipal use, the hydraulic gradient may change such that water levels in a wetland, river, stream, or lake may decline (see Figure 6.12).

Other processes can affect the discharge and recharge of groundwater adjacent to a stream, wetland, or lake. Well pumping can intercept groundwater moving toward a waterbody and capture such flows before they discharge. Pumping of a well near a waterbody can induce the flow of water from a stream into a hydraulically connected aquifer [12].

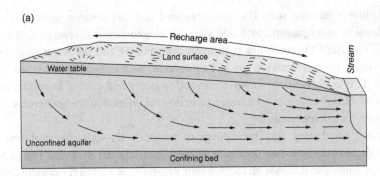

(a)

Fig 6.12 Groundwater movement influenced by well pumping. (a) Groundwater discharges into the stream. (b) A well with a flow rate of Q_1, is installed. The well operation intercepts groundwater that would have gone to the stream. (c) The well flow rate is increased to Q_2, causing not only groundwater interception but also taking water from the stream itself
(Images courtesy of USGS)

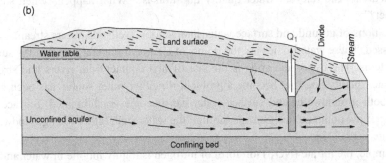

(b)

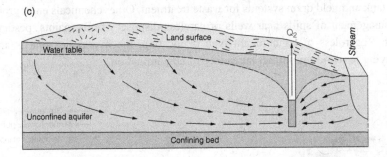

(c)

6.4 The Chemical and Aquatic Environment

Water chemistry is a theme that runs throughout this book. We discussed the role of the watershed in determining the quality of surface water. What determines water quality in groundwater?

Do you remember that soil type, and the length of time water is in contact with the soil, determine what will be dissolved into solution? Water interactions in a watershed start the process of changing water quality in a rather short period. Geologic or parent materials are present in groundwater, and over time, water quality changes. The types of chemical reactions, however, remain the same for soil water and groundwater. These are the microbial-mediated oxidation–reduction reactions and biodegradation, precipitation and dissolution of minerals, acid–base (pH) reactions, ion adsorption and exchange, and gaseous exchanges. What is different are the relative roles and importance of the reactions.

Water infiltrates into the soil. The soil is aerated and has organic matter for an energy source, which allows microorganisms to work away at biogeochemical reactions. The result is carbon dioxide (CO_2). The CO_2 combines with hydrogen ions (H^+) in the water to form H_2CO_3, carbonic acid, lowering the pH (making the pH more acid) and causing other types of mineral reactions to begin. This is particularly important in soil systems and can be important in shallow groundwater systems. However, it plays a minor role in deep aquifers that are not aerobic and do not have a "food" source for microorganisms.

Geochemical weathering (the breakdown of minerals through chemical reactions not mediated by microorganisms) becomes dominant in these groundwater systems. The exact mix of ions in solution in groundwater depends on the minerals involved, as well as the duration and type of geochemical reactions. The relevant water quality question is: "What happens when surface and groundwater meet?"

The locations of ground and surface water interactions, wetlands, rivers streams, and lakes have been discussed. These waterbodies are the points of chemical transfer between the surface and groundwater systems. The hyporheic zone is the site of first contact with rivers and streams and groundwater. Water quality does not become a problem if neither water source has been contaminated. However, both groundwater and surface water can become contaminated. Surface water is more likely to pick up a pollutant, as it travels through the watershed, rather than groundwater in an aquifer (see Figure 6.13) [13].

For example, the nitrate (NO_3^-) ion form of nitrogen is highly mobile in water and is easily transported to shallow groundwater in some agricultural areas and in communities, particularly if they have septic tank and field drain systems for waste treatment. Other chemicals enter groundwater through poor management of spills near wells or careless disposal of oils, paints, pesticides, and similar products. Regardless of contaminant type or location, the hyporheic zone is an area of concern when trying to avoid the further spreading of contamination.

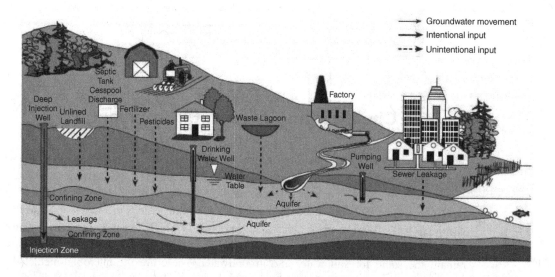

Fig 6.13 Major sources of groundwater contamination within a watershed. Notice how much groundwater quality depends on human use of the land and streams. Think of some management practices that will solve, or at least mitigate some of these inputs
(Image courtesy of USGS)

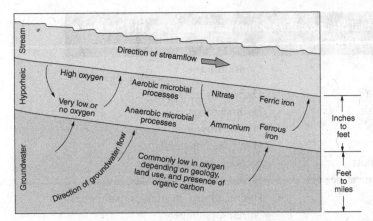

Fig 6.14 Oxygen changes found in the hyporheic zone compared to groundwater. Notice microbial and oxidation-reduction reactions at the varied oxygen levels (Image courtesy of USGS)

Fig 6.15 A piezometer is a pipe buried vertically in the ground but does not make an exciting picture; however, the information obtained from these installations can be valuable in monitoring water flow and in obtaining water samples for laboratory measurements. These were installed as part of the Wood River Wetland restoration project in the Upper Klamath River Basin, Oregon, US (Photograph courtesy of USGS)

Figure 6.14 depicts the conditions found in a hypothetical hyporheic zone. The primary things to recognize are the presence of oxygen and microorganisms. This zone is known to support algal populations and macroinvertebrates as well as microbes. Some stoneflies may also live entirely in this zone. The presence of living organisms separates this zone from deeper groundwater that is normally free of living organisms, and is very low in oxygen. In addition, notice that the hyporheic zone is usually very small compared to the total aquifer size. Even so, the reactions that occur in the hyporheic zone influence the entire aquifer. The exact nature of these reactions depends on the presence or absence of oxygen, organic matter, the minerals involved, land use, geomorphology, and aquifer type.

An important point to remember is that the hyporheic zone only exists if the waterbody, in contact with the aquifer, has a high-permeability bottom – such as coarse sand and gravel – so that water is easily exchanged between systems. A clay bottom restricts water movement so effectively that exchange between surface and groundwater is slow to non-existent. In addition, the hyporheic zone can extend into the floodplain and may exist under a wetland. One way to determine the extent of this zone is to set out **piezometer transects** (a transect is a line along which measurements are made and recorded) to monitor subsurface water flow at given depths and distances from the water source.

A piezometer is a small-diameter tube (typically 5–10 centimeters, or 2–4 inches) with perforations (slotted holes) at one end of the tube. The slotted end is placed into a borehole (a hole drilled in the ground). Groundwater will seep into the perforations and fill the piezometer to a level equal to the groundwater table depth. This depth of groundwater within the tube is called the **piezometric surface** and can be measured with a tape measure or e-tape (electronic device) (see Figure 6.15).

Table 6.2 Potential groundwater contamination sources and the major contaminants they contain

Source	Contaminant
Solid waste landfill	Hazardous materials
Auto salvage yards	Hydrocarbons
Junkyards	Hydrocarbons
Households	Hazardous materials
Agricultural feedlots	Nitrates, pathogens
Agricultural chemical dealers	Hydrocarbons, pesticides, nitrates
Urban landscapes	Hydrocarbons, pesticides, pathogens
Septic systems	Nitrates, pathogens
Weapons manufacturing	Hazardous materials
Manufacturing industries	Hazardous materials

A piezometer is not a complicated instrument but must be installed properly to obtain accurate groundwater elevation readings. After the piezometer is placed in the borehole, the area around the outside of the tube is filled with sand. Then, bentonite clay is placed to seal the borehole near the land surface to prevent potentially contaminated surface water from traveling down the outside of the piezometer. A water depth indicator is used to determine piezometric readings. Water quality samples can also be taken from the same tubes to determine chemical interactions between surface and groundwater.

Areas that do not have a definite zone of infiltration, such as the hyporheic zone, usually have contact below the soil zone and into the groundwater system. This provides the potential for contamination from many sources; Table 6.2 lists some of them. The types of hazardous materials listed are varied and have different environmental effects. They can be gasoline, metals, radioactive waste materials, any number of petroleum by-products, dry-cleaning solvents, paints, insecticides... the list goes on and on. Too few people believe that throwing the dirty oil from an oil change into the garbage, or pouring it on the ground, instead of taking it to a waste facility, can be harmful. Just imagine industries using that attitude, and the problems multiply.

Fortunately, many regions of the world have strong groundwater protection laws, and one good example is the State of Texas. Texas is very serious about groundwater protection, and has strong groundwater protection laws, programs, and monitoring. The Texas Groundwater Protection Committee (TGPC) was formed in 1989 to organize the state's groundwater efforts. Specifically, Sections 26.401 through 26.408 of the Texas Water Code state the following:

- sets out the state's groundwater protection policy
- provides legislative recognition for the TGPC
- requires the TGPC to coordinate the groundwater protection activities of state agencies
- requires the TGPC to develop and update a comprehensive groundwater protection strategy for the state
- requires the TGPC to develop the format for notices of groundwater contamination
- requires the TGPC to publish an annual report.

A TGPC report – *Joint Groundwater Monitoring and Contamination Report* – documented 156 pages of violations within the state of Texas in 2005 [14]. Petroleum products top the contaminant list followed by organic compounds, such as phenol, trichloroethylene, carbon tetra-chloride, dichloroethylene, and naphthalene (generally created by industrial processes); pesticides, including alachlor, atrazine, bromacil, dicamba, and prometon; creosote constituents (often used with wood products); solvents; heavy metals; and sodium chloride (salts from manufacturing and from winter use by road departments to melt snow and ice off roadways). The report includes actions taken to address each violation.

The fact that groundwater is closely monitored, and violations not just reported but corrected, by law speaks to a strong commitment of resources, people, time, and money by the people of the State of Texas. This is a model effort for other locations.

Think About It: Groundwater Protection

Texas is a state in one of the most developed and richest countries in the world. It has strong environmental laws, educated citizens, enforcement agencies at many levels of government, and many citizens who want to protect the environment. How, then, can governments with limited budgets be expected to deal with groundwater contamination? In many of these situations, national governments are focused on health, jobs, and other basic human needs, while environmental concerns are not high priorities. Is there a global environmental responsibility for governments or individuals to provide support in other regions of the world with limited resources? Does ground-water contamination occur even in areas with strong protection plans? If so, how could it happen?

Summary Points

- **Groundwater** comprises about 30 percent of the world's freshwater resources. It is difficult to imagine water in the ground being managed; however, groundwater management is essential to using the resource without depleting it.

- Approximately **2.5 billion people** around the world rely solely on groundwater as their only source of drinking water.

- **Groundwater** originates on the land surface as **precipitation**. Soil features determine water infiltration and percolation rates to the aquifer material. **Aquifers** are the saturated zone where groundwater is stored. They are composed of **geologic materials** ranging from sands and gravels to limestone and granite.

- Water must be able to move into and through the geologic material. **Porosity** tells how much space a material has for water movement.

- The amount of time water is held in an aquifer is called its **residence time**. Low porosity and slow movement cause long residence times.

- Use of water from an aquifer, in excess of recharge to the aquifer, is called **aquifer mining**. Mining has the potential to destroy an aquifer by collapsing its geologic material base.

- Collapsing the aquifer can cause surface land **subsidence** leaving sinkholes where soil once existed.

- **Surface water** and **groundwater** can interact in a wide range of physical, chemical, and biological processes that are all important to the health of both systems.

- The **hyporheic zone** is the location of first contact with rivers and streams and groundwater. Biological activity in this zone influences water quality in the aquifer.

- Water in a river or stream (a **losing stream**) can move into the aquifer (a **recharging aquifer**). Groundwater can move into the river (a **discharging aquifer** and a **gaining stream**). This exchange allows mixing of the two waters with possible water quality issues.

- Similarly, groundwater can also be recharged from a **wetland** or **lake** and can discharge into both.

- Maintaining the **water quality** in groundwater aquifers is a worldwide concern primarily because it is an important drinking water resource.

- It is much more difficult to clean up contaminates in groundwater than to keep them out.

- Since a groundwater aquifer is **saturated** and often deep within the Earth, there is little biological activity to influence its chemistry or to remove or transform contaminants. **Geochemical weathering**, the chemical breakdown of geologic materials, is not mediated by microorganisms.

- **Groundwater protection areas** especially in recharge areas are essential to maintaining aquifer water quality.

Questions for Analysis

1. Your firm wants to develop a housing subdivision in a major aquifer recharge area. You are a new hire fresh out of landscape architecture school, and have strong ideas regarding sustainable development. What would you suggest needs to be planned into the project for groundwater protection?
2. Why would many water professionals consider groundwater management to be the greatest water resource problem of the twenty-first century? Do you agree and why?
3. Some aquifers have high porosity and others have low porosity. Which aquifer characteristic is largely responsible for porosity, and which aquifer characteristics are influenced by porosity?
4. Groundwater discharge is an important component of water supply in a watershed. Explain how this relates to wetlands.
5. How does the direction of movement of water from a stream into an aquifer, and from an aquifer into a stream, change along the length of a river?
6. How does geochemical weathering affect groundwater quality?
7. What factors could change the elevation of the groundwater table of an aquifer?
8. What are some of the consequences of aquifer mining?
9. What is the hyporheic zone? What are some of the functions of the hyporheic zone?
10. Why would microbial reactions be more prevalent in soil water and shallow aquifers than in deep aquifers?

Further Reading

FISRWG (10/1998) (Updated 2010), *Stream Corridor Restoration: Principles, Processes, and Practices*, the Federal Interagency Stream Restoration Working Group (FISRWG) (15 Federal agencies of the US gov't). GPO Item No. 0120-A; SuDocs No. A 57.6/2:EN 3/PT.653. ISBN-0-934213-59-3, www.nrcs.usda.gov/wps/portal/nrcs/detailfull/national/water/?cid=stelprdb1043244 (accessed and downloaded 2020).

Pielou, E. C., 1998, *Fresh Water*, Chicago, Ill.: University of Chicago Press.

US Geological Survey, 1998, "The hydrologic cycle and interactions of ground water and surface water," http://pubs.usgs.gov/circ/circ1139/htdocs/natural_processes_of_ground.htm#interact

References

1 Isaiah Bowman, 1937, "Headwaters control and use: influence of vegetation on land–water relationships," *Proceedings of the Upstream Engineering Conference*, Washington, DC, pp. 76–95.

2 Geoff Brumfield, 2015, "*NASA Satellites Show World's Thirst for Groundwater*," NPR, www.npr.org/sections/thetwo-way/2015/06/17/415206378/nasa-satellites-show-worlds-thirst-for-groundwater (accessed 2020).

3 Government of Canada, 2013, "Water sources: Groundwater," www.canada.ca/en/environment-climate-change/services/water-overview/sources/groundwater.html#use (accessed 2020).

4 USGS Water Science School, 2020, *Groundwater Topics*, www.usgs.gov/special-topic/water-science-school/science/groundwater-information-topic?qt-science_center_objects=0#qt-science_center_objects (accessed 2020).

5 FISRWG (10/1998), Update 2010, *Stream Corridor Restoration: Principles, Processes, and Practices*, Federal Interagency Stream Restoration Working Group (FISRWG) (15 Federal agencies of the US gov't). GPO Item No. 0120-A; SuDocs No. A 57.6/2:EN 3/PT.653. ISBN-0-934213-59-3, www.nrcs.usda.gov/wps/portal/nrcs/detailfull/national/water/?cid=stelprdb1043244 (accessed and downloaded 2020).

6 The Joint Academies Committee on the Mexico City Water Supply Water Science and Technology Board Commission on Geosciences, Environment, and Resources National Research Council, Academia Nacional de la Investigación Científica, and Academia Nacional de Ingeniería, 1995, *Mexico City's Water Supply: Improving the Outlook for Sustainability*, Washington, DC: National Academy Press.

7 US Geological Survey, 1998, "The hydrologic cycle and interactions of ground water and surface water," http://pubs.usgs.gov/circ/circ1139/htdocs/natural_processes_of_ground.htm#interact (accessed 2020).

8 Thomas C. Winter, Judson W. Harvey, O. Lehn Franke, and William M. Alley, 1998, *Ground Water and Surface Water: A Single Resource, US Geological Survey Circular #1139*, Denver, CO: US Government Printing Office, https://pubs.usgs.gov/circ/circ1139/ (accessed 2020).

9 Andrew J. Boulton, Stuart Findlay, Pierre Marmonier, Emily H. Stanley, and H. Maurice Valet, 1998, "The functional significance of the hyporheic zone in streams and rivers," *Annual Review Ecological Systems* 29, 59–81. Copyright 1998 by Annual Reviews. All rights reserved (accessed 2020).

10 Johnathan Thompson and Sally Duncan, 2004, "An interview with Steve Wondzell, Pacific Northwest Research Station's Aquatic and Land interactions Team: Following a river wherever it goes: Beneath the surface of mountain streams," *PNW Science Findings*, Issue 67 www.fs.fed.us/pnw/lwm/aem/docs/wondzell/briefs_and_short_communications/2004_thompson_and_duncan_hyporheic_zone_science_findings_67.pdf (accessed 2020).

11 Government of Canada, 2016, "Groundwater and wetlands," www.canada.ca/en/environment-climate-change/services/water-overview/sources/wetlands.html (accessed 2020).

12 Thomas C. Winter, Judson W. Harvey, O. Lehn Franke, and William M. Alley, 1998, *Ground Water and Surface Water: A Single Resource, US Geological Survey Circular #1139*, Denver, CO: US Government Printing Office, https://pubs.usgs.gov/circ/circ1139/ (accessed 2020).

13 US Environmental Protection Agency, 2015, "Getting up to speed groundwater contamination," www.epa .gov/sites/production/files/2015-08/documents/mgwc-gwc1.pdf

14 Texas Groundwater Protection Committee, July 2006, *Texas Commission on Environmental Quality, Joint Groundwater Monitoring and Contamination Report 2005 (out of print publication)*, Austin, TX., https:// texashistory.unt.edu/ark:/67531/metapth307530/ (accessed 2020).

7 Lakes and Ponds

White Pond and Walden are great crystals on the surface of the earth, Lakes of Light. If they were permanently congealed, and small enough to be clutched, they would, perchance, be carried off by slaves, like precious stones, to adorn the heads of emperors; but being liquid, and ample, and secured to us and our successors forever, we disregard them, and run after the diamond of Kohinoor. They are too pure to have a market value; they contain no muck. How much more beautiful than our lives, how much more transparent than our characters, are they!

We never learned meanness of them. How much fairer than the pool before the farmer's door, in which his ducks swim! Hither the clean wild ducks come. Nature has no human inhabitant who appreciates her. The birds with their plumage and their notes are in harmony with the flowers, but what youth or maiden conspires with the wild luxuriant beauty of Nature? She flourishes most alone, far from the towns where they reside. Talk of heaven! Ye disgrace earth.

Henry David Thoreau, American author (1817–1862) [1]

Chapter Outline

7.1 Introduction

Henry David Thoreau is credited by many with beginning the environmental movement with his discourse on the virtues of nature and the role of man. The paragraph above ties his ideas into a few words – beauty and purity of nature were common themes in his writings. However, he also saw our inability to use and show respect for the gifts of nature without destroying the giver. One of our favorite Thoreau quotes is "Do not be too moral. You may cheat yourself out of much life. Aim above morality. Be not simply good; be good for something." What better tribute can be said of a person than "She was good for something, she accomplished something, she made a difference and mattered to something?" What better way is there to do something than to protect Earth's waters?

Lakes and ponds – how do we distinguish the two? Stock ponds are human-constructed ponds made for watering farm livestock like cattle. That use is pretty well defined, and we have no problem calling them "ponds." Beaver ponds, also constructed for a defined purpose, result in a definite inundation event. Calling them a pond seems correct. How do these compare to, say, Walden Pond – the site of Thoreau's environmental epiphany in Massachusetts? Walden Pond is a glacier pond, called a "kettle" (little more than a large depression), and formed from a remnant ice melt 10,000 to 12,000 years ago. It is 30 meters (100 feet) deep, clear, and teaming with aquatic life. Why, then, is Walden Pond not called a lake?

The answer probably lies in the original naming of a landmark. A person accustomed to, say, the Great Lakes, would see a 25-hectare (61-acre) body of water as nothing more than a pond. The fact is that ponds are much the same as lakes, in some ways. Nomenclature is largely a set of conventions rather than a formal definition; hence, a lake is known as a "pond," and something more the size of a pond gains stature when called a "lake." In this chapter, we will use the term "lake" most often, keeping in mind the following basic assumptions or conventions:

(1)   Ponds are smaller than most lakes
(2)   Ponds are shallower and therefore:
  - are non-stratified
  - are warmer in summer
  - can freeze solid in winter
  - can have plant growth across the bottom

Chemically and biologically, ponds have much the same aquatic potential as lakes, but on a smaller scale.

Freshwater lakes cover about 1,510,000 square kilometers (583,000 square miles), or less than 1 percent of the Earth's surface. Saline inland seas provide an additional 1,000,000 square kilometers (386,000 square miles) [2]. The Earth's surface area is 509,600,000 square kilometers (197,000,000 square miles). Of that, about 149,000,000 square kilometers (57,500,000 square miles), or 29 percent, are land. The water features may not account for much of the Earth's surface area but provide enormous environmental and human benefits around the world.

There are several external yet crucial factors to consider when studying lakes. These include how a lake was formed, the size, shape, topography, and chemistry of its watershed, regional climate, local biological communities, and activities of humans during the past century [3]. You will recognize these as the characteristics of the watershed or catchment studied in Chapter 5.

7.2 Lake Types

Lakes can be classified several ways depending on the needs and skills of the observer. In addition, the use to be made of the classification is often a determining factor in how it is developed. We will briefly examine three classification systems: lake-forming geologic events, hydrology-drainage characteristics, and trophic status.

7.2.1 Lake-Forming Geologic Events

Glacial Activity

Massive ice sheets gouged holes across the northern landscape of North America, Europe, and Asia during the most recent Ice Age which ended about 10,000 years ago. As climate change occurred and global warming evolved, glacial melt water filled many of these depressions. In other situations, lakes formed after chunks of ice, which were buried within the glacial till of rock, gravel, and soil melted and filled basins with water. Glacial moraines allowed lakes to form by blocking ancient valleys and forming natural dams.

Prairie potholes (sometimes called "kettle lakes") are also a result of glacial activity, and are found primarily in North Dakota, South Dakota, Wisconsin, and Minnesota in the US, and in Manitoba, Saskatchewan, and Alberta in Canada. These small lakes and wetland areas were formed by glaciers scraping over the landscape of the upper Midwest of the US, and across south-central Canada, during the Pleistocene. Unfortunately, more than half of all prairie potholes have been drained or otherwise altered for agricultural use [4]. Glaciers were the sculptors of landscapes for millennia.

In upstate New York in the US, the Finger Lakes region contains 11 long and narrow lakes that reside in glacial valleys. Generally situated in a north–south alignment, some lake bottoms are located well below sea level. Seneca and Cayuga are two of the largest Finger Lakes, and are among the deepest lakes in North America. Lake Cayuga descends about 435 feet (133 meters) at its deepest point, which puts it about 53 feet (16 meters) below sea level (Figure 7.1).

Toward the end of the last ice age, western Montana was the location of an ancient mega-lake, called Lake Missoula. It was formed in a glacial valley when the outlet of the Clark Fork River was dammed by glacial debris and a massive ice dam. The lake formed behind the dam contained more water than Lake Erie and Lake Ontario combined. Multiple floods and dam failure caused colossal flooding in Idaho, Washington, and Oregon. The remnants of this ancient flooding can still be seen today on the undulating landscape.

7.2.2 Volcanic Activity

Volcanic calderas (rimmed basin areas formed from volcanic subsidence or collapse) occasionally filled with glacial-melt water. Crater Lake, in the Oregon Cascade Mountains of North America, is an example. The lake lies inside a caldera, or volcanic basin, which was created when Mount Mazama collapsed over 7,000 years ago after a large volcanic eruption. There are no inlets or outlets to Crater Lake, so it loses water only to evaporation and seepage. More importantly for water quality, it only receives water from precipitation and snowmelt; therefore, it remains very blue and very clear. At 350 meters (1,148 feet) deep, it is the deepest lake in the US, and the seventh deepest in the world. It is located within Crater Lake National Park in southern Oregon (see Figure 7.2) [5].

Lake Taupo, on the North Island of New Zealand, is another example of a volcanic caldera that became a lake (see Figure 7.3). Lake Tauponui-a-Tia is the full Maori name. It is approximately 40 kilometers (25 miles) wide, has a maximum depth of 186 meters (610 feet), and is the largest lake in New Zealand. It is located near the active volcano, Mount Ruapehu, and south of the city of Rotorua; a popular tourist destination for viewing ongoing geologic activity of steam vents, boiling mud pots, and geysers. The most recent eruptions of Mount Ruapehu were in 1995 and 1996.

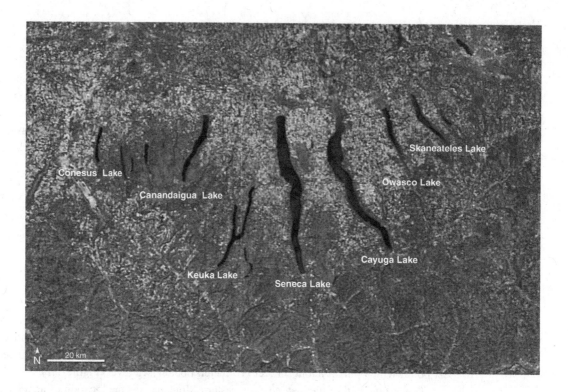

Fig 7.1 Millions of years ago, glaciers gouged deep U-shaped valleys in what is now New York, US. About 2 million years ago, during the Pleistocene glaciation, sheets of ice crept south and buried those valleys under ice. The glaciers advanced and retreated, melting for the last time about 10,000 years ago, leaving recessional moraines, large rock piles in the valleys [6]. Moraines acting as dams formed these Finger Lakes

(NASA image courtesy of the LANCE/EOSDIS MODIS Rapid Response Team at NASA GSFC. Caption by Adam Voiland)

7.2.3 Tectonic Activity

Tectonic activity (movement of the Earth's crust) can create lakes by forming rifts (fissures in rocks) that fill with water. The Red Sea in the Middle East, Lake Tanganyika in eastern Africa, and Lake Baikal in Russia are all examples of **rift lakes**. These areas are prone to the volcanic and earthquake activity associated with tectonic movement.

At 1,637 meters (5,371 feet), Lake Baikal is the deepest and oldest lake in the world and is the largest freshwater lake by volume. It holds approximately 20 percent of the world's freshwater, which is equal to all the water in the Great Lakes of North America. Lake Baikal is located in a rift valley in southern Siberia, which is slowly being pulled apart by tectonic activity (see Figure 7.4). It was listed as a UNESCO (United Nations Educational, Scientific and Cultural Organization) World Heritage Site in 1996 due to its physical and biological values [7].

Lake Tanganyika is estimated to be the second largest freshwater lake in the world, by volume, and the second deepest (see Figure 7.5). It is in the Great Rift Valley of central Africa, and is divided by Burundi, the Democratic Republic of the Congo, Tanzania, and Zambia. This is an area extensively studied for clues regarding human evolution. Since 2004, the lake has been the focus of the "Water and Nature Initiative" sponsored by the International Union for the

Fig 7.2 Crater Lake, Oregon, US, from an astronaut's viewpoint, taken July 19, 2006. The small island is Wizard's Island. The shaded areas are a deep turquoise in a lovely deep blue lake, and worth going online to view the original color image: https://eol.jsc.nasa.gov/Collections/EarthObservatory/articles/CraterLake_Oregon.htm (Image courtesy of the Image Science and Analysis Laboratory, NASA Johnson Space Center, image #ISS006-E-15238)

Fig 7.3 Lake Taupo on the North Island of New Zealand is a volcanic caldera lake. The last Taupo eruption occurred approximately 1,800 years ago and was very violent, spewing ash over all New Zealand (Getty image credit: Salvatore Alleruzzo/EyeEm)

Conservation of Nature and Natural Resources (IUCN), an international organization dedicated to natural resource conservation based in Switzerland.

7.2.4 Hydrologic Activity

Lakes can be formed by the natural process of flowing water in a river. Over time, streambank erosion can cut off an area of a river, called an *abandoned stream channel*. When this area no longer receives river flows and becomes stagnant, it becomes an **oxbow lake** (see Figure 7.6).

These U-shaped lakes are common along meandering rivers such as the Mississippi River in the US (see Figure 7.7) and the Songhua River in northeast China. In south Texas, oxbows left by the Rio Grande River are called *resacas*, while in Australia an oxbow lake is called a "billabong." The word is derived from two Indigenous Australian words: *billa* meaning creek and *bong* meaning dead.

Hydrologic activity can also create lakes when large amounts of sediment (generally sand) build up and block an area of a coastal bay. These are usually called lagoons rather than lakes.

Fig 7.4 Lake Baikal, Siberian Plateau, Russia, May 2004. It is underlain by a 25-million-year-old continental rift created by tectonic activity. The top part of the lake is still ice covered often into late spring (Photograph NASA image by Jeff Schmaltz, LANCE/EOSDIS MODIS Rapid Response)

Fig 7.5 Lake Tanganyika coastline at Gombe Stream National Park (Photograph by Andrew Cohen, University of Arizona at Tucson; National Science Foundation)

(A lagoon can also be created by a coral reef or other similar feature.) Coastal **lagoons** are often called sounds, bays, rivers, or lakes in English-speaking countries, while in Mexico, the term *laguna*, meaning "lake," is sometimes used. **Barrier lakes** are formed by sandbars usually in coastal areas and have brackish water.

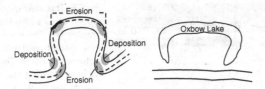

Fig 7.6 Oxbow lakes form as deposition and erosion widen and cut off the oxbow from the river. Over time, the oxbow can become silted in and gradually form a wetland and then, finally, convert into land

Fig 7.7 This image of Mississippi, US, shows oxbow formations created by years of river hydrology. The oxbow is Lake Chicot in Arkansas delta farmland. The large river to the right is the Mississippi River
(Image courtesy of Google Earth)

Sinkhole lakes form any time ground sinks and then fills with water (see Figure 7.8). These lakes can form in areas high in calcium carbonate or other salts (such as karst topography) that dissolve. This process leaves void space for the overlying land surface to sink into (hence, the name "sinkhole"). A sinkhole can also form gradually or suddenly when clays in the soil become dispersed due to sodium in the water. This causes soils to become unstable and sink. Another sinkhole process is when geologic materials slip away from each other, such as glacial sediments slipping into rock fractures. Finally, the drying and subsequent oxidation loss of organic matter in the soil can also result in a sinkhole. Sinkhole lakes can be found around the world because their mechanism for formation is so varied. Size of sinkholes can vary from a few meters (feet) to over 100 meters (several hundred feet) – both horizontally and in depth.

7.2.5 Human and Other Animal Activity

Constructed lakes, called **reservoirs**, are usually the result of dam construction. Although these are not naturally occurring lakes, their ecology can be just as complex and interesting as that of natural

Well pipe

Person

Fig 7.8 This sinkhole developed around an oil well in Barton County, Kansas, US. The rapid subsidence was cone-shaped and filled with water within hours. The vertical water well casing can be seen as a straight line on the left center portion of the sinkhole, and a person is standing outside the sinkhole toward the lower left of the picture
(Photograph by Larry Panning, April 24, 1959, Kansas Geological Survey)

Fig 7.9 Flaming Gorge Reservoir, Utah, US. Humans, too, can make a lake (reservoir actually). The Green River, a major tributary of the Colorado River, was dammed as part of the Colorado River Storage Project. The Flaming Gorge dam was completed in 1962, and created a 91-mile (145-km)-long reservoir with 350 miles (560 km) of shoreline. Hydroelectric power and water storage are project features. It is a beautiful sight to see amongst the canyons of Utah and the sagebrush grasslands of Wyoming
(Photograph by Karrie Pennington)

lakes. Reservoirs are usually managed rigorously because they were built to perform functions such as water storage, flood control, or recreation. Human management adds versatility to lake functions, and another layer of complexity to lake ecology. It can also limit some functions, such as fish movement, in any associated streams. Most often, a reservoir becomes a major source of fishing, swimming, camping, and other water activities for the local community. However, even these uses can become controversial when public use of the water conflicts with the original purposes of the reservoir (see Figure 7.9).

Another dam construction expert, not without controversy and detractors, is the beaver. These semi-aquatic rodents are native to Europe and North America. Beavers make ponds or lakes (see Figure 7.10) by damming running water in rivers and streams, often creating habitats different from the original, especially if they flood out an area of trees. Large areas of bottomland hardwood

(a)

Fig 7.10a The Google Earth image is of the largest known beaver dam in the world, found on Google Earth by a researcher in 2007. Currently 2,790 feet long, it is located on the southern edge of Wood Buffalo National Park in northern Alberta, Canada with no roads and few trails nearby
(Image courtesy of Google Earth)

(b)

Fig 7.10b This image shows part of the beaver dam. It is a mix of stones, branches, and twigs covered with grass. The grass indicates time has passed since the beavers started work around the 1970s. They have built this half-mile structure, and according to park representatives, are still adding to it. In isolation in a park larger than Switzerland, beaver find plenty of natural resources and few disturbances [8]
(Photograph courtesy of Parks Canada, Wood Buffalo National Park staff)

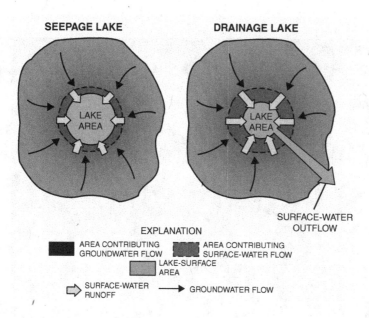

Fig 7.11 Seepage and drainage lake hydrology. The large outer circle represents groundwater inflow. Drainage lakes receive little to no groundwater when compared to seepage lakes. The inner 2nd circle shows input from surface water in the seepage lake. The seepage lake has outflow, and its surface area is smaller (Modified image from original courtesy of USGS)

wetlands in the South of the US have been converted to ponded wetlands by beavers. Beaver dams are often removed because they back water onto farmland or inundate trees that someone wants to protect.

One of the other largest known beaver dams was found near Three Forks, Montana, and was 652 meters (2,140 feet) long, 4 meters (14 feet) high, and 7 meters (23 feet) thick at the base [9]. The average beaver weighs about 14 kilograms (30 pounds). One of the largest was found in Wisconsin in 1960, and it weighed 42 kilograms (93 pounds) [10].

7.3 Lake Hydrology: Drainage Characteristics

Hydrologically, lakes can be described by their water source and outflow. Water sources for **seepage lakes** are groundwater and precipitation. Therefore, seepage lake levels closely reflect groundwater levels beneath the lakebed. Seepage lakes have no outlet but lose water through evaporation (as do all lakes). Residence time (the length of time water remains in a given location) is relatively long when compared to drainage lakes.

Drainage lakes have both surface inflow and outflow, but little, if any, groundwater input. The lake surface area is smaller than the drainage area and collects surface water runoff from storm events. Lake levels are dominated by the size of the outlet. Reservoirs usually mimic drainage lakes since they have both inlets and outlets. Drainage lakes can have groundwater as the primary inflow but that is not widely common (see Figure 7.11) [11].

7.4 Trophic Status or Classification

Lakes can be divided into categories based on their level of nutrients and subsequent growth of biological organisms – called their **trophic state**. **Trophic status** is a measure of a lake's productivity. **Eutrophic lakes** are nutrient-rich, such as a lake teeming with bass, whereas **oligotrophic lakes** are nutrient-poor. An example would be a high mountain lake with only a few small trout. **Mesotrophic lakes** are somewhere in between the two nutrient extremes. Lakes can naturally progress from oligotrophic to eutrophic states as they age, or could have altered rates of change (faster or slower) due to changes in their watersheds. **Eutrophication** results when nutrients, sediments, silt, and organic matter from the surrounding watershed accumulate in a lake. This process can take thousands of years, but human activities usually accelerate the process. This anthropogenic (human) factor can be seen in all of our discussions of water resources.

Anthropogenic eutrophication causes increased algal and rooted aquatic plant growth. Since plants produce oxygen, this would seem to be a good thing. Plants also respire, i.e., use oxygen at night. However, and more importantly, plants die. The dead plants become organic material (known as **detritus**) and are decomposed by microorganisms for food through oxidative processes. This depletes dissolved oxygen (DO) levels in the lake. This can be bad news for fish and other organisms that depend on dissolved oxygen in the water to breathe. Fish kills are all too common during hot summer months in shallow lakes.

Lakes sometimes have zones of oxygenation where mobile, free-roaming aquatic animals can move to until dissolved oxygen levels improve. Non-mobile animals do not fare as well (see Table 7.1) [12]. Aerobic microorganisms (oxygen users) consume additional amounts of the limited dissolved oxygen supplies to convert nitrogen, sulfur, and carbon compounds present in lakes into odorless, and relatively harmless, oxygenated forms like nitrates, sulfates, and carbonates. In contrast, anaerobic microorganisms (do not use oxygen) produce toxic and smelly ammonia, amines, and sulfides, and even flammable methane (swamp gas). You can see why large amounts of biodegradable materials, growing in or entering lakes and ponds, can create serious water quality problems.

Total phosphorus (TP) levels have been used to designate lake classification because it is most often the limiting growth nutrient in freshwater systems (see discussion in the lake chemistry section later in this chapter). The total phosphorus concentration is usually given in parts per million

Table 7.1 Lake organisms, how they move, and where they live

Free Roaming		
Fish	Amphibians, turtles	Larger zooplankton, insects
Carried by Water		
Living things = Plankton	Dead things = detritus	
Animals – zooplankton	Internal – produced in	
Plants – phytoplankton	lake	
Bacteria – bacterioplankton	External – carried into lake from watershed	
Bottom Dwellers		
Benthos = Animals	Plants	Bacteria & Fungi
Aquatic insects, mollusks, clams, snails, worms, other macroinvertebrates, and their larvae	Higher plants = macrophytes	sewage sludge = mixture of algae, fungi, and bacteria
	Attached algae = periphyton	

Table 7.2 Lake productivity relative to total phosphorus concentration for some lakes. Lake productivity in warm areas can be very different from productivity in cold areas due to biological activity

Productivity	Total phosphorus	
	ppb	ppm
Ultra-oligotrophic	<5	<0.005
Oligotrophic	5–10	0.005–0.01
Mesotrophic	10–30	0.01–0.03
Eutrophic	30–100	0.03–0.1
Hyper-eutrophic	>100	>0.1

(ppm) or parts per billion (ppb). Table 7.2 shows approximate lake productivity related to an average TP value from water collected in the epilimnion (upper layer) of lakes. These values can change for lakes in warmer climates.

7.5 Lake Structure

The size, shape, and depth of lakes are extremely varied. Some can be measured in square kilometers (square miles) of surface area, while others may be only a few hectares (acres) in size [13]. Depths are infinitely different. Any measurable physical, chemical, or biological characteristics can vary seasonally, or with climate change, from lake to lake and even within a lake. Physical differences can include sunlight levels, temperature, and water currents. Chemically, differences can include nutrients, major ions, oxygen levels, salinity, and contaminants.

Biologically, the structure and function of lakes, both in terms of static and dynamic variables, are different. The growth rate, type, and amount of biological organisms (called **biomass**) vary seasonally and with changing inputs to a lake. There is a great deal of locational (spatial) difference in any of these variables. These differences can occur at given times (temporal) and with a variability of scale from minutes, hours, **diel** (a 24-hour cycle), seasons, decades, or geological time.

Lakes are highly structured bodies of water. Large lakes usually stratify vertically into three physical layers, based on water density, as influenced by temperature [14]. (Remember the physical properties of water discussed in Chapter 4.) These stratified layers vary primarily in light penetration, temperature, and biological activity (see Figure 7.12).

The top layer of a lake is called the **epilimnion**. This is the shallow, warm, well-mixed layer where light can penetrate, and plants and phytoplankton thrive. The **metalimnion** is the next layer, located at deeper levels, where temperature varies rapidly with depth, the thermocline area. The actual **thermocline** is the area within the metalimnion layer where the maximum temperature change between the top and bottom of the lake occurs. Anyone swimming in a lake has experienced the delightful feel of warm water at your chest and colder water at your feet. This thermocline area varies in temperature, dissolved oxygen content, and biological activity. The depth at which this occurs can vary with lake depth and, in shallow lakes, the coolness of the metalimnion layer can disappear – leaving the hapless swimmer still longing for relief from the summer heat.

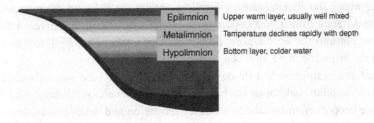

Fig 7.12 Layers found in most lakes. Lake stratification occurs with temperature influences on water density
(Image modified after original, courtesy of University of Minnesota, Minnesota Shoreline Management Resource Guide)

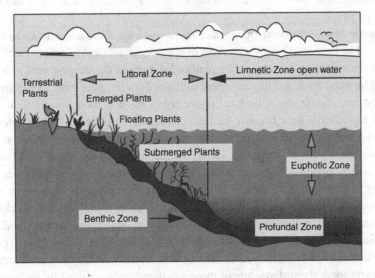

Fig 7.13 Zones in a lake and their relationship to light penetration and biological community
(Image modified after original courtesy of University of Minnesota, Minnesota Shoreline Management Resource Guide)

The deepest layer of a lake is called the **hypolimnion**. It is an area of cold water isolated by cold water's higher density and the warmer metalimnion above. It is sometimes anoxic (lacking oxygen) if sufficient mixing has not occurred to distribute oxygen to this depth. The hypolimnion has little to no light penetration and low biological activity.

These three vertical, stratified layers of a lake can be "mixed" if conditions are right. Spring "turnover" or mixing can occur as the epilimnion warms from frozen to the same temperature as the hypolimnion, very cold but not frozen. At this point, winds can mix the two layers because there is no water density barrier to overcome. A fall "turnover" is the reverse process – the epilimnion cools to the same temperature as the hypolimnion, and mixing can again occur. Lake turnover may not occur if climatic changes occur too rapidly, setting up a density change between the top and bottom, e.g., rapid freezing in winter or thawing in spring. The "normal" mixing pattern is spring turnover, summer stratification, fall turnover, and then winter stratification. Did you suspect that water density, discussed in Chapter 4, could be so important to the ecology of a lake?

7.5.1 Lake Zones

A typical lake has distinct zones of biological communities linked to its physical structure (see Figure 7.13). The **littoral zone** is the near-shore area of the lake where sunlight reaches the bottom sediment and allows aquatic plants to grow. Aquatic life is richest and most abundant in these shallow water areas. Aquatic plants provide a food source and substrate (the base on which an organism lives) for algae and invertebrates, as well as habitat for fish and other organisms. The littoral zone is quite different from a lake's open water environment. Sunlight can reach all depths of the littoral zone, but not into further offshore, deeper parts of a lake. The depth to which light can penetrate is called the **euphotic zone**, and is a vertical area within a lake.

The littoral zone can provide a diversity of habitats that include wetlands, drowned river mouths, shallow water habitats, oak savannas, beaches, dunes, relict coastal features, and abandoned dune fields. Because people often use lake beaches for recreation and development, littoral habitats are usually the most impacted. Water quality sampling at these areas provides data that allow inferences to be made about human impacts on habitat health and indicate if it has been disturbed either chemically or physically.

The **limnetic zone** is the deep, open water area away from shore and extends as far down as light can penetrate. The primary photosynthesizers in open water are phytoplankton, with larger fish spending most of their time in this zone. The **profundal zone** is the deepest part of a lake. It is located below the limnetic zone and begins where sunlight does not penetrate. Organisms that live here depend on organic matter that falls from upper layers as their food source. Some shallower lakes do not have a profundal zone because sunlight can reach the bottom of these waters.

The bottom sediments make up the **benthic zone**. It can be composed of clays, sand, gravel, detritus, etc. Its surface layer can support a variety of macroinvertebrates and small crustaceans. Larger and more diverse populations of macroinvertebrates live in the littoral (nearshore area) lake bottom where there is increased light, oxygen, and food. Biological diversity depends on the bottom substrate (rock, sand, clay) for habitat and protection, and on organic matter for food. Rocks generally provide the necessary habitat and protection, places for algae to attach, and organic food. Clays are usually low in habitat value unless logs etc., provide structure. However, clays are high in organic and mineral food. Perhaps the least diversity is seen in benthic zones with unstable sand substrate that lacks both habitat and food. Regardless, any of these situations can be altered by the natural falling and decay of trees to provide missing habitat, or by an unexpected food source washed in from the watershed into the lake.

7.6 Lake Chemistry

Water chemistry was explored in Chapter 4. Lake chemistry builds on what we have previously learned about water. This chapter places our basic water chemistry knowledge into the more complex ecological system of a lake. Oxygen is fundamental to life for higher animals, such as humans, and to the smallest microbes, called aerobes. Aquatic systems are no exception. Oxygen gets into water by **diffusion**, the random motion of molecules from an area of higher concentration to an area of lower concentration (i.e., air to water). This process strives for a state of equilibrium. However, oxygen diffuses from the atmosphere into water slowly. The molecules in water are more closely packed than those of air and make it difficult for the atmospheric oxygen molecules to move into water. This process is highly temperature dependent. Heat makes molecules vibrate and move, causing oxygen diffusion rates to increase as temperature rises. However, cold water holds more oxygen than warm water for much the same reason; it is more difficult to diffuse out once it gets in. Oxygen is more soluble in cold water since it fits into the loosely hydrogen-bonded network of water molecules, and is held there.

Chemically, the process of oxygen going into water, called **dissolution**, is **exothermic** because heat is given off. Exothermic means that cooling causes the reaction to move toward the dissolved form. Hence, colder water holds more oxygen than warm water. Diffusion rates increase with warmer temperatures. By contrast, the dissolved oxygen content increases with colder temperatures [15].

Another key factor in oxygen dissolution is water movement, called **hydrodynamics**. Water flowing in a mountain stream, splashing off rocks, or speeding downward in rapids is aerated (obtains dissolved oxygen) more easily than water sitting in a lake, primarily because there is more water surface exposed to the air.

Think About It: Dissolved Oxygen

Lakes obtain dissolved oxygen levels in various ways. Go back to the discussion on lake turnover to review the effects of lake-water movement. How does this natural process change dissolved oxygen levels in a lake?

Actual dissolved oxygen content also depends on plant populations releasing and using oxygen, aquatic organism population density, and biological and chemical oxidations losses (see Chapter 4). Dissolved oxygen levels in natural waters have diel (24-hour) fluctuations as photosynthesis adds oxygen during sunlight hours and respiration uses oxygen at night. The diffusion from the air is too slow to equalize the system during these diel fluctuations.

Nitrogen is just as important to aquatic plant growth as it is to terrestrial plant growth. Nitrogen cycles (changes form) in water by the same mechanisms that allow it to cycle on land. It enters a lake through atmospheric deposition, surface water runoff, point source releases, and in some cases from groundwater. The common nitrogen forms are nitrogen gas (N_2), ammonia (NH_3), ammonium (NH_4^+), nitrates (NO_3^-), and nitrites (NO_2^-). The form of nitrogen varies with oxygen levels and bacterial activity. Bacteria in the genus *Nitrosomonas* (important players in wastewater treatment plants) oxidize ammonia to nitrite, while bacteria in the genus *Nitrobacter* (common in locations where organic matter is being mineralized) oxidize nitrite to nitrate.

These processes are called **nitrification** and are extremely important for plant growth because nitrates are the most plant-available form of nitrogen. In addition, an aerated system (a lake with high levels of dissolved oxygen, for example) has low concentrations of potentially toxic ammonia and nitrites. **Denitrification** – the opposite process – occurs anaerobically (without oxygen) and ultimately produces nitrogen gas that is released to the atmosphere. Additions and deletions of nitrogen help balance the system; however, anthropogenic increases in nitrogen, through point sources of pollution or surface runoff of fertilizers, cause an imbalance. This causes nitrogen to accumulate in lakes, increasing plant growth, and potentially accelerating eutrophication.

Phosphorus is usually the growth-limiting nutrient in freshwater aquatic systems for phytoplankton (see Table 7.2). Plants – including phytoplankton – use phosphorus in the (PO_4^{-3}) form. This form is also bound by positive cations held on soil clays. Unlike with nitrogen, there are no microbial pathways (processes) that end in phosphorus becoming more available to plants. In a natural system, spring algal growth depletes the readily available phosphorus, which prevents eutrophication. This makes the plant-available form of phosphorus relatively scarce unless enrichment has occurred from manufactured fertilizers, or from livestock and poultry manures. This phosphorus imbalance, created by the inflow of fertilizers into a lake, can also accelerate eutrophication.

Accelerated eutrophication prematurely "ages" lake systems and can result in toxic blooms of cyanobacteria, blue-green algae, which can harm humans, wildlife, livestock, and domestic animals when consumed. Another recent concern is the dinoflagellate *Pfiesteria piscicida*, a diverse group of organisms that sometimes exist as non-toxic photo-synthesizing algae, and sometimes as a toxin-producing omnivore (a species that eats, or otherwise consumes, both plants and animals). This highly toxic chemical can cause neurological damage to humans if ingested. Fish appear almost sedated and swim abnormally, or even beach themselves, while the dinoflagellate organisms begin to digest skin and feed on exposed muscle, blood, and tissues. It's a nasty death, hence the name "fish killers" (*piscicida*).

In Depth	**Red Tide**

Even though this discussion is devoted primarily to freshwater issues, we'd be remiss not to mention the ultimate algae bloom event – the "Red Tide" phenomenon. This is a massive infestation of both saltwater and freshwater that creates dense, visible patches of phytoplankton on the water's surface, primarily in shades of brown and red. The causes of the event, common along the west coast of Florida and the Gulf of Maine on the northeast US coast, are not clear. In some locations, it appears to be a natural event, while in others it may be an anthropogenic (human-caused) problem due to high levels of nutrients in the water. Ingestion of the algae by marine organisms, such as fish, can make them potentially dangerous for human consumption due to high toxin levels. At a minimum, Red Tide can cause eye, nose, and throat irritation to humans that breathe in airborne brevetoxins (neurotoxins found in cases of shellfish poisoning).

Organic carbon accumulation as biomass can lead to decreased dissolved oxygen levels as microorganisms digest (oxidize) their feast. However, **dissolved organic carbon** (a general term, sometimes called **humic acid** or **tannin**, applied to the myriad of products produced by decay of organic materials), is naturally accumulated in lakes from surface runoff, and in some cases from groundwater. Dissolved organic carbon gives water the dark color sometimes seen in swamps and other wetlands. The blackwater swamps of South Carolina, for example, have dark, tannin, or **lignin** (constituent of wood), colored water from decomposed leaves of cypress and tupelo trees. When organic carbon is exposed to sunlight, it can lose color. In addition, calcium precipitates organic carbon causing it to drop out of solution. Sedimentary lakes (those that have lakebeds primarily composed of sand, gravel, and other sedimentary materials) high in carbonates are not usually as dark in color as lakes with bottoms of igneous rocks (normally found in mountainous regions).

7.7 Food Webs

Food webs, discussed in Chapter 5, are an important ecological concept since every organism is predator or prey, producer or consumer, depending upon its place in the web. The interaction of photosynthesis and respiration by plants, animals, and microorganisms is a food web. Most webs need a large base of primary producers to meet the demands of all the organisms they support. This can get complicated in lake systems with high diversity where literally hundreds of species, from fish to microbes, can be involved. (Refer back to Stephen Forbes' essay on lakes in Chapter 5 for a vivid portrayal of competition for food in lake systems.)

Energy is highest at the **primary producer** level and lowest at the **tertiary consumer** level. Primary producers use sunlight and photosynthesis to make food – they are the plants of the food web. Primary consumers are the little critters. Zooplankton eat algae, larval fish eat zooplankton, and invertebrates eat attached algae and higher plants. **Secondary consumers** are the small fish that eat the primary consumers. Tertiary consumers are the large fish, birds, bears, cats, and humans that eat pretty much anything they want. There are also **decomposers** that live off the dead material and recycle nutrients in the process.

It is all pretty straightforward until one of the groups gets out-of-hand due to an unexpected event – such as an input of excess fertilizers or manures, a climatic disaster in the form of drought, extraordinary storms, unseasonably cold or hot spells, excess fishing of a species, or excessive sedimentation caused by erosion. These events can change the dynamics of the food web and result in disaster for some species or groups of organisms.

Chlorophyll *a* (used by green plants to produce food energy) is often used as an estimate of algal biomass and primary productivity. Because it is relatively easy to measure and does not require the identification of individual algal species, it is an excellent surrogate. Lake water samples are taken at a given depth or depths and filtered through glass filters that collect anything larger than a micrometer (the width of human hair is about 50–100 micrometers). This collected material is processed with a solvent (usually alcohol or acetone) to remove chlorophyll *a*.

The amount of chlorophyll *a* is quantified using a spectrophotometer-based calibration curve. Simply put, a spectrophotometer reads optical density of an object at a given wavelength of light. Standard solutions of known concentrations of chlorophyll *a*, in this case, are made. Their optical densities are read at the desired wavelength and a calibration curve (graph) is produced. The optical density of the unknown sample is then compared to the calibration curve of the chlorophyll *a* result to get a concentration value.

Spectrophotometry is usually taught as part of a beginning chemistry class and then refinements are introduced in higher-level courses. It is a technique used for examination of many materials in all sorts of biological, geological, and chemical systems. It is satisfyingly fun to use this quantitative technique because it usually provides the correct answer from the start – a big plus in chemistry class. Phytoplankton populations can vary with water conditions such as temperature, wind action, nutrient availability, and light transparency (see Chapter 4). The results can be ecologically good or bad depending on the associated changes. The reactions of organisms that must adjust to these changes, such as fish, are usually a good gage of the impact of the changing water conditions.

7.8 Two Contrasting Lake Views

7.8.1 Lake Baikal, Siberia: A World Heritage Site

Lake Baikal in Siberia, the Russian Federation's "Great Blue Eye," was chosen as a UNESCO (United Nations Educational, Scientific, and Cultural Organization) World Heritage Site in 1996 [16]. These sites are selected by the country of origin and then evaluated before being designated a site of world importance by UNESCO. Lake Baikal is the world's largest freshwater lake at 3.15 million hectares (7.8 million acres), approximately 10 times the size of the state of Rhode Island, US. It is also the deepest at 1,637 meters (5,371 feet) and holds 20 percent (23,000 cubic kilometers, 5,518 cubic miles) of the Earth's unfrozen freshwater reserve. This is more freshwater

Fig 7.14 Lake Baikal in Siberia, Russia, has a rare population of freshwater seals, the Baikal seal (*Phoca sibirica*). It is the lake's only water mammal and is endemic (lives only at the lake). Lake Baikal seals have an unusually wrinkled face. These inbred seals are highly susceptible to industrial pollutants (Photograph by Uryah Wikimedia Creative Commons Attribution – Share Alike 2.1 Japanese license)

than in all of North America's Great Lakes. The fact that it is also the world's oldest lake (at 25–30 million years) seems fitting.

Lake Baikal is a continental rift lake, a gorge where the Earth's crust is splitting. It is fed by 336 rivers and streams and drained by one. Lake Baikal is so isolated that it was not well known until the Trans-Siberian Railway was constructed between1896–1902. The lake remains geographically isolated, surrounded by mountains, but regional population is estimated at 2.5 million.

Ecologically, Lake Baikal is unique. Its geological location, huge water volume, and temperature (average 4 °C, or 39°F) create a non-stratified lake. Its waters remain well oxygenated (9 ppm is the low) throughout the profile, making it a lake that supports aquatic life in its entirety. The lake water is so clear (visibility to a depth of 40–50 meters, or 130–60 feet) that susceptible people looking into the water from a boat often experience vertigo.

Biodiversity is another important feature of the lake. The Baikal seal is unique to the lake (see Figure 7.14). Large mammals in the area include brown bear, moose, elk, and deer. There are some 1,000 plus species of plants and over 1,500 species and varieties of animals. Eighty percent of the animals are endemic (found only at this one place). How about a unique flatworm that can grow to 40 cm (16 inches) and eats fish, for biodiversity?

The Baikal seal has suffered heavy declines for many reasons, one is the legal and illegal hunting that kills mostly pups for fur hats and meat that is fed to domestic animals. It has been suggested that seal hunts be offered as a tourism draw. Commercial exploitation of these seals has a long history. There is also a serious problem of pollution where organochlorines and other chemical pollutants enter the food web and accumulate in the seals as top-level predators. These pollutants can cause the same problems in seals as in other mammals such as disease, reproductive problems, and lowered immunity. Sources of these pollutants are documented. Warmer seasons have eliminated ice holds where the seals prefer to breed; increased global temperatures pose another serious threat [17] [18].

One feeder river, the Selenga River, is the major source of pollution. The river watershed has land uses including timber and agriculture, but more problematic are gold, molybdenum, tungsten, and uranium mines, and associated wastes. Industries in the watershed include a cellulose plant, concrete plants, and paper mills. "Progress" saw the development of about 100 industrial plants on the lake. Unfortunately, the inevitable damage that comes with environmentally unregulated industry was rapid. Efforts have been under way to protect the lake for 30–40 years. However, tourism has also become a major polluter as tenting visitors too often leave their garbage behind, and the region does not have satisfactory sewage facilities. Designation as a UNESCO site can help ensure future protection but cannot stop careless human actions.

Russia's Lake Baikal is a unique ecosystem that absolutely needs protection from anthropogenic progress. The 2006 Review Committee from UNESCO is working with the State Party to protect this resource. In 2018, the issue of deforestation was raised with the World Heritage Center, as well as the course of a proposed oil pipeline near the lake [19]. It is a challenge to preserve something so large and remote, but just knowing that a lake as biologically, geographically, and environmentally unique as this exists in good condition is uplifting. Helping to keep it that way would be even better.

Think About It: Is it Only Seals?

Lake Baikal is a world treasure that most of us will never see, as are many World Heritage sites. Do you think these sites, particularly water resource sites, have a purpose in today's world? The Baikal seals are a symbol of not just the lake's problems but of ecosystems all around the world, many of which are close to home no matter where you live.

What do you see as the most serious problems impacting ecosystems today? Many problems are highly controversial, for instance hunting. Common phrases we hear are "Hunters travel all over the world to kill trophy animals." "It is a money-making business." "Poachers are decimating elephant populations." Another side of the conversation are hunters who cite the need for herd population control, the use of animals to fill part of their food needs, and predator control. Do you think there is a line that must be drawn in hunting practices? How does this relate for our increasing demands for domestic meats which require large amounts of water resources?

7.8.2 Beasley Lake, Sunflower County, Mississippi

Let us move to a small oxbow lake in the southeastern United States, specifically Mississippi, home of cotton and the Blues. One of the biggest problems facing agriculture is finding the best ways to improve environmental quality, especially water quality, while maintaining their operation's efficiency and economic viability. Beasley Lake was chosen, along with two other oxbow lakes, to be part of a US Department of Agriculture Research Service (USDA ARS) study to evaluate the effectiveness of combinations of agricultural best management practices (BMPs; see Chapter 4) to eliminate degradation of water quality. The research was a multi-agency effort requiring local involvement from the landowners.

Beasley Lake is in the Delta region of northwest Mississippi, near the Mississippi River. It is 25 hectares (62 acres) in size and has an 850-hectare (2,100-acre) watershed, mostly agriculture, with a 125-hectare (309-acre) forested riparian area. The topography is flat, with a 1–2 percent slope. Crops usually grown in this watershed are cotton, corn, soybeans, and sorghum.

Scientist chose to use a combination of BMPs in the watershed for evaluation of their effectiveness (see Figure 7.15). Preventing erosion and sediment entry into the lake was a primary goal, so BMPs to stop erosion or sediment movement were chosen. BMPs included the following list [20]:

Fig 7.15 In studies to evaluate the effects of conservation buffers on nutrient retention and water quality in Beasley Lake, soil scientist Martin Locke (front left) and biologist Wade Steinriede examine a soil sample collected from a row crop area adjacent to the lake. In the background, technician John Massey operates a survey-grade GPS system to document sampling locations in the buffer area (Photograph courtesy of USDA Agricultural Research Service, National Sedimentation Laboratory Oxford, MS)

- Putting some acres into the NRCS Conservation Reserve Program (CRP) designed to promote improved water quality by planting farmland to grass or trees [21].
- Planting stiff grass hedges and grass riparian areas to slow surface water runoff.
- Creating vegetated agricultural drainage ditches to slow water and give plants time to tie up nutrients and other agricultural chemicals.
- Wetlands and riparian zones were restored to slow, filter, and process chemicals and sediment in surface runoff before it went into the lake.
- Installing slotted pipe inlets and slotted board outlets to control water.

Other measures used to restore Beasley Lake's fish population included using the chemical rotenone in the lake to remove the current population. This was used because of an unbalanced, non-native fish population heavily influenced by carp, gar, and other aggressive breeders (Figure 7.16). The lake was then stocked with blue gill, crappie, bass, and other indigenous fish. Today, the lake is said to be an excellent fishing lake by local anglers – a good, unscientific method of indicator species populations, but a true sign of success.

Scientists reporting on the progress of their research at Beasley Lake conclude the following.

- Changes in lake water quality and fisheries characteristics were used as measures of management success. Analyses of water quality, prior to the implementation of BMPs, indicated lakes that were stressed and ecologically damaged due to excessive in-flowing sediments. Significant improvements in water quality were realized using cultural and structural BMPs. Sediments were decreased 34–59 percent, while Secchi visibility and chlorophyll generally increased. The most dramatic improvements in water quality occurred in the two watersheds that featured cultural practices and combinations of cultural and structural practices, respectively.
- Reducing suspended sediment concentrations in these oxbow lakes resulted in conditions favorable for phytoplankton production. Increases in phytoplankton production resulted in increased chlorophyll concentrations and higher concentrations of dissolved oxygen, leading to improved secondary productivity. Results further indicated that cultural BMPS may play the more vital role in improving lake water quality and may be needed in addition

Fig 7.16 USDA ARS ecologist Scott Knight inspects and weighs a common carp while biologist Terry Welch records data. A wide range of fish species were collected from Beasley Lake to determine the overall health of lake ecology
(Photograph courtesy of USDA Agricultural Research Service, National Sedimentation Laboratory Oxford, MS)

to structural measures to ensure improved water quality in oxbow lakes receiving agricultural runoff [22].

Beasley Lake is still being monitored by Agriculture Research Service scientists. The improvements are significant because they offer solutions to water quality problems in one of the most intensely farmed areas in the US. Restoring habitat and function to lakes in watersheds, with intense agricultural land use, is as Thoreau said, "being good for something."

Summary Points

- **Freshwater lakes** cover about 151 million hectares (371 million acres) of the Earth's surface. Reservoirs (human-made lakes) add to this number.

- There are several **external factors** to consider when studying lakes. These include how a lake was formed, the size and shape, topography, and chemistry of its watershed, regional climate, local biological communities, and activities of humans during the past century.

- Three common **lake classification systems** are lake-forming geologic events, hydrology-drainage characteristics, and trophic status.

- Glacial activity formed many holes which later became lakes, and others that became potholes. Both are important in their own landscape.

- **Volcanic activity left craters, both large and small,** that became lakes.

- **Plate tectonic activity** (movement of the Earth's crust) can form **rift lakes**.

- An example of **hydrologic activity** forming a lake is the break-off of a meander in a river. The meander becomes an **oxbow lake**.

- **Coastal lagoons** and **sinkhole lakes** also result from hydrologic activity.

- **Trophic status** is a measure of a lake's productivity and indicates **water quality** but is not a measure of water quality. Lakes can be **oligotrophic** (low productivity), **mesotrophic** (medium productivity), or **eutrophic** (highly productive).

- Lake structure is typically layered, especially large deep lakes. The three layers of a lake from top to bottom are the **epilimnion**, **metalimnion**, and **hypolimnion**.

- Biological lake structure is also zoned in approximate layers. These zones are the **littoral (euphotic), limnetic, profundal**, and **benthic**.

- **Lake chemistry** is especially important to aquatic life because it determines water quality.

- **Dissolved oxygen** is a primary indicator of water quality. It gets into water by **diffusion**, the random motion of molecules from an area of higher concentration to an area of lower concentration.

- **Dissolved oxygen** levels in natural waters have **diel** (24-hour) fluctuations as photosynthesis adds oxygen during sunlight hours and respiration uses oxygen at night.

- **Nitrogen and phosphorus chemistry** are important to both aquatic plants and animals. They are nutrients whose effects can be positive or negative depending on how much is in the water. Not enough, and aquatic organisms do not thrive; too much, and growth can be excessive, causing water quality problems.

- **Food webs** are an important ecological concept. Every organism is predator or prey, producer or consumer, depending upon its place in the web. This is the "eat or be eaten" scenario that keeps these systems running smoothly. If a food web is disrupted, populations can become skewed and the system breaks down.

- There are **best management practices (BMPs)** to help manage lakes and what enters our lakes. These are an important tool in planning land use in a lake watershed.

Questions for Analysis

1. What are the primary differences between lakes and ponds?
2. How does turnover alter dissolved oxygen levels in a lake?
3. Why is watershed land use an important component of lake health?
4. How can healthy wetlands influence lake health?
5. Why do food webs depend upon aquatic diversity?
6. Explain the significance of geologic events in lake formation.

Further Reading

Pielou, E. C., 1998, *Fresh Water*, Chicago, Ill.: University of Chicago Press.
Thoreau, Henry David, 1854, *Walden, On Life in the Woods*, Boston, Mass.: Ticknor and Fields.

References

1 Henry David Thoreau, 1854, *Walden, On Life in the Woods*, Boston, MA.: Ticknor and Fields.
2 I. A. Shiklomanov, 1993, "World freshwater resources," in *Water in Crisis*, P. H. Glick, ed., Oxford: Oxford University Press, pp. 13–24.
3 A.J. Horne and C.R. Goldman, 1994, *Limnology*, 2nd ed, New York: McGraw-Hill.
4 US Environmental Protection Agency, "Prairie potholes," www.epa.gov/wetlands/prairie-potholes (accessed 2020).
5 US National Park Service, "Crater Lake National Park," http://www.nps.gov/crla/(accessed 2020).
6 NASA Earth Observatory, "The icy origins of the Finger Lakes," https://earthobservatory.nasa.gov/images/82448/the-icy-origins-of-the-finger-lakes (accessed 2020).
7 UNESCO, "World Heritage sites: Lake Baikal," https://whc.unesco.org/en/list/754 (accessed 2020).
8 Parks Canada, Government of Canada, 2020, "Wood Buffalo National Park," www.pc.gc.ca/en/pn-np/nt/woodbuffalo/decouvrir-discover/beaver_gallery (site updated sept. 16, 2020) (accessed 2020).
9 E. R. Warren, 1927, *The Beaver: Its Works and Its Ways*, Baltimore, MD.: Williams & Wilkins
10 North Dakota Fur Takers Association, 1997, *North Dakota Fur Takers Educational Manual*, North Dakota Game and Fish Department, Northern Prairie Wildlife Research Center, www.npwrc.usgs.gov/resource/mammals/furtake/furtake.htm (Version 16MAR98) (accessed 2020).
11 Donna M. Schiffer, 1998, *Hydrology of Central Florida Lakes: A Primer, US Geological Survey Circular #1137*, Washington, DC: US Government Printing Office.
12 North American Lake Management Society, 1990, *Lake and Reservoir Restoration Guidance Manual*, 2nd ed, Madison, WI.: NALMS.
13 University of Minnesota, Minnesota Shoreline Management Resource Guide, "A primer on limnology," www.d.umn.edu/~seawww/depth/limnology.pdf (accessed and downloaded 2020) also available as an online version: www.d.umn.edu/~seawww/depth/limnology/index.html
14 Ibid.
15 J. P. Michaud, 1991, *A Citizen's Guide to Understanding and Monitoring Lakes and Streams, Publication 94–149*, Olympia, Wash.: Washington State Department of Ecology, Publications Office.
16 UNESCO, "World Heritage sites: Lake Baikal."
17 The Seal Conservation Society, "Baikal seals," www.pinnipeds.org/seal-information/species-information-pages/the-phocid-seals/baikal-seal (accessed 2020).
18 T. A. Jefferson, S. Leatherwood, and M. A. Webber, 1993, "Marine mammals of the world," *FAO Species Identification Guide for Fishery Purposes*, Rome: UNDP/FAO, Part of *FAO Species Identification Guide for Fishery Purposes*, Rome: UNDP/FAO. Available online: *The World Registry of Marine Species (WORMS)*, www.marinespecies.org/imis.php?module=ref&refid=5627 (accessed 2020).
19 UNESCO, "State of conservation: Lake Baikal," https://whc.unesco.org/en/news/1863#:~:text=State%20of%20Conservation%20of%20Lake%20Baikal.%20On%201,informed%20of%20the%20concerns%20raised%20by%20the%20campaign (accessed 2020).

20 M.A. Locke, S.S. Knight, C.M. Cooper et al., 2006, *Beasley Lake Watershed*, Oxford, Miss.: National Sedimentation Laboratory, Water Quality and Ecology Research Unit, US Department of Agriculture, Agriculture Research Service.

21 Natural Resources Conservation Service (NRCS) Conservation Reserve Program, www.nrcs.usda.gov/wps/portal/nrcs/detail/national/programs/?cid=stelprdb1041269#:~:text=CRP%20is%20administered%20by%20the%20Farm%20Service%20Agency%2C,land%20eligibility%20determinations%2C%20conservation%20planning%20and%20practice%20implementation (accessed 2020).

22 R.F. Cullum, S.S. Knight, R.E. Lizotte, and C.M. Cooper, 2002, *Water Quality from Oxbow Lakes within the Mississippi Delta Management Evaluation System*, Proceedings of the March 11–13, 2002 Conference, Fort Worth, TX., US, pp. 83–91.

8 Rivers and Streams

Any river is really the summation of a whole valley. To think of it as nothing but water is to ignore the greater part.

Hal Borland, author (1900–1978) [1]

Chapter Outline

8.1 Introduction

Satellite photos show our Earth covered with mile after mile of rivers and streams. Yet, they contain only about 0.0001 percent of the Earth's water. That small fraction of water, as we all know, is vital to life on Earth. Our bodies have a similar system of veins and arteries essential to human survival – carrying nutrients, oxygen, salts, and disease-fighting white blood cells throughout our bodies. Rivers and streams provide similar protective functions for our Earth.

Rivers have provided water and food for as long as rivers and humans have existed together. Rivers supply irrigation water for crops and are used for navigation, to transport goods, for power production, and waste disposal. Historically, rivers have provided political boundaries and been used as natural defense barriers against rivals. Streams, bayous, and creeks are, for the most part, small rivers. We will use the term "river" to represent all flowing waters.

Think About It: Do Floods Have Value?

Remember the scene in the classic movie *The Lord of the Rings* where Arwen calls up the forces of the river for protection against the Ringwraiths (the nine dark riders)? OK, hardly a natural event, but it is a great literary and cinematic example of river power and its use for protection. The scene is a good example of the type of flood that caused early humans to attribute god-like powers to nature. Are floods of any real value to humans or an ecosystem?

Unfortunately, our efforts to control rivers to serve humans – with dams, diversions, and levees – have destroyed many river functions that are vital to humans. Dr. Luna Leopold (Aldo Leopold's son) was perhaps the leading authority on rivers. Dr. Leopold was described as an expert engineer, meteorologist, hydrologist, environmentalist, fluvial geomorphologist, geologist, author, and above all, was showered with the well-earned accolades as a true "lover of rivers." Leopold speaks to a subject that you have read repeatedly in this text – be informed, know as much as you can about your world so that you will do something that is good. The following is from his book *Water, Rivers, and Creeks* [2]:

> "Flood control, irrigation, water supply, and pollution control are examples of water projects whose merits should be hammered out in public discussion; unfortunately, such discussions often proceed with cavalier disregard for the available knowledge in the field of hydrology.
>
> Hydrologic principles are not controversial. The more that is known about hydrology, the easier it is to judge alternative proposals and to compare their benefits and costs. Sound decisions require an informed citizenry."

Major cities have developed along rivers with some disastrous environmental conse-quences but that could have been reduced with proper planning. Rivers too often became dumping grounds for human and industrial waste. America had the examples of nineteenth-century Europe as lessons, but safe waste disposal was not even considered in America's early history. We learned in Chapter 4 that river systems have a finite capacity to absorb our messes. For example, the Cuyahoga River that flows into Lake Erie from Ohio has a long history of misuse. In 1969, *Time Magazine* described the Cuyahoga as "Chocolate-brown, oily, bubbling with subsurface gases, it oozes rather than flows." Locals joked about the river: "A person does not drown (in the Cuyahoga) but decays" [3]. It was a harsh description; however, the river had been plagued by fires since 1936, and a 1952 fire caused over US$1 million in damage. Just to be clear, we are discussing a "river on fire" (see Figure 8.1).

The cities of Cleveland and Akron, Ohio, used the Cuyahoga River as a waste dump for debris, oils, sludge, industrial wastes, and sewage starting in the 1800s. Fortunately, a century later the nation was ready for change in environmental policy, and another fire on June 22, 1969, was a nationwide catalyst. Lake Erie was already considered "biologically dead" since sewage had turned it into a massive cesspool. Only three of its 62 beaches along the US shores were rated safe for swimming. In 1961, the average sludge worm population (often found in polluted waters) that carpeted the bottom of Lake Erie was 627 per square meter (750 per square yard); however, one sampling station had 27,375 sludge worms per square meter (32,740 per square yard). That made even wading unthinkably creepy.

The 1969 Cuyahoga River fire was the last straw for many American citizens. The fire itself was not large and was extinguished in about 30 minutes, but ignited a spark that resulted in the federal Clean Water Act and the Great Lakes Water Quality Agreement. It also led to the development of state and federal environmental protection agencies. These two federal actions, along with plans already in place from state and local agencies, provided the legislative tools and financial backing to improve America's water quality. Sometimes it takes a tragedy to create the momentum for positive change.

Fig 8.1 River on fire: the Cuyahoga River fire of 1952 caused over US$1 million in damage (Photograph courtesy of James Thomas, Special Collections, Cleveland State University Library)

Fig 8.2 The Cuyahoga River today. Work on the river that began in the 1990s is still ongoing. The vision of recreating a natural system, aided by government agencies and private partnerships with community volunteers, has worked to improve the river watershed by creating and enhancing wetlands, planting native vegetation, dam removal, and education. Wildlife, including great blue herons, river otters, and bald eagles, has returned, and fisheries continue to improve. There is still progress to be made in water quality, but the difference between a river on fire and one used by families in kayaks is considerable. Individuals have and can make a difference (Photograph courtesy of National Park System (NPS))

Today, the lower Cuyahoga River still has water quality problems, but it is recovering. In addition, there are citizen groups and government agencies monitoring the river to make certain that it continues to improve (see Figure 8.2). Today, the Cuyahoga River is an American Heritage River and a Great Lakes Area of Concern – both federal designations.

Think About It: River Restoration

When people have the opportunity to enjoy a summer weekend, they often head toward water. It may be a beach; a stream in the mountains to camp; a lake for fishing or water sports; it could even be the fountains of a city. However, what will we pay to keep clean rivers? Putting a price tag on such needs has become a major goal to help justify river restoration. What do you think will motivate people to restore rivers back to a healthy status? What are some of the obstacles to river restoration efforts? Is it possible to reverse the damage that has already been done to many of our rivers? Prevention is usually cheaper than a cure. What steps would you propose to stop the harm before it starts?

8.2 River System Functions

Food, water, recreation, and aesthetics are easy to understand and appreciate. Scientists have shown, as has experience, that healthy river system functions are vital – not just to our environment, but also to sustain human lives. The following examples are some river functions necessary for ecosystem health [4]:

- purification of water and wastes as they are moved through wetlands and riparian areas in the floodplain;
- mitigation of flood flows by providing low places for water to fill, slowing down the flow, and promoting infiltration and percolation of the water to groundwater;
- providing habitat and breeding sites for all kinds of animals;
- mitigating drought by capturing runoff and increasing groundwater storage;
- maintaining soil fertility with sediment deposits during flooding, and by carrying sediments to deltas and estuaries – keeping them both healthy and fertile;
- maintaining the salinity gradients of estuaries and deltas that are necessary to promote the rich diversity and productivity of life in these areas;
- preserving biodiversity, or genetic diversity.

These river functions are vital for healthy rivers, but first, what are the basic physical features of a river system? How do floods, sediment, organisms, and habitat affect river functions? What are the physical features of a river system that impact river functions? Let's find out.

8.3 Physical Features of a River System

8.3.1 River System Scale

Thinking of a river – in its ecosystem environment – helps visualize the interaction between land and water. The scale (geographic extent) of a river system provides the framework to see how various river functions are achieved. Inputs to, and impacts on, a river can originate at almost any scale. One classification of river system scale is a region – a broad, geographical area with similar macroclimate and anthropogenic (human-caused) alterations [5]. A region could be an African savanna or the Sawtooth Mountains in Idaho. The regional scale, however, is usually too geographically large for most practical applications regarding rivers or streams.

A **landscape** is the next smaller river system scale, and is extensively used in soil mapping, wetland delineation, and river evaluation. Landscapes are areas with patterns of similar component features. For example, a forest can be a landscape, but so could a city. Both contain component features (the forest has trees, acidic soils, sloped terrain, for example, while a city has impervious streets and rooftops, museums, schools, automobile pollution, etc.). These two differing landscapes would have different inputs to and impacts on river or stream ecosystems.

A smaller river system scale is the area that includes a river channel, its floodplain, and the associated upland transition zone or fringe area. As we look for influences on a river, however, we need to remember that its watershed is the controlling land link. A **watershed** provides nutrients and other inputs from surface water runoff that enter the river channel (see Chapter 5). Relief (the slope of the land) and gravity define the scale of a watershed since all precipitation eventually moves downhill. This occurs when surface water runoff moves toward a river channel or floodplain, or through the upland transition zone. Remember that the river system landscape just described is only part of the total watershed.

8.3.2 River System Parts

River systems are composed of three parts – a **channel**, a **floodplain**, and an **upland transition zone**. These three landscape features constitute the river landscape. A river needs three different things to develop – water, a channel, and some slope so that the water can flow. Rivers and streams start with their **headwaters**, also called their **source**. Headwaters are as diverse as the rivers they create. They can be mountain streams, springs, bayous, wet meadows, melting glaciers, outflow from a lake or even seeps. Headwater flows, plus precipitation from rain or melted snow or ice, provide the basic starting processes of a river. Water at the source begins the downhill journey of becoming a river – sometimes in a rush of white-capped, rolling mountain water, sometimes in the slow meandering of a bayou. Either way, downhill it is.

The river landscape is the land most closely associated with the river. The riverbank is the **riparian** area. Events within the river riparian area, floodplain, and upland transition zone – such as erosion, floods, and land use changes – have almost immediate effects on the functions of a river.

The **floodplain** areas and **riparian buffers** (vegetated land on the edge of the river) can be forested, grassed, shrubby, sparsely vegetated, or some combination of vegetative types. They may also include wetlands. Human elements such as parks, homes, trails, campsites, or even railroads and roads are common along rivers (see Figure 8.3). Land use in a floodplain is limited by flooding events even when such structural measures as **levees, channel modification**, land buildup, etc., have been constructed to protect the area. Despite human efforts, storms occur, levees break, and floods inundate adjacent lands. The result can be a drastically changed floodplain structure. Once a river is separated from its floodplain by human intervention, it clearly becomes a different waterway. However, the physical forces of a river will try to reclaim that floodplain, sometimes with enormous impacts.

Fig 8.3 View of the Animas River from the Durango–Silverton Railroad, Colorado, US. Railroads are common structures near rivers and have definite impacts on the watershed
(Photograph by Karrie Pennington)

Think About It: Flood Protection?

It is paradoxical that floodplains become fertile through repeated flood cycles. These natural events deposit sediments and organic material onto the floodplain which can produce prime farmland between the river channel and the fringe upland. In our current world order, this "prime" farmland now needs to be "protected" from the very events that formed it. What are some of the other unintended consequences of "flood protection?" What or who is being protected? Who decides the greater need?

Figures 8.4 and 8.5 are photos of two different river systems, shown for comparison. The differences between the two river landscapes are dramatic. They differ in landscape scale, watershed type and size, channel type, floodplain landscapes, and fringe upland features. Yet each system has characteristics, functions, and processes that can be traced to basic river and stream features, such as the river gradient, sedimentation, erosion, and discharge (water flow rate and volume).

8.3.3 Channels

We mentioned earlier that river systems are composed of three parts – a channel, floodplain, and upland transition zone. River channels form as water and sediments flow through them; therefore, their size and shape vary with these elements. Some rivers cut into geologic material, causing them to incise, and tend to get deeper over time (think of the Grand Canyon in Arizona). Others do not cut into the Earth but broaden out and become wider. A "typical" river channel is depicted in Figure 8.6, and shows a low flow period when much of the channel flow might be provided by groundwater (called baseflow). Notice that the scarp is the steep sloped bank section, and the thalweg is the deepest part of the channel.

Fig 8.4 Snake River canyon in southern Idaho, US. This river corridor has steep sides carved in basalt layered over rhyolite. Grasses and shrubs are characteristic of the sagebrush grassland
(Photograph by Karrie Pennington)

8.3.4 Floodplain

The floodplain bordering a river channel can be complex, and may include floodplain lakes, islands, wetlands, and many other physical features. The components of a floodplain are created by river meandering. River water carries a sediment load modified by topographic relief, land use, climate, and all the other processes that form watersheds and floodplains.

Figure 8.7 illustrates some possible river features in the depositional zone. In the example, an **oxbow lake** is shown – an old river meander that has been cut off from the main channel. **Clay plugs** can form between an oxbow lake and the river, effectively cutting off water movement between the two features. Clay plugs are highly effective moisture barriers because many clay types swell when wet. This creates an environment that both holds water within the clay and prevents water from moving through it. Figures 8.8 and 8.9 show two different floodplains, one at river flood stage.

A **chute** is a relatively straight portion of a river channel. **Splays** are created when a natural levee breaks and releases water onto the floodplain. **Natural levees** are created when floods deposit sands and gravels that form high ridges along a riverbank. **Backwater swamps** (also called back

Fig 8.5 This riparian corridor along Pole Creek in the Stanley Basin of Idaho, US, is essential to maintaining Pole Creek as crucial habitat for Chinook and sockeye salmon, steelhead, and bull trout. Pole Creek salmon and steelhead must journey up to 900 miles and through eight dams to reach their spawning grounds, one of the longest fish migrations in the world.

This tributary of the Salmon River is providing passage to improve that journey. The corridor is full of native willows with brush common to sagebrush grasslands. Lodgepole pines and Douglas firs increase as the basin gives way to the mountains
(Photograph courtesy of Dean Pennington, drone photo)

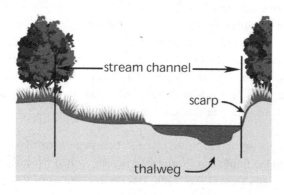

Fig 8.6 Typical river or stream channel with terraces. Note that flat terraces can flood, but higher, older terraces are usually no longer influenced by the river. The thalweg is the deepest part of the river or stream
(Image courtesy of FISRWG)

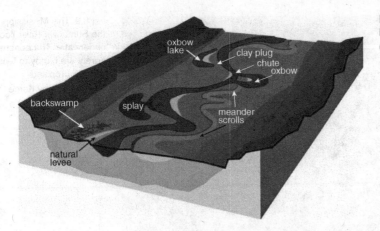

Fig 8.7 Floodplain features formed by river meandering (Image courtesy of USEPA)

Fig 8.8 Animas Valley above Animas, Colorado, US, October 2019 (Photograph by Dean Pennington)

swamps) are wetlands formed when floodwaters recede, and some water is retained behind the natural levee (away from the river channel). Backwater swamps, splays, and oxbow lakes are quite common along the Mississippi River channel in the US, the Yellow River in China, and other locations around the globe.

Fig 8.9 This Mississippi cornfield in the Sunflower River floodplain is underwater. The congregation of egrets are happy to feast on flooded cropland
(Photograph by Karrie Pennington)

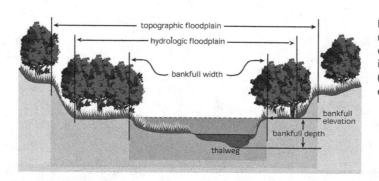

Fig 8.10 This diagram shows mature terrace development. The older terraces are no longer influenced by the river
(Modified from USEPA Office of Water)

8.3.5 Upland Transition Zone

The **upland transition zone**, or *transitional fringe* of a river channel, is the area leading into uplands or higher ground. It can be flat, sloped at any angle, even vertical, depending on how this transitional area was formed. You only need to think of the Grand Canyon (or any other river-cut canyon) to recognize one type of transitional fringe with a dramatic terrace. All transitional fringe areas are defined by the fact that they are more influenced by a river than areas further away. For example, the only trees of any size in sagebrush grassland will be near a river or stream.

This floodplain shows some of the features in Figure 8.7. It is the same valley as Figure 8.3 which was taken in 2019.

Fluvial geomorphologists use a numbering system for floodplain terraces that form, as the river changes the landscape by cutting deeply into the surface. The oldest terraces have the highest number, while the youngest is usually the channel bottom. Old terraces can become isolated from the floodplain and are sometimes used for farmland, grazing areas, or parts of urban areas. Figure 8.10 shows well-formed terraces.

Think About It: Arid and Semi-arid Landscapes

Rivers in arid or semi-arid regions are particularly welcome sites for wildlife and humans. However, a person accustomed to lush river corridors may view them as disappointingly sparse in vegetation. On closer inspection, surprisingly, these unique ecosystems provide an abundance of habitat types and animals. What role would climate change play along the transitional fringe of a river in an arid or semi-arid environment?

8.3.6 Longitudinal Zones

Longitudinally (directionally), a river has three zones. The steepest area is located in the **headwaters zone** (remember, no matter how little slope is involved, the headwaters are the highest spot of a river or stream system). Next is the **transition zone** where the slope decreases, and tributary streams can enter a main channel. The valley area gets wider during this transition, and a river starts to develop a meander pattern.

The **depositional zone** is the third longitudinal zone of a river. It has the flattest slope, the slowest water flow (called **current**), and the widest bank-to-bank area covered by water. The slower current allows sediment and other smaller materials to settle out (be deposited) in this zone. Water from the depositional zone flows into a lake, ocean, or whatever outlet ends the river's journey. Soils, geology, topographic relief, and vegetation all play a role in shaping the river along its longitudinal zones.

8.3.7 Stream Order

Stream order is a hydrology algorithm (a procedure for solving a problem) that describes how a river forms from its headwater zone to the depositional zone [6]. Figure 8.11 shows a numbered stream as an example of stream order. The number 1 represents a headwater stream and has no tributaries. The confluence (place where two streams meet and join together) of the first-order stream forms a second-order stream, designated by the number 2. Second-order streams can have first-order tributaries. The confluence of the second-order streams forms the third-order stream, designated by the number 3, and so on. Third- and fourth-order streams can also have tributaries, but do not raise the stream order. A confluence of two streams of the same order number is necessary to raise a stream to a higher order number.

Order number correlates to drainage basin size, length, and other descriptive features. Knowing a stream's order number allows you to make assumptions about which longitudinal zone it is in. It can also help in estimating how large or deep a stream might be, if you also have some knowledge of the geomorphology and soils of the area. It is accurate to say that a 12th-order stream is very large, but without knowledge of location, landscape, and watershed characteristics you cannot provide a full description by itself.

Arthur Newell Strahler (1918–2002) was a geosciences professor at Columbia University in New York City, and developed the stream order numbering system used today. Strahler first proposed the hierarchy in 1952 [7], and helped move the field of fluvial geomorphology from a qualitative to a quantitative discipline.

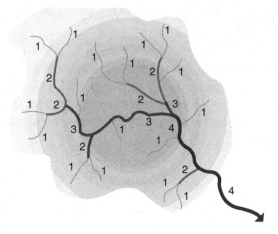

Fig 8.11 Strahler Stream Order numbering. Number 1 is a headwater stream and number 4 is the largest river or stream in this example. The Amazon, the world's largest river, is a 12th-order river while the Mississippi is a 10th-order river
(Image courtesy of FISRWG)

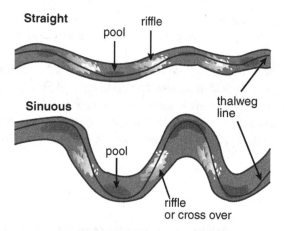

Fig 8.12 Pool and riffle structure in low and high sinuosity streams
(Image courtesy of FISRWG)

Two more important river channel characteristics are **thread** and **sinuosity**. *Thread* refers to how many channels there are – single or multiple. Multiple thread streams are called **braided** or **anastomosis**. Braided streams form from erosion in coarse sediments caused by rapidly fluctuating flows. River flows build up sandbars, gravels, and other sediments that block the flow path of water. This causes river water to form a second channel to get around the blockage. This process can continue longitudinally down the channel and forms wide shallow braids. A second condition that creates braided streams occurs in flatter land with fine sediments. These deep and narrow channels form when downstream sediment buildup causes water in a river to back up and divide into multiple channels.

Sinuosity refers to the straightness (or lack thereof) of its channel. First- and second-order rivers (in the Strahler Stream Order System described earlier) have low to moderate sinuosity. A third-order river, generally found in the flat valleys, can be very sinuous. Within a river channel, sinuosity allows for somewhat regularly spaced shallow and deep areas called pools and riffles (see Figure 8.12). Both are created by the thalweg, the deepest part of the channel. **Pools** form as the thalweg meanders to the outside, or banks, of the river, while **riffles** form as the thalweg crosses over from one riverbank to the other. Rivers with coarse-sediment bottoms, such as gravels and cobbles, have noticeable pools and riffles. Finer-sediment river bottoms are composed of sand and silts, and are more likely to have regularly spaced pools. Very fine mud bottoms can be pooled the entire length of a river segment, called a river **reach**.

8.4 Streamflow

Rivers and streams are studied separately from lakes because the differences between them are significant. Two major differences are water flows and the location of energy fixation (where the waterbody gets most of its energy or "food"). Flowing aquatic systems are called **lotic**, while still water systems, such as in lakes, are called **lentic**.

The surface runoff from a watershed plays a major role in controlling river flow and function. Our discussion of the hydrologic cycle (see Chapter 3) provides the background for understanding runoff. As we discussed earlier, there are three types of runoff:

(1)   surface or overland flow,
(2)   shallow subsurface flow, also called throughflow, and
(3)   saturated surface or overland flow.

These can occur simultaneously or individually during and after a precipitation event (see Figure 8.13). Overland flow is sometimes called **Horton overland flow**, named after Robert E. Horton, who is accredited with first describing it in scientific literature. Horton (1875–1945) was an American ecologist and soil scientist, considered by many to be the father of modern hydrology. He received his training from Albion College in Michigan, and later worked for the US Geological Survey in New York state. In 1945, after decades of work, he published his concepts of the hydrologic cycle as related to infiltration, evaporation, transpiration, and overland flow in the *Bulletin of the Geological Society of America*, just one month before his death [8].

Overland flow occurs when precipitation exceeds infiltration. It varies with the intensity of the storm event, the soil infiltration rate and permeability, and surface roughness [9]. A rapid, heavy rainfall causes overland flow almost immediately because there is no time for the water to infiltrate into the soil. In addition, during a heavy rainfall event, plant interception is not significant unless the vegetation canopy is extreme, e.g., some tropical rainforests. A gentler rain is slowed by canopy interception and soil infiltration. Slow rains gather in depressions where water can infiltrate, evaporate, or fill and overflow to continue as overland flow. Depressions are part of the surface roughness of a landscape.

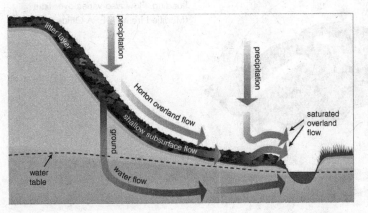

Fig 8.13 Precipitation flow paths (Image courtesy of USEPA)

Think About It: Overland Flow

How would overland flow occur in urban and municipal settings? What factors would influence the rate of flow on the land surface? What would influence water movement below the land surface, that could create shallow, subsurface flow?

Shallow subsurface flow can go into the ground to become groundwater, or if a retarding layer is present, becomes runoff into a river. Either way, more water enters the river during, or after, such a precipitation event. If the water comes from groundwater, it is part of the river's **baseflow** – the flow in the river supplied by groundwater. **Saturated overland flow** occurs when the soil has reached the point where the pores are full, and water no longer infiltrates into it. Precipitation can now move more rapidly toward the river.

Rivers react to weather, specifically in response to precipitation events, with changes in flow. Rivers can be **perennial** (contain water throughout the year except during periods of severe drought) or **permanent** (flow continuously). The term "perennial" is often used in place of "permanent." A river can be **intermittent** (meaning that it contains water only part of the year) or **ephemeral** (flows only in response to a precipitation event; this type usually exists in very arid climates).

The flowing nature of rivers means that they are constantly changing. Slow-moving rivers do not change as rapidly as a swift-moving system. This gives water and sediments an opportunity to be processed (moved, cleansed, etc.) by the natural activities of moving water. Land and river sometimes interact quite quickly and dramatically during flooding events, but more often through the subtle action of precipitation events in the watershed.

One way to look at lotic (flowing) river systems is the four-dimensional concept of J.V. Ward [10]. Figure 8.14, based on Ward's multidimensional concept, helps illustrate the variability of river flow characteristics longitudinally, laterally, vertically, and temporally. The **upstream/downstream** direction of a river is the **longitudinal dimension**. The headwater, or beginning of a stream, is the area of lowest flow volume, but usually with a fast current and high erosion rates. Little sediment deposition occurs in this section of a river.

Fig 8.14 Rivers flow in three directions, vertically into an aquifer, longitudinally in channel, and laterally during flooding. Flow also varies over time
(Modified from USEPA Office of Water)

Midway, the stream picks up volume but the speed of the current decreases. Erosion and sediment deposition are both important factors in this section, called the **transfer zone**. Furthest downstream is the **depositional zone** where flow volume is highest, current speed the lowest, and deposition of rock, gravel, and sands usually exceeds erosion. By contrast, rivers in low-slope areas usually start with low volume and low flow speed. However, erosion and deposition can be equally important. The transfer and depositional zones are pretty much the same longitudinally – they increase in flow volume, with erosion and deposition being equally important.

Laterally (the dimension across a river between its riverbanks), a river can vary significantly within a watershed. A river generally has a deepest part in its channel, called the **thalweg**. Away from that location are the low **floodplains**, which have frequent flooding, higher floodplains that are rarely flooded, a **terrace** or old floodplain, and the **uplands**. Most rivers have at least some of these features.

Vertical direction can be extremely important within a river system. Rivers often interact with the shallow groundwater aquifers beneath them. A river can either gain or lose water from an aquifer, depending on local conditions. This surface/groundwater exchange also contains dissolved components and can include organisms linked by river and aquifer water quality.

Temporal (time) variation – whether measured in minutes or millennia – is especially important because rivers constantly change, as do their watersheds. Land changes can be caused by the river and river meanders, erosional flows, and other processes including human changes. Seasonal variation can be extreme. A river may freeze in winter, pick up speed with the spring thaw, then slow to a calmer state in summer and fall. Flow duration (how long a flow lasts) and flow frequency (how often it occurs) are important influences on river function.

8.5 Fluvial Geomorphology: Forming a River

Fluvial geomorphology is the study of sediment, and how it affects river channel morphology, function, and maintenance. Luna Leopold said, "Flowing water carves a channel, but sediment forms it."[11] The study of sediment includes where sediment originates, where it goes, how it is stored, what moves it, when it moves, and what happens to the river channel and floodplain as a result of all this sediment activity over time. More formally, fluvial geomorphology is a study of process and form; the processes involved with sediment movement and deposition and the channels and floodplains formed by those processes.

8.5.1 Geomorphic Processes Associated with Flow

Erosion, sediment transport, and sediment deposition are the geomorphic processes that form river channel and floodplain structure. Watershed condition, land use, vegetative communities, soil type, relief, and flowing water all play important roles in soil erosion the starting point for soil sediments. For example, look at Figure 8.15. There are many things to notice that might influence water and sediment movement. Some of these observations are as follows:

- land use is predominantly agricultural,
- the many groups of small, rectangular, areas that appear on the east (right) side of the photo are catfish ponds,
- the river meanders are very sinuous,

Fig 8.15 A satellite image showing watershed land use
(Photograph courtesy of Google Earth)

- the southwest corner has an old oxbow area,
- a small town is located in the northwest section, and
- there are a few isolated areas of forest.

Each of these items will have an effect on surface water flow and sediment transport.

Think About It: River and Landscape

Look again, carefully, at Figure 8.15. Do you see other features that are helpful in determining how this river system will function? What kind of general land slope would you expect based on what you see of the river? What longitudinal zone best describes the river? Would this landscape have had wetlands before it was developed by timbering and agriculture? Can you make any assumptions about the soils based on the large number of ponds in the landscape?

Table 8.1 Erosion-causing factors and the resulting processes

Eroding factor	Erosion process
Surface water runoff	Sheet, rill, ephemeral gully, classic gully
Channelized flow	Rill, ephemeral gully, classic gully, wind, streambank
Wind	Wind
Ice	Streambank, lake shore
Chemical reactions	Solution, dispersion
Raindrop impact	Sheet
Gravity	Classic gully, streambank, landslide, mass wasting

Fig 8.16 Heavy rains on saturated soil have caused sheet erosion, forming an ephemeral gully midfield as the flow concentrates in a cut-water furrow. Cutting water furrows is supposed to help drain the field but in erosive soils such as this one, can result in gully formation
(Photograph by Dean Pennington)

Erosion starts with the detachment of soil particles, usually caused by the impact of rain or wind. We are concentrating here on erosion by water, but other agents can be working within the watershed. It is important to recognize other factors, such as wind, because their impacts can be significant in determining the location of erosion processes (see Table 8.1).

There are two types of land erosion. **Sheet** and **rill erosion** move soil sediments down the land surface slope. This is the dominant form of erosion, although it is not as noticeable as gullies (see Figure 8.16). **Gully erosion** forms channels as water flows concentrate in one area. These flows can be **ephemeral** (short-term), or if not stopped can become larger and turn into a classic gully (see Figure 8.17). Once sediment reaches a river, it becomes the sediment load. There are three types of sediment load:

(1) total dissolved solids (TDS),
(2) total suspended solids (TSS), and
(3) bedload.

Total dissolved solids (TDS) are the water-soluble constituents in soil water. Water causes soluble sediment to be dissolved and transported, in the form of ions in solution. The major effects

Fig 8.17 A classic gully formed by concentrated flow in erosive soil on agricultural land
(Photograph by Pam Morris, USDA NRCS)

of TDS are on vegetation. Plant roots cannot take in water with a high salt concentration (most dissolved solids are salts). In addition, if sodium ions (Na^+) are in large enough quantities, they can cause the dispersion of clays destroying soil structure. TDS can include many salts, primarily those of calcium, magnesium, potassium, sodium, chlorides, and sulfates. It can also include soluble organic matter (SOM).

Total suspended sediment (TSS) is primarily silt and clay, which are light and stay in solution longer than heavier sands and gravels. However, with high turbulence, total suspended solids can include sands. Dissolved and suspended solids float in the water column.

Bedload is composed of coarser materials that are pulled and pushed along the bottom of a stream on river. Bedload can be anything from boulders to rocks and gravels. As it moves, bedload shapes the channel morphology and creates pools and riffles. Sands, silts, and clays are soil particles formed by the chemical and physical weathering on parent materials. Bedload is often still (non-moving) rock.

Mechanical weathering is different from chemical weathering. Mechanical weathering is a series of physical processes resulting in increasingly smaller rocks that eventually become soil. Mechanical weathering of rock, or parent materials, often starts from the freezing and thawing of water trapped within the rock. This creates crevices, fractures, and surfaces for more weathering to occur. Wind is also an effective mechanical weathering agent. Wind picks up sands that act as sandpaper – abrading (scraping) rock surfaces. Thermal expansion and contraction will also weaken rock. Bedload is the shaping agent for most river channel structure, and only moves when a river is near or at bankfull stage.

Streambank erosion occurs when soil on a riverbank falls or sloughs into a river. **Mass wasting** is similar in that a chunk of soil or rockslides down the slope of the riverbank, due to gravity, as water undercuts the area. This usually results in a wider river channel being created. Once soil enters a stream, we call it **sediment**. An important point to remember is that materials the size of boulders can be eroded from a hillside, e.g., in a landslide, and washed into water. Eroded material size in rivers ranges from boulders to clay particles; sediment includes sand and the smaller particles of soil silts and clays.

8.5.2 Flow Hydrographs

River flow is highly variable because it responds to inputs from precipitation events, both from water falling directly into the stream and from watershed runoff. Unlike lakes, rivers can be flashy (rapid changes in flow) when it rains or with rapid snowmelt. For example, the Rillito River in Tucson, Arizona, is usually a dry gulch. However, a heavy rain can create a raging river, which you can watch develop in minutes. This phenomenon can occur in any area but tends to increase with a greater number of impermeable surfaces, roads, parking lots, roofs, etc. in the river's watershed.

Variability of water flows plays a major role in the ecology of a river. Luna Leopold said one must understand the hydrology of the river to understand its ecology [12]. This is done by studying a river's hydrograph. A **hydrograph** is a graph of water flow over time (see Figure 8.18). Gages measure changing river flows over a period of seasonal changes. One year is insufficient data to understand the hydrology because flow will vary with weather patterns. A drought year will produce much different flows than a very wet or normal flow year.

Many major rivers in the US have had gages installed by the US Geological Survey or US Army Corps of Engineers. Some sites have been providing data for many decades, and a few for over a century. These hydrographs provide information about past and current river flows. Land use changes can be seen in hydrographs. For example, the expansion of irrigation can cause many riverflow changes – both from surface water runoff increases and from groundwater, or baseflow, decreases. Urbanization causes a hydrograph shift, due to an increase in impervious surfaces, and causes the surface water runoff to occur much faster than normal. If there are several years of record, a pattern of natural river flow variations can be determined. Drastic changes in the natural pattern should serve as warnings that something is interfering with the natural functioning of the river system.

All water flow levels, from low flows to flood, are an essential and natural part of the river's functions. Many aquatic species have adapted their survival and reproductive strategies to natural flows. When these flows no longer occur, species can be, at the least, harmed and, at the worst, eliminated. Invasive species often gain a strong foothold in these situations because, to be invasive, the species must be adaptable. Flows also determine where aquatic organisms can travel. Many need to be able to travel both upstream and downstream. Some must also be in contact with the floodplain, especially for reproduction. Insufficient river flows stop these necessary migrations. Flow variability is therefore an important part of aquatic species survival, due to the varying habitats created for different parts of the species life cycle. Species that are able to adapt can survive the extremes of flood and drought [13].

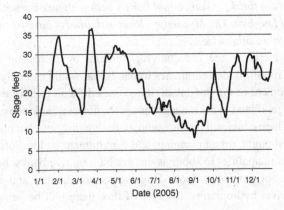

Fig 8.18 USACE gage data on the Mississippi River at Greenville, Mississippi, US. Notice that flow varies over time, with highs and lows, but very few periods of stable flow. The Mississippi River has an altered hydrology due to channel modifications made for flood control, levee building, and floodplain removal, in addition to normal precipitation events

(Image US Army Corp of Engineers USACE)

Flood flows connect the river and its floodplain, and shape the physical habitat of the river system. This process creates wetlands that serve as spawning grounds and nurseries for young fish, as well as feeding habitat for birds and other wildlife. Species diversity and health depends on the availability of these flood-produced habitats. During floods, gravels and finer sediments are deposited – fertilizing the land, creating spawning habitat, cleaning, and enriching the water with organic materials that provide food for the young fish. **Normal high flows** shape the river channel by removing and depositing sediments. They are also the time for many aquatic organisms to move from place to place. High flows help keep water oxygenated.

Normal low flows allow different aquatic species to travel in and from tributaries to new riparian areas. Low flows also expose streambanks that provide mudflats for migrating birds to feed on macroinvertebrates. They allow tree seedlings to develop to the height needed to survive high water. This is critical to bald cypress and gum tupelo, for example. Low flows concentrate fish into pools where feeding is easier for some species (predators) and population numbers are regulated on others (prey). Inadequate dissolved oxygen levels are often a problem during low flow periods.

8.5.3 Causes of Floods

Perhaps the most obvious cause of floods in large watersheds is a prolonged period (or an unusually intense period) of rain or snowmelt when the ground is already saturated or frozen. This has less to do with land use and more to do with the total amount and speed of delivery of water into the river.

The 1993 Mississippi River flood in the American Midwest was caused by an unusually long and intense rainfall coupled with the timing of flood peaks coming from its major tributaries. Circumstances were just right (or just wrong) so that the tributaries released record high discharges at their convergence with the Mississippi [14]. The rains were the second highest recorded in 121 years of records. The 1993 flood covered millions of hectares (acres) and caused unprecedented dollar damages (US$15 billion), 50 people died, hundreds of levees failed, and thousands of people were evacuated, some for months [15]. At least 10,000 homes were destroyed, hundreds of towns were impacted, with at least 75 towns completely under floodwaters. Over 6 million hectares (15 million acres) of farmland were inundated, some of which may not be usable for years to come. Transportation became a nightmare, barge traffic stopped for two months, interstate highways and bridges were closed, local roads were destroyed, 10 airports were flooded, and all railroad traffic in the Midwest was stopped. Water quality was threatened by the destruction of numerous wastewater treatment plants [16].

This was not a normal flood. Mark Twain said a hundred years ago [17] that the Mississippi River *"cannot be tamed, curbed or confined . . . you cannot bar its path with an obstruction which it will not tear down, dance over and laugh at. The Mississippi River will always have its own way; no engineering skill can persuade it to do otherwise . . ."*

Floodplain development increases the potential for flooding by increasing impervious surfaces and increasing the rate of runoff. It also eliminates wetlands that would naturally slow these floodwaters. When there is development in the floodplain, it becomes difficult to have even a small flood event without property damage. If land use is managed properly, however, a floodplain could be designated as a no-building zone. There could still be recreation, trails, hunting, etc., but no permanent structures. That would require forethought, urban planning, and commitment by local officials.

Everything from major municipalities to subdivisions are built on riverbanks, but we seem surprised every time humans cannot control a flood. Huge areas are drained and cleared for agriculture without regard to the river hydrographs. Because of this, there will be years when the land cannot be farmed because it remains too wet. This should tell us something about the wisdom of clearing that piece of ground. It is also an excellent predictor of where wetlands should be restored in that particular landscape.

8.6 River and Stream Ecology

8.6.1 Energy Source

Rivers and streams get most of their energy from the watershed. The watershed provides the necessary nutrients via sediment transport: however, the nutrients move with the flow. Water entering a river is immediately in transit, making residence times (the time water remains in one location) short. The primary producers and consumers live in the same river riparian environment. Organisms in rivers are mostly energy consumers, such as fish and frogs. The primary energy producers tend to live in the river substrate, and these are benthic organism or macroinvertebrates. Rivers with excess vegetation have usually received too many nutrients. Nutrient or energy accumulation can be complicated in low flow because there is no flowing water to carry materials to the depositional zone. The entire system becomes, in effect, a depositional zone accumulating excess energy.

8.6.2 Organisms and Habitats

Rivers and streams provide habitat on the surface, in the water column, on the bottom (the **benthic zone**), and in rivers with porous riverbeds – the hyporheic zone. Notice in Figure 8.19 that food and habitat, for macroinvertebrates and microorganisms, is at the smallest scale. Crayfish, clams, mollusks, dragonfly, mayfly, and stonefly larvae all live at this same small scale. Rocks, branches, and other surfaces provide niches for these organisms that form the bottom of the food chain.

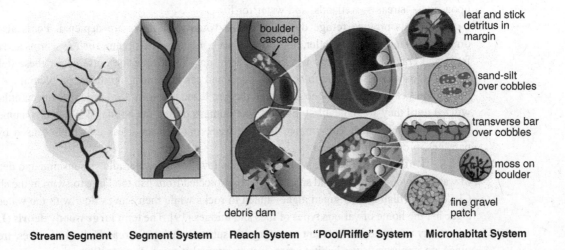

Fig 8.19 Spatial scale of a stream, focusing from large to small
(Image courtesy of FISRWG)

(a)   (b)

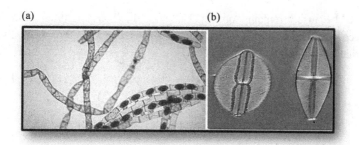

Fig 8.20 These are examples of autotrophs: (a) blue-green algae, (b) diatoms

Autotrophs are the primary food source for waterfowl, fish, turtles, and amphibians. Autotrophs are green plants, diatoms (silica-walled microscopic algae), filamentous algae, and protists (usually single-celled organisms including algae and protozoans) that use energy from the Sun and nutrients from the water to make food (see Figure 8.20). Autotrophs are the periphyton, or phytoplankton in rivers and lakes. **Heterotrophs** (fish, amphibians, and waterfowl) use organic matter produced by autotrophs, or that has been washed into the river. Autotrophs (green plants) can be limited by a lack of sunlight in turbid, deep, or turbulent streams.

Macrophytes (large aquatic plants) can be emergents with their roots in soil and their leaves coming out of the water. They may have floating leaves that are attached to a substrate or just free-floating. Some form mats which can be free-floating, submerged, or rooted in substrate. All macrophytes will only grow with adequate sunlight. Most macrophytes do not adapt well to flowing water, and are generally found in the still areas of backwaters, deltas, and contiguous wetlands.

There are many animals that do not live in rivers or streams but are dependent on stream-flow and the riparian area for habitat, food, and nesting habitat – the stuff of life. A myriad of small furbearers, such as otters, beaver, nutria, raccoons, and muskrats, also have homes in riparian areas. Alligators, crocodiles, hippopotami, countless amphibians, and birds, thousands of birds, call rivers and streams their home as well. For example: predator birds, from eagle to osprey; shorebirds, gull and tern; birds that feast on mudflats, stilts and sandpipers; the fish-eaters, pelicans, egrets, and herons; waterfowl, mallard and wood duck, swan and goose – all and many more are dependent upon rivers, streams, wetlands, and water for life.

Pools provide refuge during droughts when river flows are depleted. Pools also house organisms that prefer the gentler, non-flowing water regions of a stream, such as worms and midges, frogs and crayfish [18]. Large fish come to hunt prey or browse the bottoms for these delicacies. Dissolved oxygen levels tend to be low in pools and can become a significant problem if the aquatic population gets too large. When dissolved oxygen levels get low, dead organisms accumulate. Fish will eat until they are unhealthily bloated, like humans during the holidays! When this happens, it is not unusual to see fish kills immediately, or weeks later, from stress infections caused by fungi, parasites, protozoa, or bacteria.

Riffles tend to be rich with life. The shallow flowing water allows sunshine into deep areas of water. Diatoms flourish, and smaller fish are protected from fish too large to swim in the shallows. Cladophora (filamentous green algae) attach to rocks while their leaves flow with the water. These algae are the home of various types of flies and midges [19]. The term **large woody debris (LWD)** is commonly used and is important as an organic substrate habitat. Areas of dead branches, trees, and twigs are important river habitat areas, just as important as pools and riffles.

The inorganic substrate consists of boulders, >30 centimeters (12 inches), cobbles, 8–30 centimeters (3–12 inches), gravel 0.6–8 centimeters (0.25–3 inches), and fines, sand and silt, <0.6 centimeters (0.25 inches). These types of substrates are particularly important for spawning fish, such as salmon. In many areas, clays would be included because they are a part of muddy bottoms.

The chapter on water quality (Chapter 4) applies to rivers and streams as well as lakes and wetlands. To learn more, check out the reference websites, and take a class in limnology, fluvial geomorphology, fisheries, biology, etc., or just read and study what you find around you. You may want to keep a nature journal to see what interests you. Take walks. Take pictures, sketch (no artistic talent is needed, as they will be your own). However, you may find hidden talent and become a naturalist of note.

Guest Essay By Carolyn J. Schott

Carolyn J. Schott is principal scientist and owner of Lifezone Ecological, Inc. She specializes in wildlife ecology, watershed science, and habitat restoration. Her qualifications include a Bachelor of Science in Natural Resource Management from Colorado State University and a Master of Resource Science from the Department of Ecosystem Management at the University of New England in Australia.

Stream Restoration and Flood Mitigation at Hecla Junction, Arkansas Headwaters Recreation Area, Colorado, US

Background

Heavy rains during August 11–12, 2006, in Chaffee County, Colorado, caused several ephemeral tributaries in the upper Arkansas River valley to experience unusual flash flooding, which damaged facilities and created an alluvial fan at the mouth of a small tributary within the Hecla Junction recreation area (Figure 8.21). Hecla Junction is within a Wild and Scenic designated reach of the river known as Brown's Canyon, and is an integral component of the Arkansas Headwaters Recreation Area (AHRA).

Fortunately, there were no injuries or fatalities related to the flooding, but damages to the fabricated infrastructure, as well as the natural environment were quite significant. Immediately following the flood, Colorado Division of Parks and Outdoor Recreation (State Parks) personnel worked feverishly to repair the most critically damaged areas in order to restore basic operations at the site; however, the band-aid solutions were intended to be temporary in nature. A sustainable restoration and flood mitigation project was sought to reduce future vulnerability and to preserve ecological integrity of the site.

The AHRA is an extremely popular destination for whitewater enthusiasts. The Brown's Canyon reach of the Arkansas River is one of the most commercially rafted in the United States, and Hecla Junction serves as one of the most highly utilized raft take-out sites in the world. Brown's Canyon offers Class III and IV rapids during the normal rafting season with an average vertical drop of about 6 meters per kilometer (30 feet per mile). A Voluntary Flow Management Program offers no less than 20 cubic meters per second (700 cubic feet per second) of flow through August 15 each year. Other activities include fishing, private boating, camping, hiking, and wildlife viewing. The area has a vital impact on the valley economy because of its recreational opportunities, natural beauty, and biological productivity.

Fig 8.21 Hecla Junction maintenance showing the alluvial fan going into the Arkansas River (Photograph by Rob White, Arkansas River Headwaters Recreation Area Park Manager)

Environmental Description

The Arkansas River corridor within the AHRA is one of the most diverse in Colorado. It descends from high mountain peaks and flows through diverse montane ecosystems and ends in a mature river system. The corridor has remained in outstanding condition despite modifications including a railroad, busy highway, substantial agricultural activities, and development.

The Arkansas River at Hecla Junction is situated at approximately 2,286 meters (7,500 feet) in elevation in the upper montane life zone. Slopes surrounding the project area are dominated by pinyon juniper woodland and include areas of sage shrub steppe and open grassland. Sediments come from the Dry Union Formation and include shallow coarse soils with high permeability, which lead to highly incised silty clays, head cutting, and gullying (Figure 8.22). The Arkansas River below the project area includes a riffle–pool complex and the riparian corridor includes tall willow and alder shrub land with associated areas of grassy marsh.

Terrestrial and aquatic habitat values vary within the project area. The environment is suitable for various species including fish, amphibians, reptiles, birds, and mammals. Brown trout make up over 90 percent of the fishery and rainbow trout also inhabit the area. Macroinvertebrate populations were impacted by the flood, but according to the Colorado Division of Wildlife (CDOW), their populations rebounded quickly. Federal and State Listed Species in the AHRA require special management under the Endangered Species Act of 1973 (Table 8.2).

Field surveys revealed two federally listed species. A bald eagle was observed in December 2006, and a piping plover was observed in April 2007. According to local CDOW experts, juvenile ospreys are known to use the corridor and various migratory songbirds and

Table 8.2 Listed species in the AHRA

Common name	Species	Status[a]
Peregrine falcon	*Falco peregrinus*	SC
Bald eagle	*Haliaeetus leucocephalus*	FT, ST
Piping plover	*Charadrius melodus*	FT
Mexican spotted owl	*Strix occidentalis lucida*	FT, ST
Osprey	*Pandion haliaetus*	SC
Northern river otter	*Lutra canadensis*	ST
Canada lynx	*Lynx canadensis*	FT, ST
Townsend's big-eared bat	*Corynorhinus townsendii*	SC
Northern leopard frog	*Rana pipiens*	SC
Uncompahgre fritillary butterfly	*Boloria acrocnema*	FE

[a] SC – Special Concern, FT – Federal Threatened, FE – Federally Endangered, ST – State Threatened, SE – State Endangered.
Source: Smith, R. E. and Hill, L. M., eds., 2000, *Arkansas River Water Needs Assessment*, Denver, Colo.: USDI BLM, US BOR, and USDA Forest Service

Fig 8.22 Hecla Junction area showing erosion (Photograph by Jason Carey, RiverRestoration.Org, PE)

waterfowl species occur at different times of the year. Secluded rocky cliffs provide suitable habitat for the American peregrine falcon and Mexican spotted owl. In addition, Canada lynx, northern river otter, and Townsend's big-eared bat occur along the Arkansas River. Communication with the US Fish and Wildlife Service and the CDOW ensures protection of these biological resources, and an Environmental Impact Statement will be initiated as the project develops.

Stream Restoration and Flood Mitigation

In early 2007, experts from State Parks, the Bureau of Land Management (BLM), the Colorado Water Conservation Board (CWCB), CDOW, and the AHRA Citizens Task Force worked with environmental consulting firms Lifezone Ecological, Inc. and RiverRestoration.Org to evaluate site history, and to develop a stream restoration and flood mitigation concept plan. To uncover opportunities and constraints for long-term restoration success, studies were performed concerning hydrology, hydraulics, geomorphology, ecology, and socio-economic importance.

The project team used a holistic approach for the concept plan. Alluvial fans are a dynamic force of nature and serve an ecological purpose, refreshing sediments in the river, and therefore the concept plan anticipated sediment transport along the tributary. Respecting this force of nature allows for recreational use adjacent to a natural and functioning riparian zone.

A forensic post-flood study was completed to estimate the recurrence interval of the flood event through hydraulic modeling. A design flow was established for site mitigation based on the reconstructed flood event. Conclusions revealed the following:

- An estimated 9 centimeters (3.5 inches) of rain fell on the watershed over short duration.
- The Arkansas River gage at Hecla Junction jumped from 20 to 113 cubic meters per second (700 to 4,000 cubic feet per second) during the event.
- The flood deposited sediments and created an alluvial fan that extended over 9 meters (30 feet) into the river.
- The peak flood flow in the tributary was estimated to be (1,500 cubic feet per second, and greater than a 100-year event).

The concept plan aimed to:

- Restore full function of boating operations at the highly utilized AHRA site;
- Reduce the need for costly maintenance when flooding and sediment occurs;
- Decrease the risk of site closures and reduced user access;
- Increase the long-term sustainability for ecological and human benefits;
- Enhance the natural environment by using appropriate bio-stabilization techniques;
- Add native vegetation for buffer zones and habitat;
- Identify permitting and final design needs; and
- Prepare cost estimates and list potential funding sources for implementation.

An important part of the concept plan includes bio-stabilization techniques such as biodegradable fabrics, erosion control plantings, and coarse woody material to reduce flow velocities. Bio-stabilization techniques can provide habitat to a variety of resident and migrating wildlife species by offering security and cover, as well as food-chain support. Structures based on bio-stabilization provide pockets where small mammals can reside or where water can collect for insect larvae development, in turn offering a prey base for insectivores and carnivores associated with the riparian zone.

Conclusion

Following a significant flood event at the highly utilized Hecla Junction site, state and federal interests worked with a qualified consulting team to develop a concept plan for a sustainable raft take-out in a way that is compatible with the surrounding environment. The concept plan assisted State Parks in framing opportunities and constraints at the site. Additional hydraulic, sediment transport modeling, and final engineering design will be required prior to implementation of the plan.

Other opportunities being explored for Hecla Junction include interpretive signage for visitor education, fishing access points, and bridges or engineered pans to allow for sediment passage. Most importantly, Colorado State Parks seeks to reduce long-term maintenance expenditures while maximizing site usability.

Environmental site planning adheres to guidance and policies in accordance with stream management goals consistent with BLM Brown's Canyon Wild and Scenic Area prescriptions and with US Army Corps of Engineers Clean Water Act Section 404 regulations. Planning is also consistent with the National Environmental Policy Act, the Endangered Species Act, and the Migratory Bird Act.

Anticipated costs for construction of a sustainable project based on the concept plan are about US$1 million. The concept plan was funded by the Flood Protection Section of the Colorado Water Conservation Board. Special thanks to Tom Browning, CWCB Section Chief, for guidance and support.

Summary Points

Major river functions include **purification of water and wastes, recreation, water supply, esthetics, flood control, habitat, drought mitigation, maintaining soil, estuary and wetland health, preservation of biodiversity, and as a food source.**

- River functions can be eliminated by overuse. This has happened throughout time wherever large civilizations have developed as humankind used rivers as dumping places for our wastes. Many rivers were so overused that they were virtually dead.

- **Fluvial morphologists** study the formation and structure of rivers.

- **Erosion, sediment transport**, and **sediment deposition** are the geomorphic processes that form both the river channel and the floodplain structure.

- A **river system** is composed of three parts – a **channel**, a **floodplain**, and an **upland transition** or **fringe area**.

- Along the river is a **riparian area**, often called a **riparian buffer** because it intercepts water before it enters the river.

- Longitudinally, a river has a **headwater**, a **transition zone**, and finally a **depositional zone**.

- Laterally, a river's deepest part is the **thalweg**.

- River flow can be **perennial, intermittent**, or **ephemeral**. Since rivers flow and the flows are not constant, river characteristics are not constant.

- **Flow hydrographs** show flow over time. These can be especially useful in managing a river.

- **Flood flows** and **low flows** are natural parts of a river's regime. Both can be altered by human intervention with varying consequences.

- We have previously defined a **watershed (catchment** or **catch basin)** as the land that drains into a body of water (stream, river, pond, lake, or ocean). It is important to a river because much of the **energy** for a river comes from the land.

- There are three types of runoff: **surface** or overland flow, shallow **subsurface** flow (throughflow), and saturated surface or **overland** flow.

- River water has a short **residence time** because it flows downstream.

- **Food energy** must be intercepted by aquatic organisms.

- **Autotrophs** are green plants, diatoms, filamentous algae, and protists that use energy from the Sun and nutrients from the water to make food. **Heterotrophs** (fish, amphibians, and waterfowl) use organic matter produced by autotrophs, or that was washed into the river.

- **River system restoration** is a complicated process that **must** begin with a clear picture of what exactly is to be restored.

Questions for Analysis

1. Which American ecological disaster is credited with sparking the US's ecological movement?
2. What are some of the unintended consequences of flood protection from rivers?
3. You are a city planner in a drought area. You are directed to maximize return flow from the city to a river system for water supply. What are your major concerns?
4. Are there any circumstances under which a river could change from perennial flow to intermittent? What are some of the potential ecologic consequences of a major change in river flow?
5. Is river system restoration possible? What is the first step in the restoration process?

Further Reading

Federal Interagency Stream Restoration Working Group, 1998, *Stream Corridor Restoration: Principles, Processes, and Practices*, Washington, DC: FISRWG.
Leopold, Luna B., 1997, *Water, Rivers, and Creeks*, Sausalito, Calif.: University Science Books.
Pielou, E. C., 1998, *Fresh Water*, Chicago, Ill.: University of Chicago Press.
Postel, Sandra and Brian Richter, 2003, *Rivers for Life: Managing Water for People and Nature*, Washington, DC: Island Press.

References

1 Hal Borland, 1957, *This Hill, This Valley*, New York: Simon and Shuster.
2 Luna B. Leopold, 1997, *Water, Rivers, and Creeks*, Sausalito, Calif.: University Science Books, p. 175.
3 *Time Magazine*, 1969, "The cities: The price of optimism," August issue, http://content.time.com/time/subscriber/article/0,33009,901182-1,00.html (accessed 2020 from *Time* archived stories).
4 Sandra Postel and Brian Richter, 2003, *Rivers for Life: Managing Water for People and Nature*, Washington, DC: Island Press, p. 8.
5 R. T. T. Forman, 1995, *Land Mosaics: The Ecology of Landscapes and Regions*, Cambridge: Cambridge University Press.

6 A. N. Strahler, 1957, "Quantitative analysis of watershed geomorphology," *Transactions of the American Geophysical Union* 38, 913–920.

7 A. N. Strahler, 1952, "Dynamic basis of geomorphology," *Bulletin of the Geological Society of America* 63, 923–938.

8 A. N. Strahler, "Dynamic basis of geomorphology."

9 Federal Interagency Stream Restoration Working Group, 1998, *Stream Corridor Restoration: Principles, Processes, and Practices*, Washington, DC: FISRWG.

10 J. V. Ward, 1989, "The four-dimensional nature of lotic ecosystems," *Journal of the North American Benthological Society* 8, 2–8; and J. V. Ward and J. A. Stanford, 1989, "Groundwater animals of alluvial river systems: a potential management tool," in *Proceedings of the Colorado Water Engineering and Management Conference*, Colorado Water Resources Research Institute, Fort Collins, Colorado.

11 Luna B. Leopold, 1997, *Water, Rivers, and Creeks*, Sausalito, Calif.: University Science Books, p. 175.

12 Ibid.

13 N. Poff, 1997, "The natural flow regime: a paradigm for river conservation and restoration," *BioScience* 47, 769–784.

14 Luna B. Leopold, 1997, *Water, Rivers, and Creeks*, Sausalito, Calif.: University Science Books, pp. 100–102.

15 Lee W. Larson, 1996, "The Great US Flood of 1993," presented at IAHS Conference Destructive Water: Water-Caused Natural Disasters – Their Abatement and Control, Anaheim, California.

16 Lee W. Larson, 1993, *The Great Midwest Flood of 1993, Natural Disaster Survey Report*, Kansas City, Mo.: National Weather Service.

17 Mark Twain, 1883, *Life on the Mississippi*, Montreal, Quebec: Dawson Brothers.

18 Cristi Cave, 1998, *School of Fisheries*, University of Washington, http://chamisa.freeshell.org/habitat.htm (accessed July 2007).

19 Luna B. Leopold, 1997, *Water, Rivers, and Creeks*, Sausalito, Calif.: University Science Books.

9 Wetlands

I was horrified, drying up the springs, drying up the streams, and lowering the lake meant to exterminate the growth by running water, meant to kill the great trees which had flourished since the beginning of time around the borders of the lakes, meant to kill the vines and shrubs and bushes, the ferns and the iris and the water hyacinths, the arrowhead lilies and the rosemary and the orchids, and it meant, too, that men were madly and recklessly doing an insane thing without really understanding what they were doing.

Gene Stratton-Porter (1863–1924),
naturalist, author, comments on the destruction of Indiana's Limberlost Swamp [1]

Chapter Outline

9.1 Introduction

The word wetland conjures up different visions for all of us, depending on what we have seen, know, or think we know about these watery places. Scary things – from swamp monster creatures to giant killer alligators, and pits of quicksand for the inattentive – are common media images. The reality is that wetlands come in wondrous variety and can be forested, grassed, herbaceous, or peat bogs, but none are known to contain swamp monsters; just the alligators that prefer to be left alone. Wetlands occur worldwide, on every continent except Antarctica, and in every climate including deserts and frozen tundras.

Boreal (also called Taiga) are coniferous forests with wetlands, lakes, valleys, tundra, and peatland that cover 16 million square kilometers (6 million square miles) in Russia (primarily Siberia), Canada, and Alaska (see Figure 9.1). The Russian portion is the largest forest in the world at 12 million square kilometers (4.6 million square miles). Canada has the next largest area at 3.6 million square kilometers (1.4 million square miles); about 500,000 square miles are wetlands – called *muskegs* by the Algonquian people. The Alaskan area makes up another 810,000 square kilometers (312,000 square miles). The North American Boreal Forest is 30 percent wetlands.

Excluding Alaska, the remainder of the US has about 6 percent of its surface area in wetlands about 1.8 million acres about the size of the state of Texas. Wetlands can readily form in areas with poorly developed drainage systems. For example, glaciated areas in Alaska and Canada

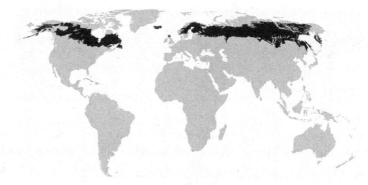

Fig 9.1 The Boreal forest exists only in the northern latitudes between about 50° and 70°N, covering about 17 percent of the world's land surface area. At these latitudes, the Sun's rays must travel long distances through Earth's atmosphere, weakening them, so there is less of the Sun's energy (heat) reaching the ground (Image developed by Mark Baldwin-Smith – own work, CC BY-SA 3.0, https://commons.wikimedia.org/w/index.php?curid=18250691)

Fig 9.2 This lake in Alaska formed in the summer when the active layer of permafrost thawed, and allowed the liquid water to pool in a thin layer of topsoil close to the surface. Plants use the water and the long summer days to grow very quickly, and bogs can form. Trees thrive in the rich, moist soil. In the Arctic, huge boreal forests cover areas of seasonally frozen ground
(Photograph courtesy of Tingjun Zhang, Principal Scientist, All About Frozen Ground Project, National Snow and Ice Data Center, University of Colorado, Boulder)

were ice covered only about 10,000 years ago. This is not long enough in geologic time for the landscape to form well-developed drainage networks of streams and rivers, that carry surface water runoff to the sea, thus allowing estuarine wetlands to develop (see Figure 9.2) [2].

Worldwide, there are 254 million hectares (628 million acres) of wetlands protected under the international agreement from the Ramsar Convention on Wetlands, held in Ramsar, Iran, in

1971. This convention was called to help develop an international agreement defining "internationally significant wetlands," and selecting criteria for countries to use in designating and protecting these wetlands. Today, 171 countries participate in this agreement [3].

The mission of the group, stated in their Strategic Framework, was [4]: "To develop and maintain an international network of wetlands which are important for the conservation of global biological diversity and for sustaining human life through the ecological and hydrological functions they perform." The fourth Strategic Framework states the Ramsar mission is the: *Conservation and wise use of all wetlands through local and national actions and international cooperation, as a contribution towards achieving sustainable development throughout the world* [5]. This was and continues to be a strong show of support for the value of wetlands and encourages the preservation of more wetlands to develop a continuity of wetland ecosystems. Protection of local wetlands is no less important in trying to reach this continuity.

We have all probably seen wetlands. Unfortunately, many people do not pay much attention to these wet areas that seem to occur in odd places. In Nebraska, for example, we may see a depression in part of a farmed field. It may not look particularly interesting, and we wonder why the farmer did not fill it in with soil and then plant a crop in that location. If we walk closer, however, we will probably see that the depression, a low spot in the ground, is teeming with life. A blue heron flies gracefully away, not wanting company. Shorebirds are feeding in the mudflats around the water. Ducks bob up and down, searching for macroinvertebrates living in the water, on rocks, branches, or in loose sediment. Dragonflies flit about while the chorus frogs trill like running fingers over a comb. Chirping noises come from backup singing insects. Cautious deer and other mammals wander down in the evening, anticipating water and food.

These prairie playa wetlands, surrounded by predominantly perennial grasses, are vital to wildlife. They provide food, habitat, and refuge for many species including insects, reptiles, amphibians, birds, and mammals. Even a small prairie depression in western Nebraska is home to many species and serves as a combination filling station and rest area to many more (see Figure 9.3).

9.2 Wetland Features

A wetland has three defining features – soil, vegetation, and hydrology. The first feature is **hydric soils**. These are soils with characteristics developed over extended periods of wetting and drying. The second defining feature of a wetland is a preponderance of **hydrophytic** (water-tolerant) vegetation. The third feature is **hydrology** that maintains a wet condition during a significant portion of the growing season.

There are many wetland definitions. Some are used for regulatory purposes while others fit a particular science – for example, botany and plant communities, or wildlife and wetland functions. The following (US Fish and Wildlife Service – USFWS) definition uses the three components that make up a wetland [6].

Wetlands are lands transitional between terrestrial and aquatic systems where the water table is usually at or near the surface or the land is covered by shallow water . . . wetlands must have one or more of the following three attributes: (1) at least periodically, the land supports predominantly hydrophytes; (2) the substrate is predominantly undrained hydric soil; (3) the substrate is nonsoil and is saturated with water, or covered by shallow water, at some time during the growing season of each year [7]. This definition serves as the basis for the following definition used today for regulatory purposes and is called the **Cowardian System**.

Fig 9.3 This is a Nebraska Sandhills wetland. The Sandhills are mixed grass prairies on stabilized sand dunes, one of the largest grass-stabilized dune regions in the world. Wetlands formed in depressions where groundwater intercepts the land surface are, basically, at the level of the area's groundwater. In Nebraska, this is the Ogallala Aquifer. Vegetation is highly diverse with some 720 species, of which 670 are native to the region. Because of this diversity, the plant community is called a Sandhills type
(Photograph courtesy of Robert B Swanson, USGS Nebraska Water Science Center)

The US Environmental Protection Agency (USEPA) and US Army Corps of Engineers (USACE) have another definition used since 1970 for regulation purposes [8]:

"Those areas that are inundated or saturated by surface or groundwater at a frequency and duration sufficient to support, and that under normal circumstances do support, a prevalence of vegetation typically adapted for life in saturated soil conditions. Wetlands generally include swamps, marshes, bogs, and similar areas."

Again, we see the definition incorporates hydrology, hydrophytic vegetation, and saturated soils. Our discussion of water quality in Chapter 4 included the fact that definitions are relative to the use of the water and viewpoint of the person developing the definition. The same is true for wetlands. However, Mitsch and Gosselink, in their excellent book on wetlands, caution that, although the precision of a definition is important, the consistency with which it is used is equally as important [9]. This is true because we live in a world where science and legal issues often meet.

Scientists are often called for assistance in situations that involve legal matters, such as compliance with environmental laws. However, the basic nature of their profession requires scientists to strive for both precision and accuracy in their work. These terms may be used synonymously – except in scientific work. An answer to a scientific question may be precise, but not accurate,

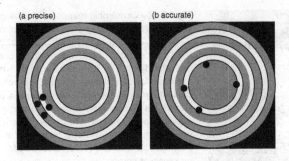

(a precise)   (b accurate)

Fig 9.4 (a) The shots are precise, reproducibly close together, but if the bull's-eye is the "answer," they are not accurate. (b) Here the shots are more accurate since they are in, or close, to the bull's-eye, but they are scattered, therefore not precise

and vice versa. The easiest way to visualize the difference is the often-used "target" example (see Figure 9.4).

Science requires that "answers" be both as precise and accurate as possible. Simply put, precision is reproducibility of the answer while accuracy is its correctness. The need for accuracy and precision is obvious – as an example, when making a chemical solution. In this situation, a pharmaceutical company must be both precise and accurate when producing thousands of vials of vaccine in accurate chemical doses.

Let's examine in more detail the three components that produce a wetland ecosystem, starting with soils.

9.2.1 Soils

Wetland soils are called **hydric soils**. These are defined as soils that formed under conditions of saturation, flooding, or ponding long enough during the growing season to develop anaerobic (without oxygen) conditions in the upper part of the soil. Several terms need to be defined to understand the definition. (This is common with scientific definitions because each field of science has its own jargon or language. These words can be very descriptive and add richness to the vocabulary but must be learned by the beginning wetland enthusiast.) The following terms are used by wetland soil scientists.

- The US Army Corps of Engineers (USACE or Corps) criteria for wetlands are the following:
    "Long enough" is 5 percent of the growing season in most years (50 percent of the time). *"Growing season"* is when the soil temperature, measured at a depth of 50 centimeters (20 inches), is greater than biological zero, 5 °C or 41 °F.
- The US Department of Agriculture's Farm Service Agency (FSA) criterion for wetlands is the following:
    "Long enough" means the soil is inundated for seven consecutive days during the growing season in most years (50 percent of the time), or saturated – at or near the surface – for 14 consecutive days during the growing season in most years.
- Both the USACE and FSA use the following definitions:
    "Near" is defined starting at the soil surface, as 15 centimeters (6 inches) below the surface for sands and 30 centimeters (12 inches) for all other soil textures. Anaerobic conditions are evidenced by **redoximorphic features**. These are morphological (physical) changes in the soil caused by oxidation and reduction conditions that occur with wetting and drying (see Figure 9.5).
    "Upper part of the soil" is the top 50 centimeters (20 inches) of the soil profile.

Fig 9.5 Redoximorphic features are striking when in color; in shades of gray, you must have a more discerning eye and some imagination. Dark gray areas in this sandy soil are actually red, which are concentrations of oxidized iron. The light gray areas are where the iron has been leached deeper into the soil profile. The darkest streak in the top center is carbon, probably following a decayed root channel. To better understand the usefulness of these differences, we strongly urge you to look up redoximorphic features of soil on the internet. Vasilas et al.'s *Field Indicators of Hydric Soils* is an excellent source
(Photograph courtesy of USDA NRCS)

Why are we interested in what the USACE or US Department of Agriculture have to say about wetlands? We have already seen that countries around the world recognize the value of wetlands and are trying to protect them. Part of that global effort involves classifying wetlands by observable characteristics. These characteristics need to be carefully defined so that classification is uniform and defensible, hence the need for agency involvement.

The United States' effort to stop wetland destruction included passing the Food Security Act of 1985, commonly called the "Swampbuster Act," requiring no net loss of wetlands [10]. This law made it necessary to develop rules to regulate how wetlands are delineated and used. The USACE became the regulating agency for wetlands in the US. However, the US Department of Agriculture Natural Resources Conservation Service (NRCS) was given responsibility for delineating wetlands on agricultural lands using FSA guidelines; therefore, we have two sets of criteria, but both generally come to the same result.

Let's return to our discussion of hydric soils: they can be mineral or organic, and their characteristics vary primarily with parent material and the duration of wetness. The identifying characteristics of hydric soils are called **indicators** [11]. These indicators are based largely on color of the soil. Why is color a logical choice? One reason is that most people can differentiate color differences. The origin of soil colors are another reason.

Soil color comes from various chemical processes acting on soil and its parent material. These include: (1) weathering of geologic material (parent material – see Figure 9.5), (2) chemistry of oxidation–reduction reactions upon the various minerals of soil, especially iron and manganese, and (3) the biochemistry of the decomposition of organic matter, primarily carbon.

Iron provides the major color component of hydric soils. Why, with all the different elements present in soil minerals, was iron chosen? For one thing, iron is ubiquitous (it is common, and occurs in many soil minerals throughout the world). It also changes color depending on its **oxidation state**. Elemental iron forms small crystals with a yellow or red color. The oxidized form Fe^{+3} (higher oxidation state, ferric) does not move with water, and is red, orange, or yellow in color. You will recognize oxidized iron as rust, which is an oxidation reaction.

The reduced form Fe^{+2} (lower oxidation state, ferrous) is mobile in water and is gray. Loss of iron from leaching produces the characteristic color left when the iron is gone from soils. These are called *gley colors;* they vary from black to blue grays or green grays. Anaerobic soils do not occur until the iron has been under reducing conditions, i.e., saturated with water. The exact time required to produce gley colors depends on the minerals involved, and so varies in different soils [12].

A quick chemistry note: elements can have different oxidation states depending on the number of electrons in their outer shell. The *neutral state* is the element at zero oxidation state. It has no charge. When an element has a charge, it is called an *ion*. Electrons can be lost, which produces an increased or positive oxidation state called the *oxidized state*. The acronym LEO (lose electrons oxidize) can be helpful as a memory aid. Electrons can also be added to produce a decreased or negative oxidation state called the *reduced state*. Remembering these facts can help you determine the oxidation state of any element in a reaction.

Therefore, iron is used as an indicator of the wetting and drying cycles that occur in wetland soils for the following reasons:

- Iron occurs almost everywhere in many mineral forms (see Figure 9.6).
- It changes color with differences in oxygen content in the soil (dependent on its oxidation state).
- It becomes mobile in water when anaerobic conditions occur, allowing it to leach from one area in the soil profile to a lower area. This results in a changing soil color in both areas.

The amount of carbon present from decomposing organic matter is also used to describe hydric soils present in wetlands. Organic matter is classified by the extent of decomposition of the vegetative materials. *Sapric organic matter* is highly decomposed without visible fibers and is black. *Hemic organic matter* is less decomposed; with 1/3–2/3 fibers visible with a hand lens in an unrubbed sample and is dark gray or reddish brown. *Fibric organic matter* is the least decomposed with > 2/3

Fig 9.6 The specifics of the geology of Yellowstone River Canyon in Yellowstone National Park, Wyoming, US, are not well known, except that its classic V-shaped formation is indicative of river-type erosion rather than glaciation. Even in black and white, the different light and dark shades in this picture show the variety of geologic or parent material colors. Rhyolite lava flows, rich in iron, formed a geyser basin before the canyon developed. Heat caused the rhyolite to become brittle and easily eroded, eventually forming the canyon seen today. The different colors are due to hydrothermal alteration of the iron in the rhyolite while the basin was active. Exposure to the elements allows oxidation reactions, with iron, producing the red and yellow colors

(Getty image credit: Brad Booth)

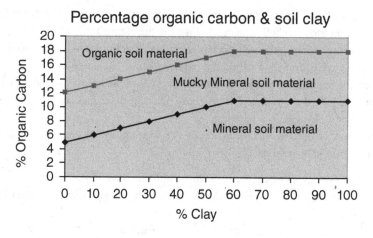

Percentage organic carbon & soil clay

Fig 9.7 Organic soils are carbon rich. An increase in clay content requires a corresponding increase in organic matter content to be defined as an organic soil. Both clay and organic materials are chemically active. For example, 70 percent clay needs > 18 percent organic carbon to dominate chemically, whereas 20 percent clay needs >14 percent carbon

fibers and is yellowish, reddish, or dark brown. Organic matter can accumulate in wet soils because it is slow to decompose in anaerobic conditions. The difference between a mineral soil and an organic soil is determined by the percent organic carbon and percent clay. Figure 9.7 illustrates this point.

Another chemical often used to describe hydric soils is manganese. Manganese forms black, immobile mineral deposits when aerobic (oxidized), but becomes clear and mobile when anaerobic (reduced). Manganese can also form hard **concretions**, or **nodules**, sometimes mixed with iron. The concretions are somewhat rounded, vary in size, are visible without a hand lens, and can be felt in a moist sample by rubbing the soil between your fingers. Sometimes the concretions feel very much like round buckshot. Their presence in some wetland soils has led to the common name "buckshot" soils. The larger concretions can be cut in half revealing concentric circles like those found in old fashion jawbreaker candy.

To summarize, the hydric soil colors are:

- The red of oxidized iron.
- The whites to grays or light colors resulting from the loss of mobile forms of iron and manganese.
- The gley colors of reduced iron.
- The darks to blacks of decomposing organic matter.

Soil is an essential component in determining how wetlands function (see Figure 9.8). Hydric soils provide the same basic functions as all soils. They are the physical foundation (structural support) for wetland ecosystems providing a place for wetland plants and animals to live and grow. Soils are the principle reservoirs (sinks) for minerals and nutrients needed by plants and other organisms. Soils are the home of microorganisms, and provide the medium for chemical transformations, i.e., minerals. Nutrient cycling starts with microorganisms in the soil.

Soils buffer and filter contaminates. Some contaminants attach to sediment and are retained when the sediments settle. Another equally important soil function is the control of water transport and storage. The physical properties of a soil determine whether water in a watershed or wetland will become surface runoff, groundwater recharge, or surface water storage. Therefore, soil properties control water movement and the hydrology of a wetland.

We have established that water is necessary for wetlands. The US Department of Agriculture Natural Resources Conservation Service (USDA NRCS) uses wetness criteria to develop hydric soils lists, from soil survey data, for each state. These criteria can be found at the

Fig 9.8 This wetland soil is a Typic Endoaquoll. The name basically means a wet **Mollisol**, or grassland soil. The left side is natural while the right side has been smoothed to better show the indicators. The soil has a dark mineral surface, about 25 centimeters (10 inches) thick, enriched with organic matter, which provides a black indicator of hydric soil. Beneath that, the light and dark areas show concentrations and depletions of minerals. Note the water at about 1 meter (3 feet) on the measuring strip. The prefix "endo" means that the soil is saturated from within. This means that water moves up from the water table through the soil profile to the land surface in the wet season (Photograph courtesy of USDA NRCS)

Fig 9.9 Hydrology in this South Carolina, US, bottomland hardwood wetland can be seen several ways. For example, standing water is present, trees are buttressed, and there are water marks on tree trunks. Can you see any other indicators? (Photograph by Karrie Pennington)

USDA NRCS soils website [13]. The system has become digital. You enter the state (US) of interest and it returns a list of hydric soils. The information you need to interpret the table is provided at a second NRCS website [14]. This site has information about soils in every state, and some worldwide features. It is a tremendous resource for anyone needing soils information. Whether you have the perspective of a gardener, engineer, architect, homeowner, farmer, or someone else, you can find soil data to make informed decisions.

9.2.2 Hydrology

Wetlands do not have to be wet all the time – they can be seasonally wet, then dry much of the year. Also, they can contain permanent or semipermanent water. Wetlands can be ponded, flooded, or have saturated soils at or near the land surface. The source of water can be from groundwater, precipitation, snowmelt, surface runoff, tidal influence, flooding, or combinations of these. The source of water has a great deal of influence over wetland formation and characteristics. Water loss from any wetland can occur from evapotranspiration or infiltration into the soil.

Field signs of hydrology (wetness) are characteristics noticed during a visit to a wetland site (see Figure 9.9). These observations can include the following:

- Visual observation of water accumulation on the surface, or as saturation near the soil surface, can be used.

- Go to the site during the growing season and see if it is wet. However, recent precipitation can influence what you observe, so consider rainfall into your observations.
- The **drift lines** present – formed by deposition of leaves and other debris from moving water – are in a line parallel to water flow. Drift lines generally indicate a minimum area of inundation. Actual water flow is usually further than debris movement.
- Flow channels cut by running water. These are more prominent in riverine wetlands after a flooding event than in other wetland types. These channels run perpendicular to or with the flow. Flow channels can be seen in debris patterns and do not have to cut into the soil. These channels can also develop in uplands, so they need to be evaluated based on landscape position. That is, if you are on a hill slope and see flow channels, they are probably not an indication of wetland hydrology. However, if you are in a floodplain depression and see channels, they can be used to confirm wetland hydrology.
- Sediment staining is present on plants or leaves after water levels decline. These mineral or organic coatings can remain on the vegetation for long periods. They also indicate a minimum level of inundation.
- High water marks on trees are dark lines which show how high water stands – long enough to develop the dark color mark. The highest mark indicates the highest water level.
- Encrusted detritus (eroded matter) such as mud-covered twigs, leaves, and debris.
- Buttressed trees show morphological adaptation to wetness.

These are primary indicators and are verified by soil characteristics (see Figure 9.8). We often look to soil characteristics because hydrology is sometimes difficult to establish visually. Records of river stage height of water surface, duration, and frequency of flooding, topographical maps, soil survey records, and climatic records all help establish a history of the hydrology in an area.

Hydrology is especially important in determining a wetland's ability to process inputs such as nutrients, contaminants, and sediments. Movements of these are all slowed by the *residence time* (discussed in Chapter 3) that water spends in a wetland. This is called the contaminant **retention time**. This delay prevents the contaminant from reaching a waterbody in a big slug (all at once) that could cause problems in water quality – a particularly important wetland function. The chemistry of the source water is also an important part of hydrology because it can make a big difference in wetland water and soil chemistry. This is especially important when the water is saline. Salt influences plant types and wildlife use. Water high in carbonates and bicarbonates can alter the pH of the wetland system, and influence plants and aquatic organisms.

Consider two organic wetlands. These are both peatlands which have a minimum of a 40 cm (16 in) layer of organic matter, the decayed remains of plants and animals that slowly accumulate over time. This usually occurs in cold climates which inhibit rapid microbial activity. Both are peatland but one becomes a fen and the other a bog. Let's examine why this occurs. Peatlands are permanently wet, with perhaps a temporarily dry surface in hot weather. Fens are fed by groundwater or slowly moving surface water. The water in the fen is slow, but still flowing, and is nutrient and mineral rich. Fens support a uniform grass, sedge type, plant population that requires nutrients and an alkaline condition. Conversely, bogs receive mostly direct precipitation, and are lentic (stagnant), nutrient poor, and acidic. However, plant life in a bog is diverse and nonuniform. Plants that can grow under adverse conditions such as peat moss, sundews, pitcher plants, black spruce and tamarack trees, and bog laurel thrive. Source water quality and hydrology (movement) are the major factors that drive these differences (Figure 9.10) [15].

Fig 9.10 This is a hillside seep bog in south Mississippi. Bogs are rarely found this far south, and the site has long been considered one of the most critical natural communities in the state. Showy orchids and carnivorous plants are all present, including sundews, butterworts, bladderworts, and pitcher plants. Rare gopher tortoises live in this type of bog. Notice that the longleaf pines have blackened stems from periodic controlled burns that keep out competing plants. Carnivorous plants are fire resilient and get a head start on other plants (Photograph by Dean Pennington)

9.2.3 Vegetation

Hydrophytic vegetation can germinate and grow in saturated conditions. The USACE 1987 Manual is used by federal agencies to determine or delineate wetlands [16]. This manual provides definitions of terms related to wetlands. The official USACE hydrophytic vegetation definition is:

> the sum total of macrophytic plant life that occurs in areas where the frequency and duration of inundation or soil saturation produce permanently or periodically saturated soils of sufficient duration to exert a controlling influence on the plant species present.

The soils remain wet enough during the growing season so the plants that survive are mainly those that best tolerate wet conditions.

Perhaps the most comprehensive list of wetland plants in the US was provided by the USFWS and is maintained by the USACE [17]. The USDA has a national plants database [18] which allows you to search for wetland indicator status of plants by state, region, scientific or common name, and to narrow the search with some plant characteristics. Once you find the plant, a photo and detailed description is provided. This site also includes plants not found in wetlands. This is immensely helpful to the non-botanist who has a plant list, but no idea what the scientific name or indicator status means. The website also provides details and pictures suitable for identification.

The term **indicator status** is defined as a probability that a plant will occur in wetlands versus non-wetlands [19]. There are five basic categories – *obligate wetland*, *facultative wetland*, *facultative*, *facultative upland*, and *obligate upland* (Table 9.1). These do not indicate the degree of wetness in the wetland, just the likelihood that a plant will occur in that environment. Some obligate wetland plants may be in permanently or semi-permanently flooded wetlands while others are in, and sometimes restricted to, wetlands that have dry periods. For example, the bald cypress tolerates flooding very well, but the young trees need a period of dryness to become established. *Facultative* refers to the plant's ability to adapt to conditions that are not optimum, it can grow in a landscape position that is not its preferred location.

Let's go back to our fen and bog examples. Both are very wet and would contain primarily *obligate* (water-tolerant) plants. The nutrient-rich, neutral, or slightly alkaline fen system produces a uniform stand of grasses, mainly sedges. The bog, on the other hand, is poor in nutrients. In addition, its acidic soils support peat moss and some tough shrubs – such as bog laurel, bog

Table 9.1 Definitions for wetland indicator status

Obligate wetland (OBL)	Contains plants that almost always occur in wetlands (99 percent of the time)
Facultative wetland (FACW)	Contains plants that usually occur in wetlands (67–99 percent of the time)
Facultative (FAC)	Includes plants that are just as likely to occur in wetland or non-wetland areas (34–66 percent chance of occurring in wetlands or non-wetlands)
Facultative upland (FACU)	Contains plants that occasionally occur in wetlands (1–33 percent of the time)
Upland (UPL)	Contains plants that almost always occur in uplands (99 percent of the time)

Fig 9.11 A fen in the state of New York, US. Notice the largely unsuccessful encroachment of trees from the nearby forest around the ponded area in the upper left of the photo. The wetland grasses carpet this fen
(Photograph courtesy of USFS)

rosemary, and cranberries. Black spruce and tamarack trees may also survive in a bog. The carnivorous pitcher and sundew plants survive by including insects in their diets. The result is that fens are less diverse and colorful than bogs. Fens also have more oxygen due to the flowing water in their system. This allows an increased decay of organic matter by microorganisms. Fen peat is finer, more decomposed, than bog peat. Neither system is better than the other since both have their own functions and values. The point is that different plants are adapted to different situations for a variety of reasons. In the case of fens and bogs, the hydrology, water source, movement, and quality controls the vegetation (see Figures 9.10 and 9.11).

9.3 Wetland Types

There are many different names for wetlands. Some have regional connotations and too many are used in contradictory ways in different countries. Standardized terminology will be necessary for wetland science to develop as an international discipline. Mitsch and Gosselink [20] propose seven major wetland types – tidal salt marshes, tidal freshwater marshes, mangrove wetlands, northern peatlands, inland marshes, southern deep-water swamps, and riparian wetlands. These categories do not cover all wetland types, and not everyone studying wetlands has agreed on a standard list. However, this list may provide a starting place for international consideration in developing a common terminology.

We've presented a great deal of physical information regarding wetlands, swamps, bogs, and fens – wonderful, mysterious water features of the world. These have been around for centuries, and their dark nature has not been lost on novelists around the world. It would be a terrible loss to English literature if we did not have the moors for Emily Bronte's darkly cruel (or was it romantic?) hero, Heathcliff, in *Wuthering Heights*, or Conan Doyle's scary dog from *The Hound of the Baskervilles*.

I (Karrie Pennington) remember reading *A Girl of the Limberlost* [21] by Gene Stratton-Porter, a bestselling early-twentieth-century author, and thinking that life in that Indiana swamp must have been both exciting and a bit scary. I did not know at the time that the author was a self-taught naturalist writing many years before I was born. Stratton-Porter loved sloughing through the swamp, taking pictures and documenting everything that she saw. Her camera used bulky and awkward glass photographic plates, not the lightweight digital wonders of today. Lest we picture a tame, manicured forest, remember that her walks involved wading murky waters, cutting through tangles of vines, and climbing trees while carrying a pistol to ward off snakes. In the nineteenth century, a woman wearing boots and men's britches was scandal enough. The fact that she travelled, often alone, in a place described as a steaming, fetid, treacherous swamp and quagmire, filled with danger, was inspiring or crazy – depending on your viewpoint. I just knew that her descriptions of butterfly and moth-collecting in that Indiana swamp – mixed with the life of a science-minded girl – were a magic combination for me. The Limberlost Swamp was drained, around 1913, to produce timber, gas, and oil, and finally, cropland. Stratton-Porter's response was scandalous for the time.

"If men do not take active conservation measures soon," she warned,, "I shall be forced to enter politics to plead for the conservation of the forests, wildflowers, the birds, and over and above everything else, the precious water on which our comfort, fertility, and life itself depend [22]."

The men in charge did nothing, and the swamp was completely lost.

Long after her death, Stratton-Porter's stories inspired the restoration of parts of the Limberlost, a work in progress. Award-winning nature writer Scott Russell Sanders, who lives in Indiana, said in an essay for *Audubon Magazine* "Her photographs and words lured me to Loblolly Marsh – a 428-acre restoration project in Jay County, Indiana – to see a remnant of the vast, magnificent, vanished wetland that Stratton-Porter made known to millions of readers around the world: the Limberlost Swamp." So, what is in a name? "Sometimes a name reflects the soul of a place."

9.4 Wetland Classification

As you read about wetland classifications, notice that no matter which terminology is used, certain things are repeated – landscape position, hydrology as water source, depth, duration, and energy, hydrophilic plant community, mineral or organic soil, and organic matter accumulation (see Table 9.2). These are all traits used to place wetlands into groups. Determining wetland type can be done several ways, using vegetative community, water source, or any other characteristic or set of characteristics. One method, developed by the USACE Waterways Experiment Station scientists, is called **Hydrogeomorphic Classification (HGM)** [23]. It uses hydrology and geomorphology as they relate to wetland function. So, how do we define wetland function?

Table 9.2 Terms commonly used to describe wetlands

Term	Definition
Bog	Peat accumulating wetland, with no significant inflow or outflow of water, that supports acidophilic mosses, particularly sphagnum
Bottomland	Periodically flooded lowlands along rivers and streams usually on alluvial floodplains. Usually forested, called bottomland hardwood forest in the American southeast
Depressional	Wetlands lower than surrounding landscape, with closed contour edges on three sides, multiple water sources possible
Estuarine emergent	Saltwater marsh
Fen	Peat-accumulating wetland that can have some overland flow inputs from the surrounding mineral soils; supports marsh-like vegetation
Flats, mineral	Usually precipitation-driven, hydrologically isolated flat areas of mineral soil
Flats, organic	Usually precipitation-driven, hydrologically isolated flat areas of organic soil
Lacustrine fringe	Wetlands formed at the edge of lakes; may have emergent or forested vegetation
Marsh	Frequently or continually inundated, characterized by hydrophilic emergent vegetation; in European terminology, marshes are mineral and do not accumulate peat
Mire	European for any peat-accumulating wetland
Moor [a]	European for peatland; a high moor is a raised bog and a low moor is a depressional bog [a]
Muskeg	Large expanses of peatlands or bogs; term used mainly in Canada and Alaska
Peatland	Generic term meaning decayed vegetation accumulating wetland
Playa [a]	Southwestern US term for potholes [a]
Pothole [a]	Shallow marsh-like [a] pond in the Dakotas, Nebraska, and Canada
Reed swamp	Eastern European wetland dominated by *Phragmites*, common reed
Riverine	Occurs in floodplains or riparian corridors, in association with streams or rivers, vegetation varies, usually frequently flooded
Slope	Occurs on hill or valley slopes from slight to steep slopes, primary water source groundwater and precipitation
Slough [a]	Swamp [a] or shallow lake north and Midwest US, or slowly flowing shallow swamp or marsh southeastern US
Swamp	US wetland dominated by trees and or shrubs
Tidal fringe	Occurs on continental edges; hydrology is influenced by tides; brackish to saline
Vernal pool [a]	Shallow intermittently flooded wet meadow,[a] generally dry most of the summer and fall
Wet meadow	Grasslands with saturated soil near the surface but without ponding most of the year
Wet prairie [a]	Like a marsh [a] but with water levels between a marsh and a wet meadow [a]

[a] You can see how confusing terminology can get when terms are used to define other terms.

Source: Modified from Mitsch and Gosselink (1993)

9.5 Wetland Functions and Values

Wetlands have a variety of functions because there are a variety of wetland types, found in different locations within many different watersheds. Wetland functions can be divided into biological, habitat, biogeochemical, physical, hydrologic, and recreational values. Functions are things that a wetland does. Values are anthropogenic concepts, things that people like or value. The following is a partial list of examples of biological-habitat function:

- Provide a diversity of habitats and habitat structure for wildlife, breeding, nesting, and foraging;
- Maintain a plant community characteristic of the wetland type;
- Serve as nurseries and feeding areas for young fish; and
- Provide the nutrients that make up the basis of food webs.

Biogeochemical function examples:
- Process organic matter and nutrients through nutrient cycling and biological uptake;
- Help maintain water quality by filtering sediments and contaminates; and
- Serve as natural buffer and transition areas, for streams, lakes, and rivers.

Physical–hydrologic function examples:
- Attenuate floodwaters and slow the rate of flow of waters going into streams and rivers;
- Replenish groundwater by providing a place where water can have time to permeate through the soil into an aquifer; and
- Protect coasts and other areas from excessive erosion.

Recreational value examples:
- Provide open space and aesthetic value;
- Provide special areas for recreation, birdwatching, hunting, fishing, walking, photography, etc.; and
- Serve as educational and research areas to study the widely diverse wildlife, plants, and animals.

How well a wetland performs its expected functions is an indication of wetland health. This brings us back to HGM and functional assessment. HGM can be an effective tool for determining wetland health and the potential to improve wetland function. Saving wetlands can involve protecting healthy wetlands, but often involves restoration of disturbed sites. (Think about the Limberlost Swamp in Indiana discussed earlier.)

Using HGM is a field exercise; by contrast, one could gather data from a desk using books, databases, computer geographic information system (GIS) programs, aerial photographs, any available tool, but actual assessment requires a field visit. Wetland field work usually involves rubber boots, repellent for mosquitoes, redbugs, ticks, etc., sun protection, rain protection, and an assortment of tools – from soil color books and shovels to binoculars, nets, measuring tapes, field guides for plants, amphibians, birds, etc., and hand lenses. A set of maps and aerial photos comes in handy, and drinking water is necessary. The best tools are an eye for observation – from the landscape scale to hand lens scale – the ability to take accurate field notes, knowledge of what you are seeing, and perhaps an appreciation of the work.

Once in the field, the first step is to assess a reference wetland in the same geographic area, and of the same type, as other wetlands you will be evaluating. The reference wetland should be well developed, i.e., old and preferably pristine. Figure 9.12 shows a rare, incredibly old bottomland hardwood (BLH) wetland. The main reason it has survived is its very wet location.

It should be functioning at its maximum potential because other wetlands will be assessed compared to the reference. Sometimes a true pristine reference wetland cannot be found due to human development or natural disaster. Generally, a team of local experts picks the reference wetlands in an area. This helps ensure consistency of assessments and the best possible sites.

Fig 9.12 This bottomland hardwood wetland would make a good reference site. An ancient bald cypress, *Taxodium distichum*, in Sky Lake Wildlife Management Area, a Mississippi Delta wetland. The oldest trees in this area are approaching 2000 years. Think about it. The taller man in the photo is 6 foot 6 inches (201 cm). Notice the hydrology indicator, high water marks (dark lines) above his head at about 2.7–3.0 meters (9–10 feet). This means that water in this wetland periodically gets about 9–10 feet deep
(Photograph courtesy of Yazoo Mississippi Delta Joint Water Management District)

Fig 9.13 Restored prairie wetland in Iowa, US. Notice the attention to birding in this restoration. The landowner was an avid birdwatcher. The prairie grass areas were carefully checked for nesting birds, and many migratory birds depend on these wetlands for food and rest
(Photograph by Karrie Pennington)

The reference site is assessed and scored in terms of soils, vegetation, and hydrology – the three components of wetlands. Wetlands to be evaluated are assessed using the same parameters, and scored, based on how well they mimic the reference. The result is an indicator of wetland health (an evaluation of weak points and strengths) that can be used in restoration of the wetland. The goal is to restore the assessed wetland to similar conditions of the reference wetland, as closely as circumstances will allow. Iowa's USDA NRCS Wetland Reserve Program (WRP) helped restore a prairie wetland (depressional) in Des Moines County (Figure 9.13). A functional assessment helped determine the goals and plans for this restoration.

Think About It

How does selecting local reference wetland sites help to ensure that restored wetlands will function properly?

HGM uses seven wetland types – riverine, slope, depressional, mineral soil flats, organic soil flats, tidal fringe, and lacustrine fringe. Riverine wetlands transition into a river. Slope wetlands develop on hillsides and depressional wetlands are in holes. The flats can be inland in any low area. Tidal and lacustrine fringe wetlands are a transition to an ocean or a lake, respectively. These terms are widely used in the US.

9.6 Trends in Wetlands

Wetlands have often been considered as "problem areas" with little value except to see what can be developed after they are "fixed" by draining and filling. Researchers in America and Canada have excellent wetland inventory databases but show that more than one-half of their wetlands have been lost between the 1780s to the 1980s. There is no worldwide database or inventory of wetlands for comparison, but any developed country has sacrificed a wetland acre or thousands to development. That is the way of humankind. Fortunately, humankind can also decide to change this trend. Wetland functions have become better understood and appreciated. In the US, for example, there is a no net loss of wetlands policy, and wetlands are being restored or enhanced in many areas. These policies are not without difficulties, but wetland loss has declined in the US.

The Department of the Interior, USFWS has the responsibility to inventory and report the status of wetlands in the US. They work in cooperation or with funding from the Environmental Protection Agency (USEPA), Department of the Army – USACE, Department of Agriculture – Natural Resources Conservation Service (NRCS), Department of Commerce, National Oceanic and Atmospheric Administration (NOAA), and the National Marine Fisheries Services (NMFS) to produce detailed reports.

Their 1986–97 report [24] summary included the following:

- There were an estimated 42.7 million hectares (105.5 million acres) of wetlands in the lower 48 States. Of this total, 40.7 million hectares (100.5 million acres), or 95 percent, are freshwater wetlands and 2 million hectares (5 million acres), or 5 percent, are saltwater wetlands. This has not significantly changed in the latest 2004–9 report [25].
- The annual estimated wetland loss was 23,700 hectares (58,500 acres). This was an 80 percent reduction in wetlands lost per year from the previous reporting period (1970s–80s). There was an increase in wetland loss (13,800 acres) in 2004–9 (Figure 9.14).

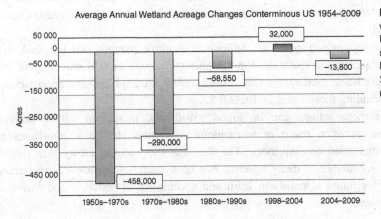

Average Annual Wetland Acreage Changes Conterminous US 1954–2009

Fig 9.14 Average annual net wetland changes for conterminous United States, 1954–2009. Data sources Frayer et al. 1983 [26], Dahl and Johnson 1991 [27], Dahl 2011 [28]

(Graph modified from Dahl 2011)

Table 9.3 Factors leading to wetland loss

Anthropogenic factors	Natural factors
Drainage	Runoff [a]
Dredging and stream channelization	Erosion [a]
Deposition of fill material	Subsidence [a]
Dikes and dams	Sea level rise [a]
Tilling for crop production	Droughts
Levees	Hurricanes and other storms
Logging	
Mining	
Construction	
Air and water pollutants	
Excess nutrients	
Toxic chemical release	
Non-native species introduction	
Grazing by domestic animals	

[a] Can be exacerbated by human activities.

Wetland losses in the US are usually attributed to urban development, increases in agriculture, changes in silviculture (timber management) practices, and rural development – all basically land use changes. The 2004–9 increase in wetland loss reversed the long-standing trend in loss reduction. We have discussed the fact that environmental issues are not simple, in several places in this text. This increased wetland loss is no different. Losses in this period were attributed to economic conditions (such as crop prices or property values), land use trends, changes to wetland regulation and enforcement measures, and possible climatic changes [29].

It is particularly important to know the types of wetlands that are lost, because each type has unique functions that it performs. If a forested wetland is lost and a ponded wetland takes its place, the habitats for plants and animals will be different. Therefore, the plants and animals will be different. When considering the functions and values involved in restoring wetlands, it makes sense to replace a forested acre with a forested acre, an emergent acre with an emergent acre, and so on. If ponded wetlands are increasing while forests and freshwater emergent wetlands continue to decrease, then the balance of wetland functions needed for ecosystem health is not being maintained.

Worldwide, wetland losses are difficult to quantify, partially due to a lack of consistent terminology and assessment methods. New technologies, such as GIS and satellite imaging, are good tools and are being used extensively. Unfortunately, wetland loss and degradation are due to many factors, including those listed in Table 9.3.

Any of these actions can be found worldwide. Increasing human populations and decreasing resources often result in tremendous ecosystem change – sometimes intentional, sometimes as unintended consequences. The fact that the Ramsar Treaty (see Introduction) is the only global environmental treaty covering one ecosystem – wetlands – is encouraging. There is always more for future scientists to learn and accomplish in restoring and maintaining these valuable ecosystems.

How does a society preserve ecosystem integrity that can potentially benefit the entire population while satisfying their immediate needs for food, fiber, and fuel? How would you balance long-term good with immediate need?

Summary Points

- **Wetlands** come in many varieties, from bogs to prairie playas, and occur on every continent except Antarctica. They are recognized globally as essential parts of the landscape with functions that must be preserved.

- A wetland has three defining features – **hydric soil, hydrophytic vegetation**, and **hydrology**.

- **Hydric soils** have redoximorphic features caused by the wetting and drying cycles in the wetland

- **Iron, organic carbon**, and **manganese** are used as an indicator of the wetting and drying cycles that occur in wetland soils because they change colors or in the case of organic matter, accumulates in wet conditions.

- **Wetland plants (hydrophytic)** are water-loving but vary in their water requirements. Many require periods of dryness to thrive and reproduce.

- A plant's **indicator status** is the probability that it will occur in a wetland.

- **Wetlands do not have to be wet all the time** – they can be seasonally wet, then dry much of the year, but they can also contain permanent or semi-permanent water.

- There is no one classification system used worldwide although most use a combination of the following features. **Landscape position, hydrology** as water source depth, duration, and energy, **hydrophilic plant community**, mineral or organic **soil**, and **organic matter accumulation** are all traits used to place wetlands into groups.

- Hydrogeomorphic classification places wetlands into one of seven groups based on landscape location; these are riverine, slope, depressional, mineral soil flats, organic soil flats, tidal fringe, and lacustrine fringe.

- How well a wetland performs its expected **functions** is an indication of wetland health. Wetland functions include the following:

- They provide nesting, breeding, and foraging habitats and structures for a variety of species from salamander to heron to large mammals. They are nurseries and feeding areas for young fish. They also provide nutrients that make up the basis of food webs.

- Biogeochemical processing of organic and mineral matter is fundamental to nutrient cycling. These processes also help maintain water quality.

- Hydrologic functions involve movement and storage of water.
- Wetlands also have recreational value.

Questions for Analysis

1. What is the difference between a wetland function and a wetland value?
2. Which soil characteristics are used to define wetlands?
3. If a wetland is not always wet, how do we know it is a wetland?
4. Wetland restoration is most successful if the wetland is restored to its original type. Do you agree or disagree? Explain your thoughts.

Further Reading

Hurt, G. W. and L. M. Vasilas, eds., 2006, *Field Indicators of Hydric Soils in the United States*, US Department of Agriculture Natural Resources Conservation Service, http://soils.usda.gov/soil_use/hydric/field_ind.pdf

Mitsch, William J. and James G. Gosselink, 1993, *Wetlands*, 2nd ed, New York: Van Nostrand Reinhold.

Pielou, E. C., 1998, *Fresh Water*, Chicago, Ill.: University of Chicago Press.

Richardson, J.L. and M.J. Vespraskas, 2001, *Wetland Soils Genesis, Hydrology, Landscapes, and Classification*, Boca Raton, Fla.: Lewis Publishers.

Stratton-Porter, Gene, 1909, *A Girl of the Limberlost*, New York: Grosset and Dunlap.

References

1 Gene Stratton-Porter, 1909, *A Girl of the Limberlost*, New York: Grosset and Dunlap.

2 E. C. Pielou, 1998, *Fresh Water*, Chicago, Ill.: University of Chicago Press.

3 Ramsar Convention on Wetlands, 2018, *Global Wetland Outlook: State of the World's Wetlands and their Services to People*. Gland, Switzerland: Ramsar Convention Secretariat, https://www.global-wetland-outlook.ramsar.org/outlook (downloaded 2020).

4 The *Strategic Framework* is available in PDF form as Handbook No. 14 at: https://www.ramsar.org/sites/default/files/documents/pdf/lib/hbk4-14.pdf (accessed 2020).

5 Ramsar Convention on Wetlands, "The Ramsar strategic plan, 2016–24," https://www.ramsar.org/sites/default/files/ramsar_convention_strategic_plan_poster_english.pdf

6 L. M. Cowardian, V. Carter, F. C. Golet, and E. T. LaRoe, 1979, *Classification of Wetlands and Deepwater Habitats of the United States*, Washington, DC: US Fish and Wildlife Service.

7 Lewis M. Cowardin and Francis C. Golet, 1995, *US Fish and Wildlife Service 1979 wetland classification: A review**, Central Plains Center for Bioassessment, Kansas Biological Survey, Vegetatio 118: 139–152, Belgium: Kluwer Academic Publishers. (*The US Government's right to retain a non-exclusive, royalty free license in and to any copyright is acknowledged.), http://cpcb.ku.edu/media/cpcb/progwg/html/assets/wetlandwg/1979Cowardin_review.pdf (PDF downloaded in 2020).

8 US Environmental Protection Agency (USEPA) Sec 404 Clean Water Act, https://www.epa.gov/cwa-404/how-wetlands-are-defined-and-identified-under-cwa-section-404

9 William J. Mitsch and James G. Gosselink, 1993, *Wetlands*, 2nd ed, New York: Van Nostrand Reinhold, p. 29.

10 Food Security Act of 1985, US Code Citation: 16 U.S.C., pp. 3801–3862.

11 L. M. Vasilas, G. W. Hurt, and J. F. Berkowitz, eds., 2006, *Field Indicators of Hydric Soils in the United States*, US Department of Agriculture (USDA) Natural Resources Conservation Service, https://www.nrcs .usda.gov/Internet/FSE_DOCUMENTS/nrcs142p2_053171.pdf (PDF downloaded 2020).

12 J. L. Richardson and M. J., Vespraskas, 2001, *Wetland Soils: Genesis, Hydrology, Landscapes, and Classification*, Boca Raton, Fla.: Lewis Publishers.

13 USDA Natural Resources Conservation Service, https://www.nrcs.usda.gov/Internet/FSE_DOCUMENTS/ nrcseprd1389479.html (accessed 2020).

14 NRCS, State Soil Data Access (SDA) "Hydric soils rating by map unit," https://www.nrcs.usda.gov/ Internet/FSE_DOCUMENTS/nrcseprd1389479.html#reportref (accessed 2020).

15 E. C. Pielou, 1998, *Fresh Water*, Chicago, Ill.: University of Chicago Press, pp. 216–218.

16 US Army Corps of Engineers Waterways Experiment Station, 1987, *US Army Corps of Engineers Wetlands Delineation Manual*, Wetlands Research Program Technical Report Y-87–1, Vicksburg, Miss.: USACE Waterways Experiment Station, https://usace.contentdm.oclc.org/digital/collection/p266001coll1/id/4532/ (This is a searchable digital edition from the USACE Digital Library, accessed 2020.)

17 Porter B. Reed, 1988, *National List of Plant Species That Occur in Wetlands: 1988 National Summary*, St. Petersburg, Fla.: US Fish and Wildlife Service in cooperation with the US Army Corps of Engineers, US Environmental Protection Agency, and US Soil Conservation Service (PDF available at https:// digitalcommons.usu.edu/govdocs/508/ or at https://efotg.sc.egov.usda.gov/references/Public/NE/1988_ National_List_of_Plant_Species.pdf).

18 USDA, 2007, "National plants database,", http://plants.usda.gov/ (accessed 2020).

19 Porter B. Reed, 1988, *National List of Plant Species That Occur in Wetlands: 1988 National Summary*, St. Petersburg, Fla.: US Fish and Wildlife Service in cooperation with the US Army Corps of Engineers, US Environmental Protection Agency, and US Soil Conservation Service, (PDF available at https:// digitalcommons.usu.edu/govdocs/508/ or at https://efotg.sc.egov.usda.gov/references/Public/NE/1988_ National_List_of_Plant_Species.pdf).

20 William J. Mitsch and James G. Gosselink, 1993, *Wetlands*, 2nd ed, New York: Van Nostrand Reinhold, p. 31.

21 Gene Stratton-Porter, 1909, *A Girl of the Limberlost*, New York: Grosset and Dunlap.

22 Ibid.

23 Mark M. Brinson, 1993, *A Hydrogeomorphic Classification for Wetlands*, Wetlands Research Program Technical Report WRP-DE-4, Vicksburg, Miss.: USACE Waterways Experiment Station.

24 T. E. Dahl, 2000, *Status and Trends of Wetlands in the Coterminous United States 1986 to 1997*, Washington, DC: US Department of the Interior, Fish and Wildlife Service.

25 T. E. Dahl, 2011, *Status and Trends of Wetlands in the Coterminous United States 2004 to 2009*, Washington, DC: US Department of the Interior; Fish and Wildlife Service (digital copies of status and trends reports are available for download at https://fws.gov/wetlands/status-and-trends/index.html, accessed 2020).

26 W. E. Frayer, T. J. Monahan, D. C. Bowden, and F. A. Graybill, 1983, *Status and Trends of Wetlands and Deepwater Habitats in the Conterminous United States, 1950's to 1970's*, Fort Collins, CO: Colorado State University.

27 T. E. Dahl and Johnson, 1991, *Status and Trends of Wetlands in the Coterminous United States, mid-1970s to mid-1980s*, Washington, DC: US Department of the Interior, US Fish and Wildlife Service.

28 T. E. Dahl, 2011, *Status and Trends of Wetlands in the Coterminous United States 2004 to 2009*, Washington, DC: US Department of the Interior; Fish and Wildlife Service

29 Ibid.

10 Dams and Reservoirs

To use the engineer's own phrase, there is no "angle of repose" (a natural and balanced resting point of material) in society's attempts to dominate nature.

Gregg R. Hennessey, editor [1]

Chapter Outline

10.1 Introduction

Other than the Great Wall of China, dams are the largest structures ever constructed by humans. Throughout history, dams have served many purposes for humanity's building of civilizations; this includes flood control, hydropower generation, irrigation, municipal water supply, recreation, and stimulation of economic growth. Without dams, innumerable cities would collapse from a lack of drinking water or adequate supplies for industry. Without dams, modern life as many of us know it would not be the same.

Think About It

Life for millions of people changed when their homesites, native lands, farm fields, and entire cities were flooded to construct dams and reservoirs. An estimated 80 million people have been displaced by dams, and too often live in poverty after being marginalized by projects that promise a better life – but for whom [2]? Continued dam building frequently increases the number of marginalized peoples. The ecology of rivers changes from free-flowing **lotic** systems to **lentic** pools downstream from reservoirs. There is always a cost for human resource changes, and with dams those costs are both human and environmental.

Dams represent human power and ingenuity in changing the Earth to suit our needs. Dam construction and operations can also represent a lack of understanding and empathy for the environmental and human needs of the people, rivers, wildlife, wetlands, and floodplains most directly influenced by such projects. In addition, it can represent a willingness to ignore the consequences of our actions in favor of economic and personal gain. The challenge for you, who are student's today and the planners of tomorrow, is to recognize that development and environmental health do not have to be mutually exclusive. There are sustainable solutions that require cooperation and a willingness to change our mindsets. You can do this.

In 1950, there were 5,700 large dams in the world (defined as being at least 15 meters (50 feet) in height or storing more than 2.8 million cubic meters (2,300 acre-feet, or 750 million gallons). Today, there are 60,000 [3]. Eighty percent of these large dams are located in just five countries – China, Spain, India, Japan, and the United States [4]. The US has approximately 6,000 large dams and 73,000 smaller dams but has slowed its construction pace significantly in the past two decades. China, on the other hand, had fewer than 100 large dams in 1949, but today has over 22,000 large and 80,000 smaller dams [5]. Among these is the world's largest dam, China's Three Gorges dam (see Figure 10.1). Brazil, India, Turkey, and Africa have all built dams with varying degrees of success in terms of energy production, human welfare, and environmental concerns.

Dams can be vital to civilization. However, onstream dams – whether large or small – are an obstacle across the natural course of a river and disrupt its normal functions. Onstream dams not only capture the flow of a river, but also its contents. Nutrient-rich silt and sediments are restrained behind such a dam and accumulate at the bottom of the reservoir. Sediment build up interferes with the water storage capacity of a reservoir. Normally, these captured sediments and nutrients would have flowed downstream and replenished fertile soils during flood events. As populations grow or move, flooding becomes more a problem than a blessing. It is a problem many dams are designed to control. The Nile River Aswan Dam (see Figure 10.2) is a classic example of mixed benefits and controversy. The dam provides flood protection, hydroelectric power, water for irrigation and stability during drought years; however, it also stops replenishing floods that built the Nile River floodplain over millennia (see Figure 10.3).

Fig 10.1 China's Three Gorges Dam is the world's largest hydroelectric dam. It is made of concrete and is about 2,309 meters (7,575 ft) long, and 101 meters (331 ft) high. Construction of this dam has been as controversial as other dam projects anywhere in the world
(Photograph by Rehman at Wikipedia File: Three_Gorges_Dam, _Yangtze_River,_China.jpg [6], converted to black and white)

Fig 10.2 Egypt's Aswan Dam was completed in 1970 to provide hydroelectric power, flood control, and water for irrigation (Getty credit: Martin Child)

Fig 10.3 This image from the International Space Station in August of 2014 helps to clearly see the impact of damming a major river. It created Lake Nasser, Egypt's largest lake. In this image, the Nile River flows to the right toward its Mediterranean delta (Photograph courtesy of NASA, www.nasa.gov/content/egypts-nile-river-and-lake-nasser/, converted to black and white, text added)

10.2 Types of Dams

10.2.1 Beaver Dams

Although the first human-constructed dam was probably built over 8,000 years ago, beavers have been constructing smaller dams for much longer. Beaver dams are nature's way of creating ponds,

Fig 10.4 This beaver dam is in Tierra del Fuego near Ushuaia, Argentina. Beavers find homes in many places around the world (Photograph by Ilya Haykinson, http://commons.wikimedia.org/wiki/File:Beaver_dam_in_Tierra_del_Fuego.jpg, converted to black and white)

wetlands, and complex ecological systems. Beaver use sticks and mud to ingeniously reduce the flow of a stream to create ponds.

Beaver do more to change their landscape, through dam construction, than any other mammal on Earth besides humans, and have been doing it for 10 million years. The range of the North American beaver (*Castor canadensis*) was wide, extending from the Arctic tundra of Canada to the deserts of northern Mexico and today even further into South America (see Figure 10.4). Its geographic limitation was the swamps of Florida and Louisiana, where alligators were a natural predator. It's been estimated that some 200 million beaver once lived in the continental US. Today, those figures are around 7–12 million, with the vast majority remaining in the northern Great Lakes region [7]. Oddly, the European beaver (*Castor fiber*) has no interest in dam construction in most regions and confines its construction activities to digging burrows in streambanks.

Beaver are a **keystone species**, meaning that they have a major influence on the structure of an ecosystem. Beavers provide homes for dozens of species in the ponds and wetland complexes that spread out behind their dams. Migrating ducks, fish, frogs, great blue herons, and moose are but a few examples of wildlife that benefit from beaver activity. Slowly and eventually, however, many of these wetland complexes fill in with silt, forcing later generations of beaver to move to a new location.

Beaver meadows, and associated wetlands, are examples of an **ecotone** – a transitional zone between ecological regions, in this case a wetland and a stream. Beaver meadows serve a valuable function between streams and dry land. These meadows contain dense food webs crowded with millions of organisms, zooplankton, and bacteria which help clean water found in wetlands. Beaver meadows also serve as a sponge for precipitation, capturing water during storm events and releasing it later during drier times. They help reduce downstream flooding and erosion and can increase groundwater recharge. Meadows will eventually replace a thinned forest, and willows may sprout at the wetland's borders.

Just as beaver dams have changed ecosystems, the loss of beavers has resulted in the loss of beaver-created wetlands. Flowing water and sediments are no longer captured behind beaver ponds. Fewer ponds mean less habitat for migrating ducks and other wildlife. The loss of beaver dams can even change the local water cycle by the reduced amount of ponded surface area.

10.2.2 Human-Constructed Dams

The earliest human-constructed dams were probably intended to supply irrigation water for small plots of grains or other agricultural crops. Larger dams and associated irrigation projects allowed civilization to develop in ancient Mesopotamia, along the Nile River in Egypt, the Indus River in Pakistan, and the Huang He (Yellow) River in China. Reliable water supplies also created agricultural surpluses; quite different from the daily chore of locating food in the hunter–gatherer communities described in Chapter 2.

Cooperation was necessary between irrigators – to remove silt from canals to shepherd water supplies to fields, to monitor water availability, and to tend crops for months at a time. These were the first steps toward an urbanized way of life. Irrigation allowed humans to develop social order, laws, and human interdependency beyond that of small clans or clustered communities. To maintain these projects, ancient rulers encouraged (and at times forced) people to create **hydraulic civilizations**. These were large-scale ancient agricultural societies dependent upon government dam and canal systems. Laborers were gathered to construct and maintain these waterworks. Later, in the Roman World and Renaissance Europe, such water systems became the basis of agricultural production, economic activity, and the power of government.

In Depth   **The Human Cost of Water Systems**

Karl August Wittfogel (1896–1988) created the somewhat controversial characterization of "hydraulic civilizations" in 1957 [8]. He was born in Germany but fled during Hitler's fascist regime during World War II. Wittfogel was an ardently outspoken antifascist and was held by the Nazi Gestapo for eight months until protests from the scientific community got him released. He came to America and later taught history and other courses at Columbia University and the University of Washington. In 1957, he published *Oriental Despotism: A Comparative Study of Total Power* [9]. His book examined the origins of complex societies and showed how water management was used by Chinese emperors to gain power over their people. The forceful use of labor, to maintain extensive water supply systems, gave power to rulers. "Those who control the (hydraulic) network are uniquely prepared to wield supreme power," according to Wittfogel.

Earthen Dams

The first dams constructed by humans were probably earthen dams on the eastern edge of Mesopotamia in the Middle East. Irrigation canals 8,000 years old have been found in this region, and it's likely that farmers used small dams of earth, sticks, and reeds to divert water from streams into irrigation canals. One of the earliest earthen dams of size was Nimrod's dam in Mesopotamia around 2000 BCE; this was constructed across the Tigris River at a location north of Baghdad, Iraq. It was probably made of earth and wood, and was constructed to reduce erosion and control floodwaters for later use as irrigation water. The dam failed around 1200 CE and allowed the Tigris River to return to its former channel.

Earthen dams are the most common because they are the easiest and cheapest to build. Approximately 80 percent of all large dams in the world today are constructed of earth and rock embankments [10]. Earthen dams rely on the massive weight of the construction material to resist the force of water held behind a dam. Since water places tremendous pressure at the base of a

Fig 10.5 Earthen dam failure due to a lack of maintenance (removal of trees) that allowed excessive erosion (Photograph courtesy of Association of State Dam Safety Officials www.damsafety.org/)

dam, construction is generally much wider at the base than at the top. A relatively waterproof core is necessary to prevent water from seeping through the material. Earthen dams are not as strong as concrete structures and are, therefore, generally much wider than a concrete dam. Modern-day engineers use compacted soils to greatly reduce water seepage through an earthen dam.

The Saint Ferreol Dam was constructed in France, between 1666 and 1675 CE, after approval of King Louis XIV. This large earthen dam was built across the River Laudot with a masonry core and had a height of 36 meters (118 feet). It could store 6.7 million cubic meters (5,400 acre-feet, or 55.9 million gallons) of water, and was the highest earthen dam in the world for 165 years.

Earthen dams had a poor success record in the US prior to 1930, and many failed due to overtopping during a flood. Such failure is caused when the water level in a reservoir becomes so great that water spills over the top of a dam. This erodes the downstream face of the dam, and ultimately can lead to failure of the structure. One of the worst examples of an earthen dam failure is the South Fork Dam upstream of Johnstown, Pennsylvania. During a torrential rainstorm in 1889, the dam overtopped and over 2,000 people lost their lives from surging floodwaters. See the US National Parks Service website (www.nps.gov/jofl/) for additional information regarding this nineteenth-century catastrophe (Figure 10.5).

Think About It: Choices

Earthen dams, like all dams, require constant monitoring and maintenance to prevent failure. As some dams age, costs may not offset the economic benefits or environmental problems created by the structure. Difficult choices are sometimes required to justify removal or continued maintenance of a dam. Can you think of a situation when dam removal would be the best choice for a dam owner and the environment? Describe what you imagine and ask yourself, is it feasible, is it safe, is it reasonable?

Gravity Dams

The second type of human-constructed dams is gravity dams – thick, triangular-shaped walls of concrete or masonry (stone blocks). Generally, gravity dams are built across relatively narrow river valleys that have firm bedrock foundations. The massive weight of the material within the structure holds back the water pressure behind a dam. The oldest known large gravity dam was constructed in ancient Egypt between 2950 and 2750 BCE, nearly 5,000 years ago. The Sadd el-Kafara (an Arabic name meaning "Dam of the Pagans") was one of the oldest civil engineering structures in the world and gives the history of dam construction an ancient beginning [11].

Remains of the Sadd el-Kafara Dam were discovered in 1885 by Georg August Schweinfurth, a German archaeologist. This ancient gravity dam was constructed across a seasonal (intermittent) stream named the Garawi ravine (Wadi el-Garawi) in Egypt, near Helwan, about 30 kilometers (20 miles) south of Cairo. It's the first known gravity dam constructed across a river anywhere in the world. The dam had a crest length of 106 meters (348 feet) and a base length of 81 meters (265 feet). It was approximately 13 meters (43 feet) wide across the top and immensely thick, 24 meters (79 feet) in width at the base of the dam, with a maximum height of 11 meters (37 feet). Storage capacity was estimated at 600,000 cubic meters (486 acre-feet, or 160 million gallons). The dam interior was filled with massive quantities of gravel and stone [12]. Although carefully placed limestone blocks covered the face of the dam, it was not watertight. No mortar was used to seal gaps between stones, and this likely allowed water to flow between the limestone blocks, a condition called **piping**.

The Romans invented concrete in the second century BCE and perfected the combination of lime and gypsum materials used to construct more permanent gravity dams. Emperor Nero (37–68 CE) had a dam constructed across the River Aniene at Ponte di San Mauro for the purpose of a pleasure lake near his villa at Subiaco near Rome. It was one of the earliest Roman dams and was the tallest constructed. The structure remained intact until 1305 CE, when two monks removed some of the stones from the dam to lower the water level in the lake. Evidently, water overtopped the dam and led to its failure.

Modern gravity dams serve many functions. For example, the US Department of Agriculture's Soil Conservation Service (the agency was renamed the Natural Resources Conservation Service [NRCS] in 1994) built 10 gravity dams between 1962 and 1987 in the Sudbury–Assabet–Concord rivers watershed (Figure 10.6). These dams still provide flood control in eastern Massachusetts; however, in 2014 the NRCS Watershed Rehabilitation Program provided a US$8.8 million grant to upgrade these dams. Development in the watershed resulted in the need to increase dam capacity for public safety. Plans also included maintenance of wetlands and wildlife habitat and recreational nature and hiking trails [13].

A note from history: Many Roman gravity dams were constructed on the Iberian Peninsula of modern-day Portugal and Spain, in North Africa, and in the Middle East. The largest was a dam located near Homs, Syria, in 284 CE. The dam height was 200 meters (650 feet) and impounded approximately 90 million cubic meters (73,000 acre-feet, or 755 million gallons) of water.

Arch Dams

Arch dams are also made of concrete but are limited to narrow river canyons with solid rock walls. An arch dam depends on its shape for strength and is relatively thin in width – to reduce the amount of material and cost of construction. The natural, curved shape of the arch helps withstand the tremendous forces of water pressure in the reservoir, and with only a fraction of the concrete

Fig 10.6 Natural Resources
Conservation Service (NRCS)
flood control dam near Tyler in the
Sudbury–Assabet–Concord rivers
watershed
(Photograph courtesy of USDA
NRCS, converted to black and
white)

required by a gravity dam. One of the first known arch dams was constructed by the Romans at
Vallon de Baume, France, for water supply to a nearby city. The dam was 12 meters (40 feet) high
and 18 meters (60 feet) long.

The next arch dam constructed was around 1280 CE across the Kebar River in modern-day
Iran. It was discovered in 1956 by Henri Goblot, a French engineer. The structure is still intact and is
regarded as the oldest known surviving arch dam. It stands 26 meters (85 feet) high, 55 meters
(180 feet) long at its crest, and 5 meters (16 feet) thick at its crest. Arch dams are generally
constructed into rock formations on either side of a valley to support the dam's arched abutments.
This construction method "keys" (attaches) the structure into rock and creates a watertight joint with
the valley walls and at the base of the dam. Only about 4 percent of all large dams in the world are
arch dams [14]. Figure 10.7 is Gordon Dam in Tasmania and is a hydroelectric facility. Note the
dramatic arch design of the dam.

Buttress Dams

Don Pedro Bernardo Villarreal de Berriz of Spain wrote the first book on dam design in 1736. It was
called Máquinas Hidráulicas de Molinos y Herrerias, y Govierno de los Árboles y Montes de
Vizcaya (translation: *Hydraulic machines for mills and blacksmiths and government [sic.] Of the
trees and mountains of Vizcaya*) and was published in Madrid. He was a Basque nobleman, and the
owner of mills and forges (both of which required water to operate) in the province of Vizcaya.
During his time, only two types of dams were constructed – arch dams in narrow canyons with solid
rock foundations and walls, and gravity dams where the site was wide and shallow.

Don Pedro's book describes how a multiple-arched dam (one that curves both horizontally
and vertically) could be constructed. In addition, he described how artificial supports or buttresses
would be required to support such a structure. This led to the invention of buttress dams, a gravity
dam that included buttresses (similar to those used in cathedrals in Spain and Europe – a design
feature used to support higher ceilings and larger openings in church walls for rose windows). Dam
buttresses provided the additional weight and support to ensure the stability of the structure.
Almendralejo Dam located near Badajoz, Spain, is one of the earliest examples of a large buttress

Fig 10.7 A double curvature arch dam on the Gordon River in Tasmania, Australia. It is 140 meters (460 ft) high, making it the tallest dam in Tasmania and the fifth tallest in Australia
(Photograph by "Noodle Snacks" at http://en.wikipedia.org/wiki/File: Gordon_Dam.jpg, converted to black and white)

dam. Also known as the dam of Albuera de Feria, it was constructed in 1747 and still survives. The original structure was approximately 20 meters (65 feet) high and 170 meters (560 feet) long. Buttresses provided support to the downstream face of the dam [15].

10.3 Purposes of Dams

Climate and changing weather patterns too often deliver unpredictable amounts of precipitation. Global climate change has caused prolonged droughts and extreme storms, too little or too much water can both create problems. Dams can provide water storage opportunities to capture surface water runoff during times of plenty for use during water shortage periods. Dams can also reduce and prevent damages caused by floods, generate electricity, and can improve navigation on rivers. However, dams cannot make it rain. Figure 10.8 is the Grand Coulee Dam in Washington state. It

Fig 10.8 Grand Coulee Dam, taken August 1986 (Photograph courtesy of US Bureau of Reclamation, http://en .wikipedia.org/wiki/File: Gordon_Dam.jpg, converted to black and white)

Fig 10.9 The Iron Gate Dam on the Klamath River outside Hornbrook, California. This dam is at the end of a series of dams on the Klamath. Its primary purpose is to regulate water flows and provide hydroelectric power. It is one of four that the Yurok Tribe is trying to have removed to help restore salmon habitat and migration patterns.

Dams on the Klamath River have generated tremendous controversy between irrigators, lakeside homeowners, and those wanting to maintain fish passage and to restore a more natural river flow. A decision to remove four dams on the Klamath River, including Iron Gate, by 2020 was delayed and a new deadline has not been finalized

was built to supply hydroelectric power and irrigation water. Interestingly, the original dam project was expanded to a higher specification, under President Franklin Delano Roosevelt, to supply more jobs, water, and power during the height of the Great Depression. Adding a human component to dam specifications was an unusual action. This dam was a major factor in the industrial development of the Pacific Northwest. Another component of the human dimension in dam construction is the issue of competing interests that may not have been fully considered when the dam was built, for example see Figure 10.9.

Dams create reservoirs to store water for drinking supplies, sanitation needs, industrial uses, and other requirements of towns and cities. Controlling water, particularly in arid climates, is a goal of most municipal water providers.

10.3.1 Flood Control, Four Examples

Around the world, rivers have created human hardship and destruction for centuries while providing many benefits such as drinking water, enriched soils from floodwaters, transportation, irrigation water, wildlife habitat, aquatic biodiversity, and fish for food and commerce. However, in exchange for these benefits, floodplain communities can fear inundation and catastrophe from floods. Governments have constructed dams and levees around the world intended to reduce the risk of flooding.

The Yellow River

China has a long and tragic history of floods. The Huang He, or "Yellow River," has created the most devastation, loss of life, and greatest floods. China's second longest river flows for over 5,464 kilometers (3,395 miles) through the Bayan Har mountain range in the northern mountain province of Qinghai, through nine provinces, ending in the Yellow Sea. The Huang He is called "China's Sorrow" because it has killed more people than any other river in the world. Flooding in 1887 killed nearly 2 million people; in 1931, nearly 4 million lives were lost from flood and ensuing famine and disease; and in 1938 another 1 million perished from its floodwaters [16].

The major problem with trying to control the Yellow River is created by the land it flows through, the Loess Plateau. Soils deposited by the wind are called **loess.** Loess deposits are composed of silt sized particles (20–50 micrometers) and can be highly erodible when disturbed. These soils also provide prime farmland due to its ease of cultivation, a high-water holding capacity, good root penetration, and good aeration. Unfortunately, farming is one of the most disruptive erosive processes used by humans. It is these silts that cause flooding problems along the Huang He – nearly 60 percent of its volume by weight is composed of silty sediment. These murky waters deposit millions of tons of mud which choke the flow of the river channel, causing it to overflow, change course, and build the river higher than the adjacent land surface. Flood control was attempted as early as 300 BCE by Yu the Great, Emperor of China. Larger and taller levees were constructed to restrain the flow of the Huang He, and dams were built to assist in flood control. However, the large volumes of silt clogged many of the new reservoirs. For example, the Sanmenxia filled with silt in only two years, causing it to require extensive reconstruction and management changes [17].

The Xiaolangdi Multipurpose Dam Project, completed in 2000, was designed to hold sediment for the first 20 years before it would reach its capacity. Silty muds are released through controlled floods using 15 specifically designed discharge tunnels [18]. Clearing the sediment has been done yearly since 2002. It is so fascinating to see the huge plumes of murky water shoot out of the tunnels that it is attracting tourists by the busload.

The Nile River

The relationship between Egyptians and the Nile River has been much different from that of Chinese residents along the Huang He. For thousands of years, Egyptians have called the annual Nile River flooding the "Gift of the Nile." Each summer, like clockwork, the Nile carries floodwaters from the Blue Mountains of Ethiopia to the lower Nile River Valley of eastern Egypt. When the waters recede, a thin layer of black, fertile mud is deposited. Crops are immediately planted in the rich soils that require no fertilizer, and was the only farmable land in Egypt. This narrow river valley represents only 3 percent of the total land in the country but has fed the region for centuries.

Construction of the Aswan High Dam blocked the Nile River, about 970 kilometers (600 miles) south of Cairo. Lake Nasser – the reservoir created behind the 111-meter (364-feet)-tall dam – was created in 1971. Many ancient Egyptian sites were removed before the area was flooded by the world's third largest reservoir, but some were permanently inundated and destroyed. Thousands of homes were also submerged, and 90,000 peasants were forced to relocate.

Because of reduced flood flows, saltwater from the Mediterranean Sea is now forcing its way up the Nile River. In addition, the lake's massive water surface area has reduced the average air temperature in the area. Since more land was placed under irrigation, the Nile now rarely floods. However, the river water is now relatively clear and free of heavy sediment loads. Since the fertile sediments are now captured behind the dam in Lake Nasser, fertilizer must be applied to crops downstream of the dam. These are consequences of dam construction that may have been anticipated but were not considered project stoppers.

The Mississippi River

The "Mighty Mississippi" begins as a trickle in northern Minnesota but finishes as a great river nearly 3,900 kilometers (2,400 miles) south in Louisiana and the Gulf of Mexico. Along the way, thousands of tributaries – from 31 states and two Canadian provinces – collect surface water runoff. The land area of the Mississippi River Valley is 20 percent greater than the Huang He of China, twice as large as Africa's Nile and the Ganges River Valley of India, and 15 times greater than Europe's Rhine River. Over 40 percent of the continental US lies within the Mississippi River Valley, one of the most productive regions in the world. However, the river causes devastating floods.

A terrible flood occurred in 1927, and soon after the US Army Corps of Engineers (USACE) was charged with protecting the people and property of the watershed. More than 1,000 people lost their lives, and over 900,000 were left homeless by the relentless floodwaters that year. The USACE took charge and created the longest levee system in the world to control future floods. Today, the Mississippi has 29 locks and dams, hundreds of runoff canals, and hundreds of miles of levees on both sides of the river.

Residents living near the levees felt relatively safe until the devastating flood of 1993 which flooded the upper Midwestern states. Fifty lives were lost and over US$15 billion dollars in damages occurred. This included failed levees, and thousands of residents were forced from their homes, some for months. Over 150 streams and tributaries affected the flooded area, making it one of the most significant floods in US history [19]. However, this tragedy was overshadowed by Hurricane Katrina's devastation of New Orleans and surrounding regions of Louisiana and Mississippi in 2005. The prospect of eradicating flooding on the Mississippi River is a difficult, if not impossible, goal to achieve. For an interesting look at the history of floods and flood control on the Mississippi River see John M. Barry's outstanding book *Rising Tide: The Great Mississippi River Flood of 1927 and How It Changed America.*

The Colorado River, Urban Water Use

Water from the Colorado River keeps 30 million people alive. Located in the western and desert southwest of the US, growing cities like Denver, Las Vegas, Phoenix, and Los Angeles need

Fig 10.10 The Hoover Dam forms Lake Mead
(Photograph by Pamela McCreight, commons. wikimedia.org/wiki/File: Hoover-Dam.jpg)

water from the Colorado for drinking, irrigation, sanitation, business and industry, and recreation. During the early 1900s, the future prosperity of these cities depended upon the construction of dams and diversion structures along the Colorado River. Water is liquid gold to these desert communities.

Las Vegas, Nevada, for example, receives little more than 10 centimeters (4 inches) of precipitation each year, making it one of the driest cities in the US. It's understandable that the Colorado River serves as the lifeblood of the community. Lake Mead, located approximately 80 kilometers (50 miles) from Las Vegas, provides water via a pipeline to the city. Lake Mead is the reservoir formed by the construction of the Hoover Dam (see Figure 10.10). The reservoir, created by the 221-meter (726-foot)-tall Hoover Dam, looks out of place in the hostile desert environment, and evaporation takes a toll on its water supply. To serve its growth, citizens of Las Vegas recently approved a US$2 billion bond to transport and treat additional water supplies from Lake Mead. The Southern Nevada Planning Authority is considering a US$5 billion, 555-kilometer (345-mile) water pipeline northward to central Nevada to solve Las Vegas' future water needs. However, some irrigators and residents in rural parts of the state oppose the "water grab" by the gambling city to the south. Some refer to the coming water battle in Nevada as "crops versus craps," referring to the gambling tables of Las Vegas in competition with irrigators north of the metropolitan region.

10.3.2 Irrigation

Reservoirs provide water for 10 percent of the cropland irrigated in the US. Thousands of jobs, and the food supply of thousands, depend upon these sources of water [20]. This situation is common around the world, and the Murray–Darling River system is presented as an example.

The Murray–Darling River Basin

The Murray–Darling River Basin dominates irrigation in southeastern Australia and contains over 70 percent or 1.5 million hectares (3.6 million acres) of all irrigated crops and pastures in the country. The basin comprises the Murray, Darling, and Murrumbidgee Rivers – the three largest on the continent. Crops grown in the area include fruits, grapes, nuts, alfalfa, rice, wheat, corn, and cotton. Irrigation in the Murray–Darling Basin began in the 1880s, and after the drought of 1895–1902, the governments of Australia, New South Wales, Victoria, and South Australia signed the River Murray Waters Agreement to construct a number of dams in the region. In 1917, the River Murray Commission met for the first time, and was a predecessor of the Murray–Darling River Basin Commission.

In its natural state, Australia's River Murray did not flow year-round. In fact, during periods of drought, the river was reduced to a series of saline watering holes. In the southern portion of the basin, seawater intruded up the river from its mouth at the Southern Ocean at Encounter Bay, southeast of Adelaide. The completion of the Hume Dam (see Figure 10.11) in 1936, however, has allowed the Murray to flow continuously. The 51-meter (167-foot)-tall dam created the 3.1 billion cubic meter (2.5 million acre-feet, or 800 billion gallon) Hume Reservoir, named in honor of Hamilton Hume, one of the first Europeans to see and cross the River Murray in 1824. The dam is earthen, with a concrete core, and is covered on its exterior with concrete to provide protection from erosion. Without Hume Dam, the droughts of 1938–9, 1944–5, 1967–8, 1982–3, 1997–8, and 2002–6 would have certainly dried up the River Murray in many locations.

Each day, River Murray Water staff "run the river" by making water releases from reservoirs to meet irrigation demands in South Australia. However, minimum streamflow requirements, channel capacity, and water quality parameters must also be met [21]. The Murray–Darling Basin Agreement in 1992 provides the current institutional framework for administration and protection of water resources in the region.

In 2004, a historic first step was taken to restore flows and address the declining health of the River Murray system as part of the Living Murray Initiative. The program requires up to 500 million cubic meters (400,000 acre-feet, or 132 billion gallons) of water each year – enough to fill Sydney Harbor. After communities were informed, consulted, and involved, the Council of Australian Governments (COAG) provided AUS$500 million. The funds were be used to recover water for environmental use and will address the over-allocation of water in the Murray–Darling River Basin. The Council purchased water entitlements from irrigators and will develop institutional water trading agreements. Water supplies were purchased from willing sellers, on-farm water use efficiencies improved, and river flows better managed. Plans include construction of some engineering works, but it is intended that there will be no adverse social or economic impacts on river communities. The goal is to restore the River Murray to an ecologically healthy working condition [22].

Fig 10.11 The Hume Dam on the Murray River in Australia
(Photograph by Grumpyoldman1959 – Own work, CC BY-SA 3.0, https://commons.wikimedia.org/w/index.php?curid=21270288)

Guest Essay By Dr. Sara Beavis

Dr. Sara Beavis has a background in engineering geology and hydrology, and undertakes both research and teaching at the Fenner School of Environment and Society, and the Department of Earth and Marine Sciences at the Australian National University, Canberra, Australia. Her research focuses on the impacts of anthropogenic activities on catchment hydrology, the flow of water/solutes through substrate materials, and the physical/engineering properties of soils and regolith (loose, heterogeneous material covering solid rock). She teaches undergraduate and graduate courses in hydrology, water resources management, and coastal environmental earth sciences.

A Drying, Dying Waterway: The Murray–Darling System, Australia

Australia has the most variable rainfall in the world, with mean annual precipitation ranging from over 3,000 mm to <200 mm (118 inches to <8 inches) per year in areas as diverse as the northern tropics, arid 'Centre' and temperate Tasmania. Moreover, there is significant temporal variability because of the El Niño–Southern Oscillation (ENSO), which imposes cyclical fluctuations in

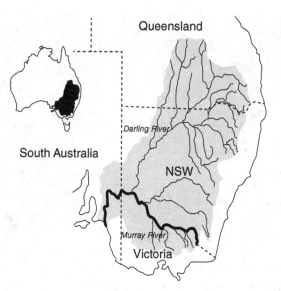

Fig 10.12 Location of the Murray–Darling Basin and the spatial distribution of the two main river systems

rainfall characterized by episodic floods and drought. A consequence of this spatial–temporal variability, in terms of water resource management, is that Australia has too much or too little water in the wrong place and at the wrong time. To mitigate the effects of this, a significant proportion of Australia's surface water resources has been impounded in dams and controlled through regulated river systems, so that users such as irrigators can be assured of secure, reliable water supply.

The Murray–Darling Basin (Figure 10.12) is the most important drainage basin in Australia, in terms of stream discharge and productivity. This drainage area occupies ~14 percent of Australia's landmass, generates 40 percent of Australia's agricultural product, and delivers water to four States and the Australian Capital Territory. Approximately 70 percent of the total water used for irrigation in Australia is applied to agricultural land within the Basin. The value of the Basin's agricultural produce is estimated at AUS$10 billion/year, with AUS$3 billion derived from irrigation [23].

Despite a mean annual water input to the basin of 508,000 gigaliters/year (411.8 million acre-feet/year, or 134.2 trillion gallons/year) as precipitation, high evapotranspiration losses result in total basin runoff to streams of only 23,850 gigaliters/year (19.3 million acre-feet/year, or 6.3 trillion gallons/year) [24]. Much of the evaporative loss (~11,000 gigaliters/year, or 8.9 million acre-feet/year, 2.9 trillion gallons/year) occurs in the extensive floodplains, floodplain lake systems, and wetlands that characterize the system. Most of the precipitation occurs in the east and southeastern parts of the basin, and hence, complex infrastructure exists to impound and regulate available water here in order to harvest it elsewhere (Figure 10.13). Some of the water is diverted for irrigation or intercepted and stored in large reservoirs and small farm dams, from which further, significant evaporative losses occur. The total volume of water in large storage dams equals ~25,000 gigaliters (20.3 million acre-feet, or 6.6 trillion gallons), equivalent to one year's runoff. Volumes impounded in small dams are ~2,200 gigaliters/year (1.8 million acre-feet/year, or 581.2 billion gallons/year) [25].

As a consequence of this broad set of conditions, catchment yield of the Murray–Darling River system is relatively low (as a means of comparison, its mean *annual* discharge is less than the mean *daily* discharge of the Amazon River). Median flows from the basin to the sea (the

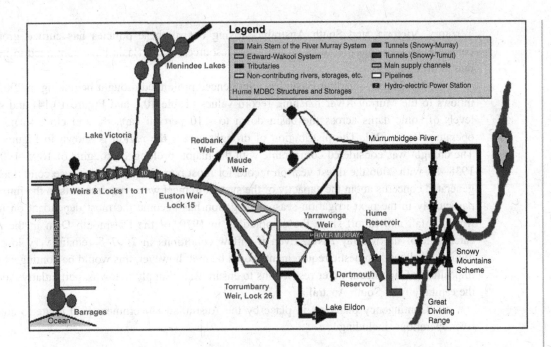

Fig 10.13 Infrastructure as control structures and reservoirs on the Murray and lower Darling Rivers (Source: http://www2.mdbc.gov.au/river_murray/river_murray_system/images/system-new_dec.GIF)

Southern Ocean) are ~27 percent of those that occurred under a natural flow regime, and the lower reaches of the Murray River experiences severe drought-like low flows in over 60 percent of years, compared to 5 percent of years predevelopment. In addition to these modifications to flow volumes, due to regulation and consumptive use, the impoundment of water within dams and its controlled release for use by irrigators has reversed the flow regime so that maximum flows no longer occur in winter–spring, but during summer.

Because of these changes to the flow regime, wetting and drying cycles in wetlands have been modified and salinity has increased. The frequency of algal blooms has also increased in response to periods of low flow when conditions for eutrophication are ideal. These changes all have profound impacts on riverine and floodplain biota. In addition, low-quality water is a serious issue for users at the end of the system, including the inhabitants of the city of Adelaide in South Australia, whose water has relatively high salt concentrations.

In 1920, the annual diversions from the Murray–Darling System totalled ~2,100 giga-liters/year (1.7 million acre-feet/year, or 555 billion gallons/year), but had increased to 10,789 gigaliters/year (8.7 million acre-feet/year, or 2.9 trillion gallons/year) in 1994, with a 7.9 percent growth in water consumption between 1988 and 1994 alone. This pattern of growth prompted decision-makers to realize that whilst unrestricted expansion of diversions and consumptive use in the Basin would support economic growth, it would come with significant environmental costs. This led to a raft of water reforms including the Council of Australian Governments (COAG) Water Reform Framework in 1994, and the Murray–Darling Cap on Diversions in 1995, which aimed to limit consumptive uses to 1994 levels of development. The process of water sharing across five jurisdictions (Queensland, New South Wales, the Australian Capital

Territory, Victoria, and South Australia) arising out of these policies has curbed growth in consumptive uses, but pressures on the water resources of the Basin have continued to increase due to drought.

The Murray–Darling Basin has experienced prolonged drought beginning in 2001, with inflows to the Murray River reaching record values (Table 10.1 and Figure 10.14) and storage levels of some dams across the basin down to <10 percent capacity and close to minimum operating levels [26]. The distribution of drought across the region is shown in Figure 10.15. The drought was considered comparable with the major, prolonged droughts of 1895–1903, and 1938–45, with 2006 the driest year on record for most parts of the basin. These conditions have generated concerns about the capacity of the system to meet water requirements in the future, and particularly in the next irrigation season. Irrigation has become the most dependent on inflows from rainfall and runoff since the completion, in 1979, of the Dartmouth Dam in the upland catchment of the Murray River. Even if inflow conditions in 2007–8 remain very low, basic human, stock, and domestic requirements could be met; however, this would be contingent on the imposition of significant water restrictions to ensure water supply to towns, particularly Adelaide, the capital city of South Australia.

Contingency plans set in place by the Australian Government [27] relate to adjusting river operations, including:

- target end-of-season reserves in Lake Victoria (a water storage that was designed in the 1920s to provide reliable water supply for the Lower Murray region in South Australia, and to mitigate and augment flood peaks);

Table 10.1 Annual River Murray system inflows (MDBC 2007)

Period	Annual total inflows (gigaliters)
Long-term average	11,200
1895–1904 drought	5,400
1938–1942 drought	6,300
2001–2007 drought	4,150
2006–2007	predicted >1,000

Source: Murray–Darling Basin Commission, *Drought Update No. 7*, April 2007

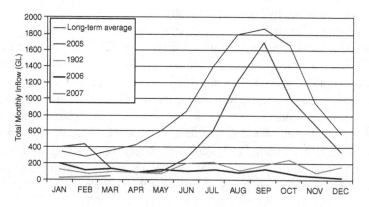

Fig 10.14 River Murray inflows showing the contrast between long-term mean and selected years during the current drought (Source: Murray–Darling Basin Commission, *Drought Update No. 7*, April 2007)

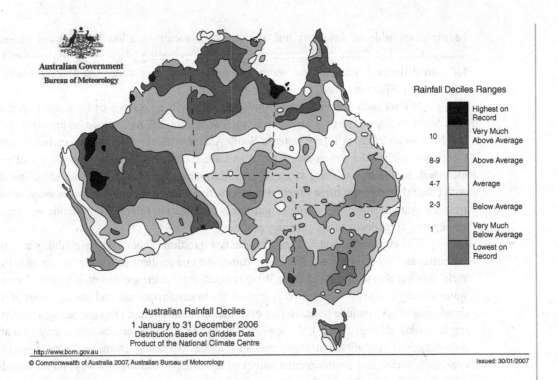

Rainfall Deciles Ranges

	Highest on Record
10	Very Much Above Average
8-9	Above Average
4-7	Average
2-3	Below Average
1	Very Much Below Average
	Lowest on Record

Australian Rainfall Deciles
1 January to 31 December 2006
Distribution Based on Griddes Data
Product of the National Climate Centre
http://www.bom.gov.au

© Commonwealth of Australia 2007, Australian Bureau of Motocrology

Issued: 30/01/2007

Fig 10.15 Australian rainfall deciles in the severe drought year 2006 (Source: Australian Bureau of Meteorology, 2007)

- minimum flow targets;
- build reserves for Adelaide through early pumping from the River Murray to urban supply storages;
- disconnect selected permanent wetlands that are artificially inundated under "normal" conditions.

Additional contingency plans for towns at risk have been formulated for implementation if necessary, including highly restrictive water use.

Conditions by April 2007 were so serious (see Figure 10.15) that planners and decision-makers considered no allocations would be possible for anything other than critical demand to meet human and stock needs at the "normal start" of the irrigation season. This generated media headlines across the country, although rains confirming the end of El Niño and the beginning of stable conditions have subsequently raised expectations of water users. By June 2007, with rainfall across the upper catchment reaching "average rates," State governments are finalizing arrangements to allow limited volumes of water to be available for irrigators. However, significant, and persistent rainfall will be essential to wet up the basin and generate runoff into the stream network. Indeed, it is estimated that even under wetter than average conditions (<15 percent probability), it would take several years for reservoirs and other storages to return to long-term average levels.

The severity and length of the drought have produced caution in planning, so that, despite the onset of rainfall, *Murray–Darling Basin Dry Inflow Contingency Planning* continues to develop. This means that even when water allocations are defined at the beginning of the irrigation season, State and Territory authorities will review water availability monthly, and information will be made available to irrigators and the livestock sector to allow for forward planning and management. The States have agreed to interim water-sharing plans above the amount allocated for critical demand, until normal water sharing arrangements under the Murray–Darling Basin Agreement [28] are resumed.

Within such arrangements, South Australia will give some of its water to Victoria and New South Wales early in the irrigation season. When conditions improve, as anticipated, Victoria and New South Wales will give water to South Australia in recognition of its earlier contributions. Contingency plans for towns are in place and are evolving as conditions change and issues are identified. In addition to these strategies, the disconnection of wetlands (not including wetlands with cultural heritage or environmental values) actively saves water by reducing evaporative losses from the system. However, ongoing community consultation relating to wetlands disconnection is an ongoing component of contingency planning.

The current drought has raised some key questions about water availability and supply in the basin, and has shown particularly how vulnerable and resilient the system is, not only to climate variability but also to climate change. Whilst contingency plans are addressing immediate issues of water scarcity and supply, longer-term approaches to sustainable use and management of water are developing. These include research (for example, drought-tolerant crop species and spatial modeling for sustainable agriculture); the application of water use efficiencies across irrigation and other industries; and institutional arrangements such as flexible water reform packages. Given the high economic, social, and environmental values of the Murray–Darling Basin, planners and decision-makers readily acknowledge the importance of using the lessons derived from the current drought to mitigate against the impacts of future extreme events and the more subtle changes of climate change. However, there are concerns by various sectors, including basin communities, that water sharing devices and strategies will not meet all the requirements in an over-allocated basin characterized by complex biophysical and economic pressures and responses.

10.3.3 Power Generation

Energy plays an important role in the socio-economic development of a country. In past centuries, small dams (called mill dams) were constructed across rivers and streams to divert water into raceways to turn waterwheels and mills. In modern times, hydropower continues to provide cheap, clean, and renewable sources of energy. In addition, hydropower provides peaking power (energy needed during high use periods of a day or season) that might otherwise be provided by gas turbines or other fossil fuels. China, Brazil, Canada, United States and Russia are the largest producers of hydropower in the world. Dams produce over ~3,002 terawatt hours (3,002 million megawatt hours) of renewable energy, which represents approximately 71 percent of all renewable electrical energy in the world. Hydropower is considered to be clean energy because it does not create air pollution, acid rain, ozone depletion, or contribute to global warming [29].

Do dams contribute to global climate change (GCC)? Research on global climate change and reservoirs has found that sediment in reservoirs is a significant source of the greenhouse gases methane (CH_4) and carbon dioxide (CO_2). This is due primarily to the decay of buried organic materials present when the reservoirs were formed, and soil nutrient runoff (erosion) from farmed fields, logging, and land development [30]. Other energies considered clean – solar, wind, and geothermal – do not currently produce as much energy as hydropower; therefore, it remains an important source of energy but is not problem-free in terms of GCC. This is an area where more research is needed and will be an important choice of research areas for students interested in becoming research scientists in many areas (e.g., soils, silviculture, chemistry, global climate modeling, biology, and more) pointing out the connectivity of human activity in all biological system changes.

An example of hydropower development is at the Itaipu Dam, which straddles the border between Brazil and Paraguay in South America. The dam altered the course of the seventh largest river in the world, the Paraná. It is a hollow gravity dam located near Iguassu Falls, and contains the largest hydropower plant in the world. Construction began in 1970 and was completed in 1991 at a cost of US$18 billion. Itaipu Dam was constructed primarily for irrigation and industrial uses. However, it also provides 25 percent of all energy used in Brazil, and 78 percent of the electrical demand in Paraguay (enough to power most of the state of California). The power plant is also a major tourist attraction in the region, receiving more than 9 million visitors from across the globe. Approximately 10,000 families had to be relocated during construction when 1,350 square kilometers (520 square miles) were inundated by the reservoir. The dam is 196 meters (643 feet) high, equivalent to a 50-story building. About 40,000 people worked on the construction project at its peak [31].

10.3.4 Inland Navigation

Dams and lock systems on rivers provide a more stable inland river transportation network for the movement of grain, coal, gravel, and other materials. These transportation systems provide more stable river flows to generate large national economic benefits, as well as options for consumers to transport commodities such as grain, minerals, and other items.

Tennessee Valley Authority

The Tennessee Valley Authority (TVA) was created by an Act of the US Congress in 1933 to develop dams along the Tennessee River. The purposes were flood control, navigation, hydropower, and the improvement of the social and economic well-being of the southeastern US. TVA's system of dams and locks has created a navigable river between Knoxville, Tennessee, and Paducah, Kentucky, a distance of 1,049 kilometers (652 miles). Since the change in elevation is over 150 meters (500 feet), nine locks and dams were necessary to allow boats to "climb" up and down the series of quiet, pooled waters created behind the nine dams. Industrial activity is extensive along the river, with transportation costs lowered by the river transportation network.

Today, over 28,000 barges carry 45 to 50 tons of goods (coal, stone, sand, gravel, grain, chemicals, iron and steel, and forest products) up and down the Tennessee River annually. It has been estimated that transportation costs are lowered by over US$400 million annually when goods are shipped by river barge rather than by truck or rail [32].

10.3.5 Recreation

Dams provide flat-water recreation opportunities to boaters, water-skiers, swimmers, and anglers. Reservoirs create camping, picnic areas, and wildlife viewing opportunities, as well as boat launch facilities. Recreation and associated health benefits are especially important as people in more-developed countries often lead sedentary lifestyles. Mental stress is also a serious concern of

Fig 10.16 Lake Tsukui in Kanagawa, Japan, is a beautiful place for recreation (Photograph courtesy of Wikipedia Commons, creator unknown, https://commons.wikimedia.org/wiki/File:Lake_Tsukui_01.jpg)

modern society, and the outdoor recreation provided by reservoirs is an important outlet for millions. The economic benefit of these activities, and associated health cost reductions, are enormous.

Figure 10.16 is Lake Tsukui in Kanagawa, Japan. In Japan, as in most other countries, reservoirs are a vital source of water supply for irrigation, industrial, household uses, and hydropower generation. However, tourism and recreation are also extremely important uses of reservoirs created by dams. Japan's Ministry of the Environment has developed programs to encourage more of its citizens to utilize reservoirs for recreation and health purposes. The Ministry states that people feel relaxed around water and encourages the use of lakeshores for taking walks and sport activities such as boating and fishing. Tens of millions participate in water-based recreation each year in Japan. School children plant wild oats as feed for swans and other waterfowl at reservoirs. Other groups organize lake water festivals, collect litter, plant trees, hold citizen's symposiums, conduct nature watches and studies, and perform observations of the lake using canoes. This is in addition to recreational boating, fishing, and other aquatic sporting activities promoted by the Ministry of Environment in Japan.

10.4 Discussion of the Impacts of Dams and Reservoirs

Even though rivers are a gift of nature, humans have altered the location and flows of these systems to better serve the needs of civilization. During modern times, water needs are becoming more in conflict with the needs of the natural world. The challenge for modern society is to continue to meet human water needs while also protecting the environment. No easy solutions exist, and men and women around the world continue to debate the need for new and existing dams and reservoirs.

As we have been discussing, few changes are all good or all bad – whether in nature or life. Change can produce unintended consequences, such that the term "the law of unintended consequences" has become a commonly used phrase. It's hardly a scientific law but expresses the problems of trying to anticipate everything that can happen when an action is taken. The effects of dams and reservoirs on water quality and wildlife habitat fall into this category.

Water releases from reservoirs help dilute harmful dissolved substances during periods of low flow. This helps maintain and preserve some aspects of water quality, within safe limits, for aquatic wildlife and humans. However, water released from reservoirs is typically low in dissolved oxygen and can create an unhealthy aquatic environment downstream. On the other hand, dams provide some environmental protection through the retention of detrimental sediments (such as pesticides and nutrients) behind the structure. In fact, sediments captured in reservoirs can bury and isolate some toxic materials from the food chain – a good consequence unless reservoir maintenance requires that sediment be removed. Proper disposal of contaminated sediments can be a problem. They can also release methane – a negative consequence. The caveat here is that water releases from reservoirs can harm downstream aquatic species because of reduced oxygen levels in the water, or higher or colder water temperatures.

The chemical, thermal, and physical changes created by dams and reservoirs – both upstream and downstream – can be great. Rivers in their natural state possess a regular cycle of disturbances. Plants and animal communities that utilize rivers and riparian habitats have evolved with particular river patterns of flood and drought, varying water temperatures, and slow and fast currents. Dams alter these regular cycles of disturbance and can alter the ecology of a river system.

Various species have developed their reproductive cycles to correspond with annual flood seasons. For some, floodwaters provide the necessary nutrients and water levels to enhance survival of offspring. Floods can provide shallow backwater areas that are conducive for young to hide for protection from predators. The reduction or elimination of flood flows can reduce needed food supplies or can also change water temperatures to ranges not tolerable by young.

10.4.1 Sediment and Nutrient Capture

Lotic (moving water) ecosystems are found along rivers and streams. These ecological systems are lost when a dam is constructed, and a site is inundated with water. Reservoirs create a **lentic** (standing water) ecosystem, and generally provide habitat for a much smaller range of species. Slow-moving or still water allows suspended sediments to settle out and build up at the bottom of reservoirs. Silt captured behind dams can damage or destroy fish spawning grounds. Impounded floodwaters will no longer provide fertile sediments to adjacent floodplain areas.

Water released through a dam can be much clearer, and carry less sediment and their associated nutrients, than water previously contained in native river flows. This may seem like a benefit; however, water released from a reservoir is "hungry" to carry sediment. This process can remove important downstream sediments and nutrients from a stream and destroy aquatic habitat that was previously used by organisms adapted to the rising and falling water levels in riverine systems.

Changes in downstream river morphology may include more inorganic substrate (gravel and cobbles) and less organic material (detritus, including deteriorating woody matter). Cleaner river water means that less sediment will be available to fill in pore spaces between sand, gravels, and other substrate (river bottom materials). Over time, the river downstream of a dam will tend to become deeper and narrower, which will reduce the diversity of plant and animal life that can be supported.

High-energy, sudden, powerful water surges – sometimes caused by reservoir releases – will flush sediments. Aquatic plants and leafy and woody debris will also be swept downstream. These complex components of habitat are lost, and too often, what remains is a scoured riverbed of larger rocks. The system is "starved" of sediment and other organic materials, which are not suitable habitat for many plants, fish, and benthic macroinvertebrates. (Freshwater benthic macroinvertebrates are animals without backbones that are larger than 0.5 millimeter – the size of a pencil dot.) They are essential components of the **food web**. Macroinvertebrates live on rocks, logs, sediment, debris, and aquatic plants during some period in their life. These can include crustaceans such as crayfish, mollusks such as clams and snails, aquatic worms, and immature forms of aquatic insects such as stonefly and mayfly nymphs [33].

For example, sediment loss is a problem along the Ebro River and its tributaries in Spain. There are 187 dams along its course that trap about 99 percent of all river sediment that would normally flow into a downstream delta region – the Ebro Delta National Park. The lack of sediment replenishment is causing the erosion of its delta coastline, making it unstable. We have already noted that a similar problem exists in Egypt, where dams along the Nile River trap about 98 percent of suspended materials no longer replenishing farmlands. Unfortunately, there is no safe or econom- ical method to remove vast amounts of sediment from behind reservoirs and deposit it in a normal river cycle.

Dams hold back important nutrients contained in the sediment and debris of flowing water. Leaves, twigs, branches, and entire trees provide organisms (including fish with food for survival. These carbon-rich materials nourish organisms at the base of the food chain. In addition, debris provides important hiding places for life as varied as phytoplankton, microorganisms, and fish. Dams can hold back a substantial portion of these materials, depriving downstream users – particularly for the bottom level of the food web. The loss of these freshwater benthic macroinverte- brates can leave larger species hungry, and ultimately will reduce their numbers.

10.4.2 Water Temperature

Another impact of dams and reservoirs is water temperature. Rivers are generally homogeneous in water temperature, but reservoirs are layered – warmer on top and colder on the bottom during the summer months, with the reverse occurring during the winter (see our discussion in Chapter 7 regarding *lake turnover*). Reservoir releases are usually made from the bottom of a dam (the typical location of outlet works). This deep water is colder in the summer than the downstream river water and can create a thermal shock to some species. For example, a stonefly *(Plecoptera)* may sense colder river water temperature and delay its metamorphosis (change in form, as from a tadpole into a frog). This can alter the life stage and cause it to live at a more susceptible stage during the winter rather than in the autumn.

The pre-dam temperatures of the Colorado River in Glen Canyon in northern Arizona, for example, varied from around 27 °C (81 °F) in the summer to around freezing in the wintertime. However, after construction of Glen Canyon Dam, the temperature of the deep reservoir water – some 70 meters (230 feet) below the reservoir surface – is almost constantly around 8 °C (46 °F). This water temperature has caused severe problems for native fish reproduction as far as 400 kilo- meters (250 miles) downstream of the dam [34]. Also, water temperatures may increase and become warmer at the reservoir surface, and can harm additional aquatic species.

10.4.3 Mercury

The issue of mercury contamination in freshwater fish has been widely publicized and has become more prevalent in recent years. Dams create large bodies of water that inundate vast areas of vegetation. The decaying plant material beneath a reservoir creates an anoxic (without oxygen) environment – a key component necessary for sulfur reducing bacteria to change inorganic mercury (Hg^{2+}) into the organic neurotoxin **methylmercury** sometimes written (MeHg) or $(CH_3Hg)^+$. Mercury is a toxicity problem when it becomes organic or is a vapor. Methylmercury affects nerve cells (a characteristic of neurotoxins) and can be fatal.

Organic forms of mercury become part of the food chain when ingested by small creatures, which are then ingested by larger creatures, and so on, up to the top predators – humans. The anoxic

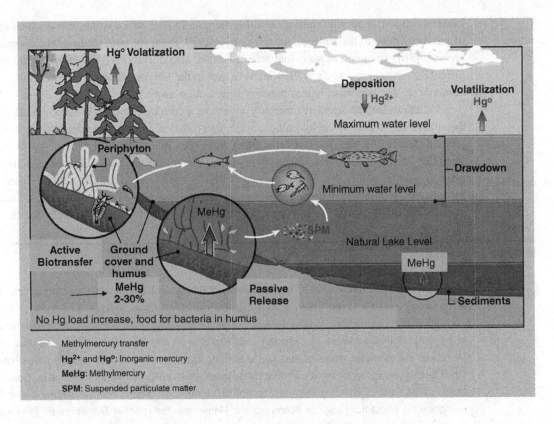

Fig 10.17 Mercury cycle in new reservoirs where organic matter as food for bacteria, that transform inorganic mercury to organic methylmercury, is plentiful (Figure modified from Hydro-Québec www.hydroquebec.com/sustainable-development/specialized-documentation/mercury.html)

environment found in new reservoirs can also exist in other aquatic systems that contain fine sediment bottoms and excess vegetative growth. A problem with reservoirs is that most are heavily fished by humans and other fish-eating predators. Dam proponents probably didn't anticipate that mercury would enter the food chain because of these projects.

A compound that is retained and moves from one level to another in the food chain – increasing in concentration – is said to **biomagnify**. A compound that occurs in higher amounts in aquatic organisms than in water is said to **bioaccumulate**. The problem with biomagnifications and bioaccumulation is that, by the time humans ingest methylmercury from fish, the methylmercury levels can be high enough to cause significant adverse health effects – particularly to the human nervous system. The effects are worse in young children and pregnant women) [35].

Rather than becoming an alarmist, it is good to remember that dose makes the poison, and that the human body can eliminate some mercury as waste. Eating a few servings of mercury-containing fish, over a period, may not be harmful. However, eating mercury-containing fish every day, for a length of time, is not a healthy idea. How much methylmercury accumulates in a fish depends on: (1) whether there is a source of mercury available, (2) whether the environmental conditions necessary to form methylmercury are present, and (3) what makes up the entire diet of the fish, since not everything consumed contains methylmercury.

Let's discuss some background regarding mercury. Mercury is a naturally occurring element and is released to the atmosphere during volcanic eruptions and from geothermal springs.

Industrialization, especially mining, coal-burning, and waste incineration, has doubled the amount of mercury added to the atmosphere since about 1950. Polar ice core studies have documented this problem [36]. The good news is that scientists have also documented the effectiveness of environmental protection laws, such as the Clean Air Act in the US, on reducing industrial pollution to the atmosphere. The bad news is that mercury does not go away.

How does atmospheric mercury enter our hydrologic cycle? We know that moisture in the atmosphere is part of the hydrologic cycle. Precipitation carries particulates, such as mercury, to the land surface, and into rivers, streams, lakes, reservoirs, and oceans. Mercury in the atmosphere can also be dry deposited onto the surface environment. The result is a ready supply of reducible inorganic mercury – both a natural and human-caused condition (see Figure 10.17). Mercury in reservoirs, both new and old, is a well-documented global water quality issue. For example, the United States has had mercury advisories across the country from New York [37] to Indiana [38] to California [39].

In Depth	**Mercury**

Mercury is normally contained in an unavailable mineral form such as cinnabar, mercury sulfide (HgS), and some mercury salts. It is present in all rock classes from sandstones to basalts.

Elemental mercury $(Hg)^0$ is a liquid at room temperature – a shiny, silver puddle of metal. The use of mercury in thermometers became an unacceptable health risk when its toxicity was recognized. (A broken thermometer releases mercury vapor into the air and is toxic if inhaled.) Most people have never seen mercury up close; it's a shiny silver metal that can be broken into droplets. Surface tension creates little balls of mercury that can be rolled together to reform into its original mass. You might recognize the effect from science fiction movies. Remember the robot in *Terminator II*? (You may need to find it under the classic Sci-Fi Section wherever you access old movies.) The robot – a high-end bad guy model – was shot into thousands of shiny metal pieces, but always flowed back together. That is exactly what "liquid" mercury does, except that you must move the silver droplets together. It isn't movie magic, just chemistry and physics.

Mercury is a fascinating metal. Even though metallic mercury appears to be poorly absorbed, even if swallowed, it is blocked from the bloodstream by the stomach and intestinal lining. However, mercury vapors are readily absorbed into the human body. Metallic mercury can be harmful if it can evaporate and then is inhaled, since it quickly moves from the lungs to the bloodstream. Mercury can become a neurotoxin, which means it acts on the nervous system (brain, spinal cord, and nerves), damaging bodily functions – from walking to thinking. Mercury poisoning symptoms can take months to appear if a person repeatedly eats contaminated food. However, symptoms can disappear if people stop eating mercury. Clearly, there is still a lot to learn about mercury's effects on humans and other living things.

A Canadian study on the increase in mercury levels in reservoirs was conducted by Hydro-Québec in 1978–1985. Fish mercury levels were monitored at the La Grande complex of reservoirs in the James Bay region of northern Québec. The purpose was to understand the recently discovered phenomena of increased mercury levels in the reservoirs. The study documented actual changes in mercury concentrations in fish, aquatic organisms, and humans [40].

Indigenous Cree residents consumed fish from the La Grande reservoirs, with pike and walleye fish accounting for 15–20 percent of the bush food consumed by locals. Families that regularly caught piscivorous fish (fish that eat other fish, such as pike and walleye) were exposed to high levels of methylmercury, sometimes as high as six times pre-reservoir levels. By 1984, six years after La Grande Number 2 Dam was completed, 64 percent of the Cree living on the estuary had blood mercury levels far exceeding health safety limits. Fifteen years after the natural lakes were impounded to make the larger reservoirs, mercury levels were still higher than in natural lakes, but were decreasing.

Mercury levels in most of the Cree are now below those considered harmful to human health, potentially because the Cree agreed to eat fewer piscivorous fish. Mercury levels in fish are also decreasing, so that probably contributed to the success. Research from these reservoirs suggests that time is necessary to bring mercury levels back to normal (10–20 years for non-piscivorous fish and 20–30 years for piscivorous fish). The Cree Board of Health and Social Services of James Bay has been actively involved in this now decades-old program, which continues under the James Bay Mercury Agreement [41].

10.4.4 Dissolved Oxygen Levels

Another impact of dams and reservoirs is changes in dissolved oxygen. Dissolved oxygen levels are reduced when water is held in a reservoir for an extended period. When a reservoir is first formed, submerged vegetation and soil decomposes, depleting oxygen levels in the reservoir water. These "young reservoirs" can contain drastically depleted levels of dissolved oxygen in the water. Deoxygenated water can be harmful and even fatal to certain fish species and for aquatic life downstream of the dam. Young reservoirs can take as long as a decade to "mature" to higher dissolved oxygen levels.

10.4.5 Fragmentation of River Ecosystems

As we have mentioned, dams fragment river ecosystems, isolating species that live upstream or downstream of a dam. Dams and reservoirs create barriers for fish migration, such as adults traveling upstream to reproduce, and the subsequent smolt (juvenile fish) working their way back downstream after hatching. Settlements around Lake Ontario (surrounded by the Province of Ontario and the state of New York) in the 1800s, for example, ultimately blocked dozens of streams flowing into the lake due to the construction of mill dams. These facilities were created to divert river water to turn mills for grain grinding, lumber milling, and other industrial activities, which depended upon waterpower generation to operate.

Think About It: Who or What Needs the Water?

The endangered Rio Grande silvery minnow (*Hybognathus amarus*) has become trapped between two dams along the middle Rio Grande River in New Mexico. This reach of the river sometimes dries up during the summer, and minnow populations have decreased. The city of Albuquerque, New Mexico, relies on water supplies from these reservoirs, but had to reduce water diversions to help maintain river flow requirements for the endangered minnow. How should society weigh human needs in a desert environment with the water needs of an aquatic species? Of what value is one minnow species? Does the ecological health of the river itself play a role in your decision?

10.4.6 Elimination of Flood Flows

Dams generally capture floodwaters and eliminate vital pulses of floodwater. These pulses historically provided ecological benefits to adjacent floodplains, wetlands complexes, and associated alluvial groundwater aquifers (groundwater hydraulically connected to adjacent surface streams, wetlands, and marshes).

Dam operations alter the historical flow of water downstream in a river system. Reservoir releases, even hourly fluctuations, generate a range of new impacts on rivers. This is because fish and wildlife species have adapted over hundreds and thousands of years to historic river flows. Human-caused changes in these flows alter the tightly linked patterns of ecosystems to the existing water flow patterns of rivers. The result is a change in the natural ebb and flow of a river system. This disrupts habitats and species that have adapted to that natural rhythm.

In Depth   **The Case of the Missouri River**

Six dams have greatly affected the Missouri River since the Lewis and Clark Expedition of 1804–6. These are Fort Peck, Garrison (the fifth largest earthen dam in the US), Oahe (which created the fourth largest reservoir in the US), Fort Randall, Big Bend, and Gavins Point Dam in Montana, North Dakota, South Dakota, and Nebraska. The Missouri River drains almost 20 percent of the US as it flows 3,767 kilometers (2,341 miles) from its headwaters in North Dakota to its confluence with the Mississippi River at St. Louis, Missouri. The basin is home to about 10 million people, including 28 Native American tribes, and other residents of 10 states in north-central US and a small part of Manitoba and Saskatchewan in Canada. Precipitation is in the range of 25–100 centimeters (10–40 inches) per year, with an elevation drop from 4,300-meter (14,000-foot) mountain peaks at its headwaters to only 120 meters (400 feet) at its confluence with the Mississippi.

Predevelopment, the Missouri River was one of North America's most diverse ecosystems, which included abundant braided river channels, extensive riparian lands, islands, sandbars, and extensive wetland complexes. Floods reshaped the river channel over centuries and provided high sediment and nutrient loads. Today, 32 percent of the Missouri River is channelized, often between levees, and over 1,300 kilometers (800 miles) of the river – an amazing 35 percent of the entire length – is impounded behind dams in a lake environment. Only 33 percent of the river remains unchannelized. Dams have transformed approximately one-third of the Missouri River into a lentic (lake) environment with a storage capacity of 91.3 billion cubic meters (74 million acre-feet, or 24.1 trillion gallons). The surface area of these lakes exceeds 400,000 hectares (1 million acres). As you can imagine, this has greatly changed the features of the Missouri River and its ecosystems, and it is no longer the free-flowing stream observed by Lewis and Clark in the early 1800s. In 2007, the US Geological Survey noted:

> These changes have significantly altered the Missouri River ecosystem. In the upper river, a new ecosystem has been created with the deep-water reservoirs replacing the free-flowing river and the inter-reservoir reaches affected by lower water temperatures and reduced sediment loads. In the lower river, channelization has eliminated sandbars, depth diversity, and river connections with off-channel side channels and backwaters. The historical flow regime has been transformed with spring high flows now captured in reservoirs and low summer and fall flows augmented with reservoir releases. All of these changes have lowered populations for many river fish and bird species, some to the extent that they are federal or state-listed as endangered, threatened, or species of special concern [42].

The Rivers and Harbors Acts of the US Congress in 1912, 1917, 1925, 1927, 1930, 1935, and 1945 each provided funds to develop the Missouri River for navigation, development, and flood control. The river was dammed, channelized, relocated, and concentrated so that flows could be utilized for human use. This work has provided thousands of acres for urban and agricultural development within the historic floodplain of the Missouri River. These projects also effectively ended the "meander belt" of the Missouri River which, prior to human intervention, allowed the river to wind back and forth across nearly one-fourth of its floodplain area. The meander belt contained wetlands, sandbars, forests, and other ecological systems that provided for a wide diversity of fish and wildlife habitats. Seasonal floods also replenished the shallow-water habitats located in the floodplain.

Channelization reduced the course of the Missouri River by 116 kilometers (72 miles) and eliminated 143,000 hectares (354,000 acres) of meander belt habitat. In 1986, this loss was recognized by the US Congress in 1986 with the authorization of the Missouri River Fish and Wildlife Mitigation Project. Its goal is to acquire and restore 11,000 hectares (28,000 acres), or about 5 percent, of the land lost over the past century to development funded by the federal actions [43].

10.4.7 Urbanization

Construction of dams, and the creation of reservoirs, often leads to increased urbanization in a region. This means that roads, buildings, parking lots, and other impervious materials cover the land surface that was previously covered by vegetation. These land use changes reduce the ability of soils to allow precipitation to infiltrate into soils and groundwater aquifers. More water may be needed to service new human needs, and river flows downstream of the dam could be reduced even further. Urban runoff can also contaminate surface water supplies.

10.5 Rivers, Dams, and Rehabilitation Efforts

10.5.1 The Rhine River

For years, the Rhine River was one of Europe's worst waste dumps. It was a river sacrificed for economic progress; a common occurrence in more-industrialized nations around the world. This process has, unfortunately, been repeated over and over again in less-developed nations as they become more industrialized.

There is a German folk tale – The Mouse Tower. It is the story of a greedy bishop who, after killing his peasants, says they were no better than mice. He then meets his own death by an invading army of rats and mice. This could also have been a cautionary tale of what greed would do to the Rhine (Figure 10.18). The Rhine River is Europe's busiest waterway and has been developed for centuries. Its riverbanks are crowded with cities and factories, dams impede its flow, and much of its channel has been straightened and lined with concrete. The Rhine begins in the Swiss Alps and flows across Switzerland, Germany, France, and ultimately to the Netherlands at the port city of Rotterdam. Dams along the Rhine generate hydropower, provide river transportation, irrigation water, and water supplies for drinking and sanitation. Approximately 50 million people live in the Rhine River watershed.

In 1816, the Rhine Commission was created to preserve the river as a navigable waterway. Remarkably, the Commission has met every year since 1831, except during World War II, to manage

Fig 10.18 An anonymous artist rendition of the Rhine River in 1840 shows early development near Bingen, Germany. The Mouse Tower is the small structure in the water on the right side of the painting (Source : http://commons.wikimedia.org/wiki/File: Rhine_river_and_Bingen_post-1840.jpg)

the shared commercial waterway for human use. In 1885, five countries signed the Salmon Treaty to protect the migration of fish, which were vanishing due to pollution and dams that were preventing their passage to spawning grounds. Coal mining, iron, and steel manufacturing changed the watershed from an agricultural society to one of industry – highlighted by Germany's Ruhr River Valley, a tributary of the Rhine. Some of the world's densest road and railway networks parallel its banks.

River pollution became so serious in the 1950s – from reregulation of river flows, human waste, fertilizers, and household products such as detergents – that between 1970 and 1990, US$38.5 billion was spent on water purification plants. These plants were designed to eliminate heavy metals and improve dissolved oxygen levels in the water. The improvements were successful, but sources of new pollution remained. It was discovered that the water quality problems could not be "fixed" by purification attempts alone.

The turning point came in 1986, in the form of the Basel disaster. A catastrophic fire struck the Swiss Sandoz agrochemical plant near Basel, Switzerland, sending tons of toxic chemicals to the nearby Rhine River, turning it red. Firefighters doused the flames with large quantities of water, and inadvertently flushed 10–30 metric tons (11–33 tons US) of insecticides and pesticides into the Rhine. Virtually all fish and plant life died along the river as the chemicals made their toxic flush to the North Sea. Witnesses reported the foul smell of rotten eggs and burning rubber throughout the area. Fortunately, the government of the Netherlands took the lead on river cleanup and created the Rhine Protection Commission. Long-term cleanup efforts expanded rapidly, and today, groups and governments along its banks have worked to rehabilitate the river. Fish ladders and other passages have been constructed at dams to allow the return of salmon migration; over 200 restocked salmon have been caught since 2002. However, the restocking, pollution prevention, cleanup, and watershed rehabilitation programs must be expanded and supported if the Rhine ecosystem is to continue to recover [44].

10.6 Is Dam Removal the Answer?

We've listed many benefits and problems created by dams and reservoirs. In some locations, however, dams are being removed because of old age and maintenance problems. In the US, it

Fig 10.19 Thai family fishing at sunrise on the Bang Phra Reservoir, Chonburi Province, Thailand, provides an excellent example of human use of natural resources for food and fun
(Getty image credit: Sarote Pruksachat)

has been estimated that 25 percent of the nation's dams are over 50 years old [45]. Older dams can have cracks if built of materials other than earth. Water could excessively pipe (leak) through the base of an aging structure. Sediment buildup behind dams is another serious problem and reduces available space for water storage. In some cases, the original purpose for the dam may no longer exist.

Dams alter the ecology of rivers. However, economics are generally the primary incentive to remove a dam. Dam maintenance costs and liability are often significant problems for the dam owner. Environmental benefits of dam removal are important but are difficult to quantify. By contrast, recreational benefits such as fishing, canoeing, rafting, and riparian activities at reservoirs – such as hiking, picnicking, biking, and bird-watching – all bring in recreational dollars that can be quantified (Figure 10.19). Debate continues over the environmental and economic impacts of dam removal.

10.6.1 A Successful Removal

Several dams have been removed from the Baraboo River, which flows approximately 160 kilometers (100 miles) across Wisconsin and through its capital city of Madison. The Baraboo River once had 11 dams across the main stem of the river, owing in part to an elevation drop of over 46 meters (150 feet) along its course. Fourteen meters (45 feet) of that gradient occurs in an 8-kilometer (5-mile) stretch. Early European settlers recognized its potential to generate power for mills, and constructed numerous dams and mills along its course. During the mid to late 1800s, Baraboo River dams drove the local economy, powering grain, lumber, and other types of milling operations.

The milldams changed the Baraboo River, from a fast-flowing stream with rapids, into a series of slow-moving impoundments. Fish populations declined or were replaced by carp and other less desirable species. Silt covered the river bottom, destroying once-productive aquatic habitat. Fish migration and spawning ceased.

Then, in 1994, the Waterworks Dam, located in Baraboo, failed a safety inspection. The cost of repairs was estimated at nearly US$700,000, while dam removal was around US$200,000. In 1998, the City of Baraboo decided to remove the dam. Backhoes and pneumatic hammers destroyed the structure, and that segment of river soon returned to its more natural conditions. Later, two additional dams – the Linen Mill (built in 1898 of timber, rock, and gravel to provide power for a linen mill) and Oak Street dams (built in 1929) – were destroyed, along with another one upstream at Lavelle, Wisconsin.

The results of the dam removal projects along the Baraboo River have been dramatic. Accumulated sediments behind the dams were flushed away with the first high water flows. Aquatic plants soon grew on the exposed riverbanks, and native fish returned to this segment of the river. Stream chemistry and ecology have been investigated by students and faculty at the University of Wisconsin–Madison. In addition, the Wisconsin Department of Natural Resources uses an "index of biotic integrity" to assess the overall environmental quality of the Baraboo River, before and after dam removal. Before removal, the river received an index score of 25–50 (poor to fair). Two years after removal, the scores reached 75–80 (excellent) in the same location. This was based on the presence of species such as smallmouth bass, which require high-quality aquatic conditions. Previously, warm water species such as carp and other suckers were present [46].

Often, the process of dam removal begins with a dam inspection procedure by a state inspector. Repair costs may be so great that removal is justified if the dam no longer provides significant economic benefits. Very often, safety and potential lawsuits are greater factors than environmental protection or enhancement. Typically, such factors do not generate direct economic benefits to the owner of the dam.

Think About It: Dam Removal

Dams are often the focus of intense controversy and debate. Depending upon your viewpoint, dams can represent economic and social salvation, and perhaps national economic development for less-developed countries. Do less-developed countries have to repeat the mistakes of more-developed countries? Are there cheaper, less environmentally harmful sources of power?

For others, dams represent negative environmental implications, human relocation problems, and societal change created by large-scale projects. Do the people most impacted by building a structure, that floods their homes and destroys their livelihoods, have a say in the process? This diversity of views illustrates how divergent attitudes and perspectives can be created based upon contrasting value systems. Is the issue of dam removal a simple right or wrong situation? What do you think about removing dams as they lose their usefulness? Spend a little time and look for examples of dam removals on the internet. What are you finding in terms of benefits, costs, and problems with the removal?

Summary Points

- **Dams** can be vital to civilization, but inevitably disrupt the ecology and normal functions of the dammed river.

- **Beavers** create the most natural of dams using sticks and mud to stop the flow of water. This technique has also been used by early humans on small streams.
  - Beavers create an **ecotone** transitioning from river to wetland.
  - Because of their ability to manipulate their structural environment, beavers are classified as a keystone species.

- There are several basic dam types from simple to highly technical.
  - Eight-thousand-year-old irrigation canals have been found in the **Mesopotamia** area. **Earthen dams** supplied the water for these canals.
  - **Gravity dams** built of concrete or masonry blocks across relatively narrow valleys came next. Concrete was not invented until the Romans developed it in the second century BCE.
  - **Arch dams** can be built of concrete in narrow canyons with solid rock walls.
  - **Buttress dams** include the best features of arch dams, with buttress structures providing additional support.

- Dams have a variety of stated purposes, including irrigation water supply, flood control, municipal and industrial water supply, hydropower, and recreation.
  - Flood control with dams has had mixed results in some locations but is one of the common desired functions. China's **Yellow River** is called "China's Sorrow" because it has killed more people than any other river. The landscape and soils of the Loess Plateau allow the river to change course and build natural levees higher than the surrounding landscape. The damming of the Yellow River is a continuing lesson in how difficult it can be to tame a river.
  - Some **reservoirs** created by dams provide water supply for municipalities and agriculture. They also provide **recreational** opportunities for boating, swimming, and fishing.
  - **Hydroelectric dams** provide energy in much of the world. The accelerated growth of dam building in many countries has consequences in terms of human welfare and river ecology. Hydroelectric dams are considered clean renewable energy producers. Studies showing methane releases from reservoirs adds a caveat to that assumption.

- Some of the consequences of dam building are unintended but not unexpected if one studies river ecology. For example:
  - Water released from reservoirs for **flow maintenance** is typically low in dissolved oxygen (DO). The water may improve flow and water quality in the river by for example diluting municipal effluents added to it, but may harm aquatic organisms by limiting DO causing fish kills.
  - **Flood control** disturbs the natural cycle of flow levels in the river.
  - Less **sediment** in released water picks up sediment from the river channel removing **habitat** and **nutrients**.

- Reservoirs are thermally layered like lakes. The released water can cause thermal shock downstream.
- Low DO levels in reservoirs also allow **methylmercury** to bioaccumulate. Biomagnification through the food chain can result in mercury levels in fish that are harmful to humans if consumed on a regular basis.

- The **removal of existing dams**, in some parts of the world, is becoming more common as age and usefulness of specific dams are examined in light of environmental quality issues and often in the case of dam owners the high cost of renovating dams to meet regulated standards.
  - The **Rhine River** restoration project provides excellent examples of the complexity of river restoration after centuries of abuse.
  - The **Baraboo River** in Wisconsin is an example of successful dam removal and river restoration.

Questions for Analysis

1. What part did dams play in the formation of early civilizations?
2. What are some of the different types of dams? Why did engineers develop different types of dams?
3. Beavers are a keystone species. What does this mean? Which species characteristics epitomize the keystone species concept, and why?
4. How are dams used to enhance inland navigation?
5. How do dams effect river ecology?
6. When is the removal of a dam a good solution to a local water resources problem? What problems does dam removal solve? What problems can dam removal cause? Who should have a say in making this type of decision and why?
7. Does ecological degradation have to be the inevitable consequence of dam development? What can be done in planning a dam project to minimize damage to the river, its inhabitants, the surrounding landscapes, and wildlife?
8. The issues of displaced peoples are complex. Worldwide, an estimated 80 million people have been displaced, and many end up living in poverty, marginalized by projects that promised a better life. What can be done or should be done for these people? Is there a social responsibility to human welfare in planning projects? Can dam builders be held responsible for unkept promises made to displaced families, and if so, how?

Further Reading

McCully, Patrick, 1996, *Silenced Rivers: The Ecology and Politics of Large Dams*, London: Zed Books.

Outwater, Alice, 1996, *Water: A Natural History*, New York: HarperCollins.

Smith, Norman, 1972, *A History of Dams*, Secaucus, N.J.: Citadel Press.

Worster, Donald, 1985, *Rivers of Empire: Water, Aridity, and the Growth of the American West*, New York: Pantheon Books.

References

1 Gregg R. Hennessey, 1987, "Book review of Rivers of Empire: Water, aridity, and the growth of the American West, author Donald Worster," *Journal of San Diego History* 33, 4.

2 World Commission on Dams, "Dams and development: A new framework for decision-making," 16 November 2000, p. 102, available at: https://pubs.iied.org/sites/default/files/pdfs/migrate/9126IIED.pdf (accessed 2021).

3 G. Grill, B. Lehner and M. Thieme et al., 2019, "Mapping the world's free-flowing rivers," *Nature* 569, 215–221, www.nature.com/articles/s41586-019-1111-9 (accessed 2021).

4 World Commission on Dams, "China," www.dams.org/kbase/studies/cn/cn_exec.htm (accessed February 2007).

5 International Rivers program report: "China," www.internationalrivers.org/programs/china (accessed 2020).

6 By Source file: Le Grand Portage Derivative work: Rehman – File:Three_Gorges_Dam,_Yangtze_River, _China.jpg, CC BY 2.0, https://commons.wikimedia.org/w/index.php?curid=11425004

7 Alice Outwater, 1996, *Water: A Natural History*, New York: HarperCollins, p. 20.

8 Donald Worster, 1985, *Rivers of Empire: Water, Aridity, and the Growth of the American West*, New York: Pantheon Books, pp. 23–24 and 29.

9 Karl A. Wittfogel, 1957, *Oriental Despotism: A Comparative Study of Total Power*, New Haven, Conn.: Yale University Press, p. xix; and WaterHistory.org, "Karl August Wittfogel," www.waterhistory.org/histories/wittfogel/ (accessed February 2007).

10 Patrick McCully, 1996, *Silenced Rivers: The Ecology and Politics of Large Dams*, London: Zed Books, p. 12.

11 Norman Smith, 1972, *A History of Dams*, Secaucus, N.J.: Citadel Press, p. 1.

12 Mohamed Bazza, 2006, "Overview of the history of water resources and irrigation management in the Near East region," in *1st International Water Association International Symposium on Water and Wastewater Technologies in Ancient Civilizations*, under the Aegis of EU, EUREAU, FAO, UNESCO, and Association of Greek Municipalities, Iraklio, Greece.

13 USDA NRCS Massachusetts, 2014, "Sudbury-Assabet-Concord (SuAsCo) rivers watershed dams," www.nrcs.usda.gov/wps/portal/nrcs/detailfull/ma/programs/planning/wpfp/?cid=nrcs144p2_015977 (accessed 2020).

14 Patrick McCully, 1996, *Silenced Rivers: The Ecology and Politics of Large Dams*, London: Zed Books, p. 12.

15 Mohamed Bazza, 2006, "Overview of the history of water resources and irrigation management in the Near "East region," in 1st International Water Association International Symposium on Water and Wastewater Technologies in Ancient Civilizations, under the Aegis of EU, EUREAU, FAO, UNESCO, and Association of Greek Municipalities, Iraklio, Greece.

16 NOVA Online, 2007, "Flood!" February, www.pbs.org/wgbh/nova/flood/deluge.html (accessed 2020).

17 Guangqian Wang, Baosheng Wu, and Zhao-Yin Wang, 2005, "Sedimentation problems and management strategies of Sanmenxia Reservoir, Yellow River, China," first published September 21, 2005, https://doi.org/10.1029/2004WR003919 (accessed 2020 at https://agupubs.onlinelibrary.wiley.com/doi/full/10.1029/2004WR003919

18 Power Technology, "Xiaolangdi multipurpose dam project," www.power-technology.com/projects/xiaolangdi/ (accessed 2020).

19 Lee W. Larson, 1996, "The Great USA Flood of 1993," in *International Association of Hydrological Sciences Conference Destructive Water: Water-Caused Natural Disasters – Their Abatement and Control*, Anaheim, California.

20 US Federal Emergency Management Agency, "Benefits of dams," www.fema.gov/hazard/damfailure/benefits.shtm (accessed November 2008).

21 Australian Government, 2007, Water Act 2007 Act No. 137 of 2007 as amended, "Murray–Darling Basin Agreement," www.environment.gov.au/minister/env/2003/mr14nov03.html (accessed 2020).

22 Ibid.

23 National Water Commission, 2007, *Australian Water Resources 2005: A Baseline Assessment of Water Resources for the National Water Initiative*, Level 1 Assessment, http://inform.regionalaustralia.org.au/industry/electricity-gas-and-water/item/australian-water-resources-2005-a-baseline-assessment-of-water-resources-for-the-national-water-initiative-level-1-assessment (accessed 2020).

24 Mac Kirby, and Murray–Darling Basin Commission (Australia), 2006, *The shared water resources of the Murray–Darling Basin / Mac Kirby ... [et al.]* Murray–Darling Basin Commission Canberra 2006, www.mdbc.gov.au/__data/page/1130/CSIRO_Part_1_shared_water_resources.pdf (accessed 2020 at https://catalogue.nla.gov.au/Record/3887833).

25 A. Van Dijk, R. Evans, P. Hairsine et al., 2006, *Risks to the Shared Water Resources of the Murray–Darling Basin*, MDBC Publication 22/06, Canberra: Murray–Darling Basin Commission.

26 New South Wales Storage Reports, http://waterinfo.nsw.gov.au/StorageSummary.html

27 Commonwealth of Australia, 2007, *Murray–Darling Basin Dry Inflow Contingency Planning Overview*, Report to First Ministers, May 2007, www.environment.gov.au/water/publications/mdb/pubs/dry-inflow-planning-may07.pdf

28 Murray–Darling Basin Commission, "The Murray–Darling Basin Agreement," http://mdbc.gov.au/about/the_mdbc_agreement/

29 US Federal Emergency Management Agency (FEMA), "Benefits of dams," www.fema.gov/hazard/damfailure/benefits.shtm, January 2007 (2019 last update), www.fema.gov/benefits-dams (accessed 2020).

30 Bridget R. Deemer, John A. Harrison, Siyue Li et al., 2016, "Greenhouse gas emissions from reservoir water surfaces: A new global synthesis," *BioScience* 66, 11, 1 November, 949–964, https://doi.org/10.1093/biosci/biw117 (accessed 2020).

31 PBS, 2007 "Wonders of the World," February, www.pbs.org/wgbh/buildingbig/wonder/structure/itaipu.html, February, www.pbs.org/wgbh/buildingbig/wonder/structure/browse.html (accessed 2020).

32 Tennessee Valley Authority, "Navigation on the Tennessee River," www.tva.com/Environment/Managing-the-River/Navigation-on-the-Tennessee-River (accessed 2020).

33 Maryland Department of Natural Resources, "Freshwater benthic macroinvertebrates," www.dnr.state.md.us/streams/pubs/freshwater.html (accessed 2020 at https://dnr.maryland.gov/streams/Pages/streamLife.aspx).

34 Patrick McCully, 1996, *Silenced Rivers: The Ecology and Politics of Large Dams*, London: Zed Books, p. 12.

35 J. G. Wiener, D. P. Krabbenhoft, G. H. Heinz, and A. M. Scheuhammer, 2002, "Ecotoxicology of mercury," in *Handbook of Ecotoxicology*, 2nd ed, D. J. Hoffman, B. A. Rattner, G. A. Burton, Jr., and J. Cairns, Jr., eds., Boca Raton, Fla.: CRC Press, pp. 407–461.

36 Ibid.

37 New York State Department of Environmental Conservation, "Mercury in fish and wildlife," www.dec.ny.gov/chemical/8517.html (accessed 2020).

38 Martin Risch and Amanda L. Fredericksen, 2015, *Mercury and Methylmercury in Reservoirs in Indiana*, professional paper 1813 https://pubs.er.usgs.gov/publication/pp1813 (accessed PDF download 2020).

39 California Water Boards, 2016, "Statewide mercury control program for reservoirs," www.waterboards.ca.gov/water_issues/programs/mercury/reservoirs/ (accessed 2020).

40 Hydro-Québec, "Research and agreements on mercury," www.hydroquebec.com/sustainable-development/specialized-documentation/mercury.html (accessed 2020).

41 Ibid.

42 US Geological Survey, "A brief history and summary of the effects of river engineering and dams on the Mississippi river system and delta," https://pubs.usgs.gov/circ/1375/C1375.pdf (accessed 2020).

43 Ibid.

44 P. H. Nienhuis, A. D. Buijse, R. S. E. W. Leuven et al., 2002, "Ecological rehabilitation of the lowland basin of the river Rhine (NW Europe)," *Hydrobiologia* 478, 1–3, 53–72.

45 University of Wisconsin, "What's the dam problem?" http://whyfiles.org/169dam_remove/index.html (accessed 2020 from archives, this truly excellent, award-winning, factual source of information from the University of Wisconsin is now only available as archived files).

46 Ibid.

11 Drinking Water and Wastewater Treatment

Water is fundamental for a life of human dignity. It is prerequisite to the realization of all other human rights.

World Health Organization [1]

Chapter Outline

11.1 Introduction

The fact that water is fundamental to a life of human dignity is so obvious that you would surmise that we would not have to repeat it so often in this text. Looking at the history, how different societies through time, approached obtaining, transporting, and disposal of water, sounds terribly dull. It is not. In fact, it is fascinating to see how far humans can develop and then fall backwards; how we can get some things very right while not seeing the total picture. Sometimes we lack the fundamental knowledge to understand complex systems, sometimes we just don't make the right connections, and sometimes current thinking gets in the way of progress. Read on.

11.2 Early Drinking Water Treatment

Even though they had no knowledge of the microbial world, as early as 2000 BCE the ancient Greeks filtered water through charcoal, sand, and gravel to improve its drinkability. Sunlight, boiling, placing hot instruments in the water before drinking, and straining were also used to improve taste. Ancient Hindu culture directed that foul-tasting water be either boiled, exposed to sunlight, or dipped with a hot piece of copper seven times, filtered, and then allowed to cool in an earthen vessel [2]. Many religious restrictions regarding food and drink actually did protect the believer's health. Ancient societies generally assumed that good-tasting water was also clean since no scientific connection existed between impure water and disease.

Water treatment methods are depicted in the tombs of the Egyptian rulers Amenhotep II (1447–1420 BCE) and Ramses II (1300–1223 BCE). Painting on the tomb walls show servants removing sediments, using a settling device or wick siphon, in an Egyptian kitchen. Later, Egyptians tried to remove suspended solid materials by adding alum (aluminum sulfate) [3]. Alum works as a flocculent, clumping particulates together and creating a larger, faster-falling mass. When this mass settles, it leaves clearer water to be decanted. As mentioned earlier, the Greeks also treated water, and created extensive water delivery systems. A famous hydraulic work of Greek engineers was the ancient aqueduct of Samos. It was constructed around 500 BCE on Samos, an island in the Aegean Sea off the coast of Turkey. The delivery structure was a 1,000-meter (3,400-feet)-long water tunnel (also known as the Tunnel of Eupalinos – named for the engineer in charge of the project).

The ancient city of Athens had a population of over 200,000 people around 500 BCE. When its population was smaller, natural springs, groundwater wells, and storm water provided adequate water supplies. But as the city grew, an extensive system of public wells, fountains, and springs were developed, and a primitive water distribution system was constructed with ceramic tiles buried beneath streets. Similar tile systems were used for sewers. A public administrator, called the Officer of Fountains, was appointed to operate, and maintain the city's water system, and to enforce the fair distribution of water. Guards were posted at public springs and fountains to ensure order. In 333 BCE, Pytheus was awarded a gold wreath for his work in restoring and maintaining numerous fountains.

Around this same time, the Greek physician Hippocrates (460–377 BCE), also known as the Father of Medicine, devised a cloth bag or strainer (called a "Hippocrates' sleeve"), usually made of flannel or linen and sewn into a cone shape. Sediment was removed by pouring boiled water through the sleeve to remove sediments. The cloth trapped sediments that caused bad taste or smell.

Ancient Rome was famous for its drinking water supply and wastewater disposal systems. Originally, Romans relied on the Tiber River, groundwater springs and wells for their water supply, but these became too polluted or inadequate to meet increasing needs [4]. Aqueducts were subsequently constructed from distant springs and rivers to deliver cleaner water to public fountains in central locations around the city. Around 300 BCE, Roman engineers constructed water aqueducts to transport daily as much as 492,000 cubic meters (130 million gallons, or 400 acre-feet) to the capital city of Rome. There, residents could fill containers with freshwater any time of the day or night.

Although direct water treatment was not provided, sunlight provided incidental treatment along the unenclosed aqueduct. Figure 11.1 is the Pont du Gard, an aqueduct built by the Romans about 19 BCE near Nîmes, France. At the end of the first century CE, Sextus Julius Frontinus (35–103 CE) was the water commissioner of Rome and wrote an account of the water system for Emperor Nerva called *De aquae ductu Urbis Romae* (The Aqueducts of Rome). In it, Frontinus describes, in proud detail, the size, length, and functions of the various Roman aqueducts.

During the time of Emperor Marcus Cocceius Nerva (30–98 CE), nine aqueducts conveyed water to Rome to 600 *lacus*, or major delivery points. Underground storage tanks, called *cisterns*, were also used to save water for later use. However, a common problem was water theft. Landowners along the aqueduct often tapped into the conduits to divert some of the public water supplies in the aqueducts to irrigate their private gardens or fields (Frontinus called these illegal acts "puncturing").

Fig 11.1 The Pont du Gard, a Roman aqueduct at Nîmes, France, built about 19 BCE is an UNESCO World Heritage site. Notice the beauty and complexity of the architectural details and functionality achieved without modern technology. One negative note, Roman's economic backbone was considered to be slavery, with about one in five individuals in the Empire, living as slaves with no human rights. They were from all walks of life – from agricultural workers to educators, and even many skilled in medicine [5] (Getty image credit BessieB)

Illegal diversions were often made from the aqueduct through hidden lead pipes which were buried beneath city streets and extended into private homes [6].

The aqueduct of Segovia, Spain, is another dramatic example of Roman engineering. Its two tiers of arches were built around 50 CE in central Spain and are remarkably well preserved today. The Segovia aqueduct is 813 meters (2,667 feet) long and 28 meters (92 feet) high, constructed of granite rocks, contains no mortar or concrete, and has 166 arches and 120 pillars set at two levels. The aqueduct obtained its water supply from the Spring Fuenfría, some 18 kilometers (11 miles) from Segovia. Sand naturally settled out of the flowing water as it made its way toward the center of town. The gently sloping yet dramatic aqueduct was declared a UNESCO World Heritage Site in 1985.

Evidence exists that the ancient Mayan civilization also created extensive aqueduct technology on the North American continent prior to 500 BCE. These systems included aqueducts, dams, reservoirs, and canals. They have been documented in the ancient cities of Varaçol, Tikal, and Palenque [7].

Aqueducts sometimes terminated in cisterns. Figure 11.2 is the Theodosius Cistern built beneath Istanbul, Turkey, in 428–443 CE. The cistern is about 25 by 45 meters (80 by 150 feet) and the roof is supported by 32 marble columns, each about 9 meters (30 feet) high.

Water taste was an important issue in these ancient water systems, and experiments on odor and taste continued. In the first century CE, the Greek agricultural writer Diophanes of Nicaea

Fig 11.2 The Theodosius Cistern is a Roman cistern built under the city of Istanbul, Turkey. The area is about 45 by 25 meters and the roof is supported by 32 marble columns about 9 meters high (Photograph by Roger W. Haworth at https://en.wikipedia.org/)

recommended placing macerated (made soft by soaking or steeping in a liquid) laurel in rainwater. (Laurel ranked high as a healing plant and was probably equivalent to a bay leaf in today's culinary and herbal medicine circles.) Diophanes' advice was sound since bay leaf contains eugenol – a chemical that breaks up cell membranes much like soap. Eugenol killed microbes in water if used in sufficient quantities. Not to be outdone, the Roman writer Paxamus suggested placing a bag of pounded barley, or bruised coral, into foul water to improve its taste.

The fall of Rome, in the fifth century CE, ended water treatment advancements in Europe for the next 1,000 years. Water purification experiments were uncommon during the Middle Ages due to the lack of scientific and cultural improvements. Sickness and plague – spread by fleas, mosquitoes, excrement, and filthy drinking water – were horrific during this bleak period of human history [8].

Fortunately, the search for clean drinking water did not completely stop during the Dark Ages of Europe. In the eighth century CE, the Arabian alchemist Geber wrote a dissertation on water distillation that suggested the use of a wick siphon to remove "spirits." In the eleventh century CE, the Persian physician and writer Avicenna recommended that water should be strained through a cloth, or boiled, before drinking to prevent illness. He suggested that minute animals lived in water – too small to be seen by the human eye – and could cause illness. Avicenna was narrowing in on the concept of bacteria and waterborne pathogens [9], but he needed a microscope to confirm his medical theory. However, that technology would not be invented until 600 years later in Holland.

Beginning in the late 1400s CE, the Middle Ages gave way to the Renaissance. Scientific and intellectual stagnation ended as new inventions and innovations sprang to life. This was the "Age of Discovery." Contaminated drinking water was still a major cause of illness and death around the world. Fortunately, two unrelated inventions – the microscope and multiple water filters – provided tools to improve human health.

11.3 Discovery of the Microscope

In 1595, two Dutch spectacle makers, Zacharias Janssen (1580–1638) and his father, Hans, experimented with glass lenses in a tube, and are believed to be the first to invent the compound microscope. Eyeglasses were becoming popular, and the Janssens made glass lenses for customers in Middleburg, Holland. Through their work, Zacharias and his father discovered that objects could be magnified if lenses were placed at opposite ends of two tubes which could slide within one another. The device was about 0.8 meters (2.5 feet) long when fully extended, and like a modern telescope could be lengthened or shortened. The handheld Janssen "drawtube microscope" could enlarge an image up to three times when fully closed, and up to nine times in size when the tubes were fully slid apart. The rudimentary microscope opened the invisible world of microbes (bacteria) and led to their discovery a century later.

In 1609, Galileo Galilei changed the curvature of the Janssen microscope and became the first person to observe the skies through a telescope. Robert Hooke (1635–1703) was Curator of Experiments of the Royal Society of London and improved the design of the Janssen compound microscope to magnify objects up to 30 times their actual size. In 1665, Hooke published *Micrographia*, with detailed drawings of objects, including the minute detail of a magnified slice of cork, which he called "cells," viewed through the microscope. Hooke also viewed tiny, living organisms in a drop of water – theorized by Avicenna in Persia 600 years earlier. Hooke's book was a bestseller and helped guide scientists to the discovery of bacteria.

After reading *Micrographia*, Antony van Leeuwenhoek (1632–1723) became fascinated with microscopy. The Dutch textile merchant created hundreds of microscopes and achieved magnifications of items over 200 times their normal size, and with clarity and brightness. His instruments had a single lens and were more like magnifying glasses than microscopes. Leeuwenhoek's talent was in his careful grinding of curved lenses, his polishing process, and excellent light manipulation. He could place a sample on the point of his instrument and increase magnification by turning two screws (see Figure 11.3). The scope was only 8–10 centimeters (3–4 inches) in length and was held close to the eye. Leeuwenhoek was able to study everything from water to blood, cork, optic nerves, seeds, mouth scrapings, even a parasite inside an aphid.

Leeuwenhoek spent 50 years observing and meticulously recording observations and experiments with his tiny microscope. He also sent letters, along with preserved samples, to the Royal Society in London to report his findings. Remarkably, in 1981, Leeuwenhoek's specimens were found in the Society's archives by Brian Ford, a Fellow of Cardiff University. Ford evaluated the samples using modern microscopy techniques and found that Leeuwenhoek's work was

Fig 11.3 The Leeuwenhoek microscope looks nothing like a modern microscope; yet with care and skill, he was able to make accurate drawings of everything from blood to seeds, and is credited as being the first to see bacteria (Getty image credit: Tetra Images Creative)

both accurate and meticulous, in execution and accuracy of description. Brian Ford praised Leeuwenhoek's 300-year-old work: "His [Leeuwenhoek's] delight in making new observations, in piecing together new levels of biological understanding, has given him a unique role in the formation of a scientific approach at the very dawn of the discipline" [10].

In his letters to the Royal Society of London, Leeuwenhoek described substances removed from his own teeth as having "many very little, very prettily a-moving animalcules." He also explained the substance's structure and movements – the first recorded discovery of bacteria [11]. Leeuwenhoek's work opened the door for others to understand the connection between water, bacteria, and disease.

11.4 Epidemics and the Microscope

Until the 1800s, most believed that illness and disease were caused either by evil spirits or by bad air known as *miasma* (a Greek word meaning "pollution"). Epidemics were attributed to poisons from bad air originating in the "bowels of the Earth." (The bad air of Paris was such a serious health concern, during the summer months, that it was one of the reasons Versailles was used as a retreat for King Louis XIV.) As another example, the miasma of Washington, DC, was cited numerous times as a major problem when the site was considered as the national capital back in the late 1700s.

In contrast, Louis Pasteur (1822–95), the famous French scientist (see Figure 11.4), and a few others, believed that germs existed and caused disease, not miasma. However, most scientists and medical professionals did not agree. In 1854, **Dr. John Snow** (1813–58) witnessed another deadly outbreak of cholera in his London neighborhood. Snow was convinced that germs in the

Fig 11.4 French microbiologist and chemist, Louis Pasteur, helped solve the mysteries of how diseases are spread

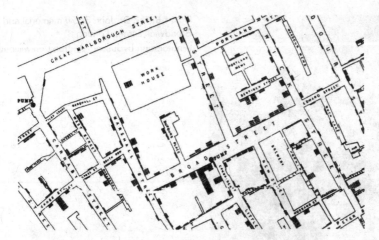

Fig 11.5 A portion of John Snow's map of the 1854 cholera epidemic in central London. He was a meticulous researcher and kept detailed records

public drinking water supply caused the disease. He was almost alone in his theory since most blamed miasma.

Snow used his skills as a doctor and a detective, and a microscope, to determine the source of the cholera epidemic. During the epidemic of 1854, Snow walked around his London neighborhood, interviewed families of the victims, and noted the location of deaths on a street map (see Figure 11.5). He discovered that most victims died within a short distance of the community water pump on Broad Street, where they obtained drinking water. By contrast, no deaths occurred among the 70 workers at the nearby Broad Street brewery. This did not surprise Snow because brewery workers never drank water from the Broad Street pump. Instead, they were given a daily allowance of free beer (made with water that was boiled during the brewing process).

Two deaths didn't follow the pattern, however, and Snow was puzzled. He travelled some distance to visit the family and found that the victims drank water from the Broad Street pump because they preferred its taste to their local water source. A servant had used a cart to haul back a large container of water, every day for many months, to the now deceased victims.

Snow used a microscope to confirm the presence of cholera bacteria in the drinking water from the Broad Street pump and found white particles. Snow plotted the deaths on a map, and their locations resembled a well-used rifle target with the bull's-eye at the Broad Street pump. Snow convinced the city to inspect the well for weaknesses or breaks that would allow contamination. City inspectors found the well to be solid, and the local sewer line to be lower than the well, and some 3 meters (10 feet) away. They again discounted his theories, but Snow did not give up.

Dr. Snow learned that a baby died in a home directly behind the Broad Street pump. The child's mother was interviewed, and it was learned that she cleaned the child's dirty diapers – and dumped the waste – into a drain going into a cesspool 1 meter (3 feet) from the well. The whole system was examined and found that the outside well bricks were cracked, and waste from the stopped-up cesspool drained into the well. The source of contamination was revealed [12].

Today, Snow's old London neighborhood contains two tributes to his work, and both are located on Lexington and Broadwick Streets (the street name was later changed from "Broad"). The first is a replica of the Broad Street pump – minus its handle (see Figure 11.6). The second is the "John Snow" pub on Broadwick Street, in the Soho neighborhood of London near the Piccadilly Circus station on the Tube (subway). It contains various photographs and is at the site of the historic pump; a marker outside the pub designates its location. All are tributes to Dr. John Snow, the Father of Epidemiology (see Figure 11.7) [13].

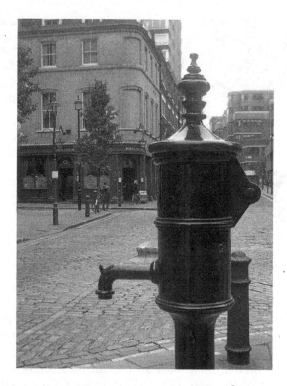

Fig 11.6 The Dr. John Snow memorial and pub, at Broadwick Street, London
(Photograph by Justinc at https://en.wikipedia.org/)

Fig 11.7 Dr. John Snow, 1857, the Father of Modern Epidemiology

In 1849, M. Gabriel Pouchet had found cholera bacteria, *Vibrio cholerae*, in the stools of cholera patients, but did not recognize their significance. An Italian anatomist, Filippo Pacini, published his "Microscopical observations and pathological deductions on cholera" in 1854, with four subsequent publications. These all correctly proved cholera was caused by bacteria and he also correctly identified treatments. These were all, unfortunately, ignored by the scientific community of that time.

Between 1849 and 1861, over a million people died from cholera in Russia and Europe [14]. The Home Secretary of the United Kingdom, Lord Palmerston, was aware of the environment that led to cholera. When he was asked to appoint a national day of fasting, humiliation, and prayer to stop the cholera epidemic, he suggested that "it would be more beneficial to feed the poor, cleanse the cesspools, ventilate the houses, and remove the sources of contagion" [15]. Lord Palmerston's words are still true today.

Robert Koch rediscovered the bacteria in 1884 and again the work was published and ridiculed by members of the scientific community. Even though different scientists had suspected living causative agents of disease before the Common Era, the germ theory of disease was not accepted until Koch published his postulates in 1890 [16]. Today's technology – such as scanning electron microscopy – allows us to see the *Vibrio cholerae* bacteria (see Figure 11.8). Snow's work and theories were vindicated, and his techniques are still used by modern epidemiologists [17]. Lord Palmerston was also correct concerning nutrition and cleanliness as basic to good health.

Parallel to the discovery of the microscope, inventors were experimenting with a wide range of water purification methods. Sir Francis Bacon (1561–1626), a British scientist and philosopher, experimented with filtration, boiling, coagulation, and percolation to purify ocean water. Scientists had long attempted to make saltwater into freshwater since ocean-going vessels spent enormous time and resources hunting for freshwater on voyages. Bacon had a simple hypothesis – if a hole was dug along an ocean shoreline, the heavier sand particles would prevent salt from passing upward into the excavated hole on the beach. If successful, sailors would be able to obtain drinkable freshwater from pools created along beaches. We know that Bacon's concept was wrong, but the idea led others to experiment with sand filtration methods on freshwater to improve taste and odor.

In 1685, one year after Leeuwenhoek created the microscope in Holland, the Italian physician, Luc Antonio Porzio, published the first known details of a multiple sand filtration method. Porzio worked on sanitation methods for soldiers fighting in the Austro-Turkish War of

Fig 11.8 Scanning electron micrograph of *Vibrio cholerae* bacteria; the cholera mystery was solved without the use of electron microscopes, but scientists today are technologically better equipped for the search (Getty image credit: Callista Images)

1685. He proposed filtering water through sand to provide clean drinking water to troops. Porzio's work was printed in the soldier's *vade mecum* (a pocket reference, from Latin meaning "go with me"). It was probably the first published work on mass sanitation and advanced the interest in filtered drinking water for communities.

About 20 years later, around 1703, the French scientist Philippe de la Hire (1640–1718) recommended to the French Academy of Sciences that every household should have a sand filter and cistern to capture rainwater. He believed that filtered water could be stored for years if it was not mixed with salts. La Hire also believed that everyone had the basic human right to clean drinking water. He wrote extensively on the subject and helped encourage the construction of municipal water filtering plants. (Sadly, over 300 years later, families in developing countries are still waiting for clean water and sanitation.) In 1746, another Frenchman, Joseph Amy, was granted the first patent for a water filter design composed of sponges and sand. In 1791, British architect James Peacock patented a three-tank backwash water filter.

In 1804, Paisley, Scotland, became the first town to deliver filtered water to everyone within its urban limits. Filtered water was distributed to households by horse and cart. Robert Thom engineered the slow sand and gravel filtering system for the town. Only three years later, the system was extended, and treated water was delivered through pipes to the residents of neighboring Glasgow. The system relied on a bed of sand and gravel to block passage of waterborne contaminants. On the downside, these filtering beds were large, and quickly filled with sediments and other constraining materials. This required extensive and frequent cleaning or replacement, which required shovels and much physical labor.

Why Scotland? It was no coincidence that the first municipal water treatment plant was designed and located in Scotland. The period of the Scottish Enlightenment, from about 1707 to 1800, was a great time in Scottish society, and generated unbelievable development in science, literature, culture, and human rights. Education has been compulsory in Scotland since 1496 and shows the value Scottish society places on education. Influential thinkers and innovators helped develop Scottish society and much of the world during this period. In addition, Scottish universities were strong, intellectual institutions. It was not by accident that Paisley and Glasgow, Scotland, became the first cities in the world to provide filtered water to all its residents. As noted by Sir Winston Churchill, Prime Minister of the United Kingdom, "Of all the small nations of this earth, perhaps only the ancient Greeks surpass the Scots in their contribution to mankind."

Two years later, in 1806, a similar filtered water treatment system began operation in Paris. Seine River water was allowed to settle in basins for 12 hours before it entered the filtering process (coarse river sand, clean sand, and pounded charcoal which was changed every six hours). In 1827, the English scientist, James Simpson, created a similar design to service a portion of London. The slow sand filtration process was quickly adopted by larger cities throughout Europe and the world.

In the 1800s, many sand filtering experiments were conducted in England, France, Germany, Russia, and the US. The first **slow sand filtration system** in the US was built in Richmond, Virginia, in 1832, and had 295 water subscribers. The next system was in Elizabeth, New Jersey, in 1855, one year after the cholera outbreak in London.

Back in Europe, a network of public health councils was created in France in 1849, the same year a cholera epidemic killed 5,000 in New York City. In 1852, the Metropolitan Water Act of London was passed, two years before the Broad Street cholera epidemic. The law required that all drinking water be filtered in the London area, and was one of the first government efforts to regulate a drinking water supply system. The need for better, more effective, and efficient water filters was rapidly increasing.

In 1856, Henry Darcy patented water filters in France and England. Darcy also worked on **rapid sand filtration systems** in America. This new filtering process allowed more water to flow through filters, and utilized backwashes, or water jets, and mechanical agitators to clean the sand and gravel filters [18]. The slow-filtering method, however, used sand and gravel beds and required extensive land areas to meet the water needs of large European cities. In 1849, London treated 166,600 cubic meters (44 million gallons, or 135 acre-feet) of water daily, and which required 5 hectares (12 acres) of land for the filtering bed area. By 1901, London's water needs exceeded 757,100 cubic meters (200 million gallons per day, or 614 acre-feet), and the land area required was not available. Rapid sand filtration soon spread to the US, with cities in Massachusetts using this type of process in the 1870s [19].

Going back, remember it was 1876 when the Germ Theory of Disease was verified when the German physician and microbiologist Robert Koch discovered the actual bacteria that caused cholera – *Vibrio cholerae*. Eighteen years later, in 1894, the American Public Health Association began work on a standardization of bacteriological testing so that results from various water laboratories could be compared. This work led to the Standard Methods of water sampling used today in the US. From 1900 to 1913, the typhoid death rate in the US dropped by more than 55 percent, along with significant drops in cholera, as more people were served by filtered drinking water systems [20].

11.5 Wastewater Treatment

In addition to the discovery of the microscope and the improvement of sand filtration methods, chlorine treatment has revolutionized the safety of drinking water. In 1774, the Swedish pharmacist, Karl Wilhelm Scheele, placed a few drops of hydrochloric acid onto a piece of manganese dioxide. A greenish yellow gas was emitted – chlorine. Several decades later, the English chemist, Sir Humphry Davy, recognized chlorine gas as an element, and in 1810 suggested the name from the Greek word *khloros*, for greenish-yellow, and called it "chloric gas" or "chlorine" [21].

Although drinking water filtration has provided enormous health benefits over the past 150 years, the use of chlorine has accounted for even more. In fact, the use of chlorine in the treatment of drinking water has been so effective that in 1997 *Life* magazine named water chlorination "probably the most significant public health advance of the millennia." In the US, municipal water treatment systems have used chlorine for over 100 years and is used by 98 percent of all water treatment facilities [22].

Chlorine was first used to purify water in England in the 1850s, where Dr. John Snow used chlorine as a water disinfectant at the Broad Street pump. In 1897, Sims Woodhead used a chlorine bleach solution as a temporary measure to sterilize the water distribution pipeline system in Maidstone, Kent (England), after an outbreak of typhoid fever. The great success of chlorine treatment – in drinking water to prevent cholera – led to its quick adoption in 1908 in Jersey City, New Jersey, and then throughout America. Chlorination, along with sand filtration systems, virtually eliminated waterborne diseases such as cholera, typhoid, hepatitis A, and dysentery (a disease of the intestinal tract, generally caused by bacteria or protozoa infection, and characterized by fever and severe diarrhea).

Chlorine also provides residual protection within water delivery system pipes and prevents growth of germs in the distribution system. In addition, chlorine not only eliminates germs that cause waterborne diseases, it also helps control odor and taste problems from naturally occurring substances, e.g., algae secretions and decaying vegetation. The filtration and disinfection of drinking

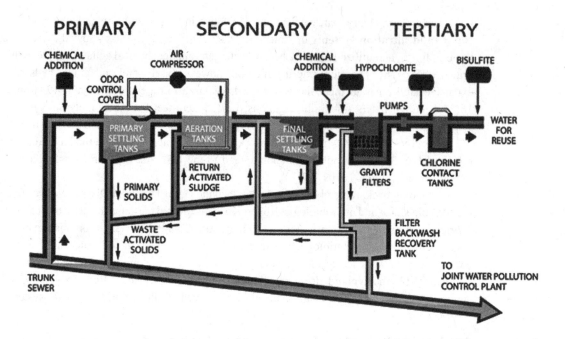

Fig 11.9 Modern wastewater treatment is a combination of the techniques discussed in this chapter. Primary treatment is a physical process that basically settles out large solids. Secondary treatment can involve biofiltration through sand filters, contact filters, or trickling filters, aeration treatment with aerobic microbes, or oxidation ponds where natural microbial reactions purify the sludge.

This diagram ends with disinfection using chlorine (sodium hypochlorite NaClO, a concentrated liquid bleach) treatment then sodium bisulfite ($NaHSO_3$) to remove chloramines (NH_2Cl, $NHCl_2$, NCl_3) formed during disinfection that can harm aquatic organisms. We will cover some alternative techniques used in many cities with technologically advanced systems in Figure 11.12

water with chlorine has been largely responsible for a 50 percent increase in life expectancy in the US, during the twentieth century. Figure 11.9 is a diagram showing the modern wastewater treatment process. Although there are difficulties with chlorination byproducts, the World Health Organization (WHO) recognizes chlorine as the best guarantee for providing microbiologically safe water which is the primary goal [23].

11.5.1 Ozone Treatment

Ozone treatment has greatly improved drinking water safety, primarily in more-developed countries of the world. Ozone (O_3) is found naturally in the atmosphere as a colorless gas, but with a very pungent odor. (The ozone layer in the Earth's upper atmosphere provides protection from dangerous solar radiation.) It is formed naturally during thunderstorms by the discharge of electricity, or any time an electrical discharge creates a spark.

In 1785, a Dutch scientist, Martin van Marum (1750–1837), who was Secretary of the Dutch Society of Sciences, is the first known scientist to record the presence of an unusual odor present in the air near an electrostatic machine (an electromagnetic device that produced "static electricity" by turning a crank by hand). When electric sparks, caused by friction, passed between metal collectors and inductor plates, a particular smell was created. Van Marum called it the "odor of electricity." In 1840, Christian Friedrich Schönbein (1799–1868), a professor of chemistry at the University of Basel in Switzerland, was conducting experiments by passing an electric charge through water. He named the gaseous substance "ozone" from the Greek word *ozein*, to smell.

The first drinking water treatment plant to use ozone as a disinfectant was in Oudshoorn, Holland, in 1893. The Bon Voyage water plant in Nice, France, has used ozone since 1906, and is often referred to as the "birthplace of ozonation for water treatment." This method of drinking water purification became more common in Western Europe, even though it was a more complex and expensive process than rapid sand filtration systems. Many preferred the ozone process over chlorine because chlorine gas was used as a chemical warfare agent in World War I. Use of ozone in the US began in the 1940s.

In Depth	Ozone

There are advantages and disadvantages to most water treatment methods. Below is a list of properties of ozone, along with advantages and disadvantages:

- Ozone is a flocculating agent (it causes small contaminants to stick together and then settle out of water). Ozone precipitates out oxidized iron, sulfur, and manganese, and other heavy metals, allowing them to be filtered. This can be very advantageous in areas with high heavy-metal concentrations.
- Ozone has a higher oxidation potential than chlorine, which makes it a more efficient disinfectant. It can oxidize all bacteria, mold, yeast spores, organic materials, and viruses. *Cryptosporidium parvum* and *Giardia lamblia* (common parasites) are very resistant to most chemical disinfectants but are effectively destroyed by ozonation.
- Ozone has a positive effect on chemical oxygen demand removal by breaking down refractory compounds (hard materials to break down) and making them biodegradable.
- Oxidizing organic materials removes both odor and color from water. Odors generally come from bacteria decomposing organic matter. Ozone oxidizes the organic material and kills the bacteria. Color from humic, fulvic, and tannic acids (all are acids from decomposed organic materials) is usually brown.
- Chlorination produces chlorine-containing organics – such as chloroform, carbon tetrachloride, chloromethane, and others, generally known as trihalomethanes (THMs). Ozone does not produce THMs, but does not remove them, either. These THMs have been implicated as carcinogens in kidney, bladder, and colon cancer.

The effectiveness – both in cost and water treatment of ozonation – has been demonstrated in many large cities around the world, including, in the US: Anaheim, Los Angeles, San Francisco, Oakland, suburban Philadelphia, New York City, Oklahoma City, Dallas, El Paso, Orlando, and Milwaukee. Mexico City, Barcelona, Singapore, Paris, Shanghai, and Zurich represent other large cities using ozone water treatment.

There are some disadvantages of ozone treatment:

- Operator skills must be high, and maintenance is critical – more so than with other methods.
- The systems must be professionally designed to maximize water exposure to the ozone.
- Byproducts must be properly handled.
- Ozone is highly corrosive and toxic; safety measures are necessary.
- Some form of chlorine residual is nearly always required in the distribution networks.
- Ozone does not last a long time at high temperatures and pH.

Los Angeles, California, had the world's largest water treatment plant using ozone (2.3 million cubic meters, 600 million gallons a day, or 1850 acre-feet). It is the most economical method to meet or exceed state and federal water quality laws for that city. Wylie, Texas, now claims one of the largest plants, sending 770 million gallons per day to 1.6 million people at an annual operating cost of less than a nickel per thousand gallons. It is run by the North Texas Municipal Water District (NTMWD) [24]. New plants are being built and old plants remodeled to include ozonation around the world.

11.6 Federal Protection of Drinking Water in the US

Modern treatment plants, such as this one (see Figure 11.10), use a combination of methods to treat wastewater. The US has had federal regulations regarding drinking water and sanitation for many years. In 1914, the US Department of the Treasury approved the first drinking water standards for bacteria. Coliform count was adopted as an indicator of fecal contamination in water, and the maximum level was 2 per 100 milliliters. By the 1920s, filtration and chlorination eliminated most waterborne diseases from US public drinking water systems. In 1925, the US bacteriological standard was lowered to 1 coliform per 100 milliliters, and standards were adopted for lead, copper, zinc, and excessive soluble minerals. In 1942, standards were adopted for lead, fluoride, arsenic, and selenium.

By 1942, the US Public Health Service implemented the first set of drinking water standards in the country. In 1974, the Safe Drinking Water Act was enacted to safeguard drinking water supplies and to protect the public health. It is the primary federal law in the US that regulates drinking water treatment. It was amended in 1986 and 1996 to provide source water protection, because contaminates can originate anywhere in a watershed.

In the US, health standards are established in a four-step process. First, a substance is determined to be harmful. This is done through recommendations from scientists and their research. The second step is to determine a maximum amount of the substance that is considered safe for human consumption. This number includes a large margin of safety to protect highly susceptible

Fig 11.10 Modern wastewater treatment plants, such as this one in Zhejiang, China, can not only be efficient but can also blend into landscapes with the careful planting of trees and shrubs (Getty image credit: sinology)

individuals, such as pregnant women, infants, and immune system-compromised individuals. The entire process takes time and is intended to be as scientifically defensible as possible. Next, an enforceable level for the substance is proposed. This may not be intuitively obvious, but unenforceable standards are meaningless. Last, after scientific review finds it effective and enforceable, the standard is adopted. The process seems straightforward, but each step requires time, money, and expertise. There are times when the public wants to move faster on some issues, but the importance of having a scientifically defensible standard cannot be over emphasized.

The US Environmental Protection Agency (USEPA) also oversees water suppliers to ensure certain standards are met. These oversight duties are usually given to individual state governments. The states, through departments involving water quality, set state standards and work in coordination with the USEPA. The USEPA retains the right to reject state recommendations that are considered to be in error, for whatever reason. The USEPA also sets health-based maximum contaminant levels (MCLs) to minimize disinfectant byproducts, and to reduce chemical and microbial contaminants in the nation's drinking water supply systems [25].

MCLs are the highest level of a contaminant that is allowed in drinking water. MCLs are set as close to maximum contamination level goals (MCLGs) as feasible – using the best available treatment technology and considering cost. MCLs are enforceable standards.

A MCLG is the level of a contaminant, in drinking water, below which there is no known or expected risk to human health. MCLGs allow for a margin of safety and are unenforceable public health goals. The margin of safety in MCLGs means the values are very conservative – and possibly not attainable – due to cost or a lack of technical knowledge. However, a margin of safety ensures that susceptible members of the population are protected.

MCLs are set for a wide range of contaminants including microorganisms, turbidity, chemicals such as disinfectants and disinfectant byproducts; inorganics ranging from arsenic, asbestos, lead, to nitrates; organic chemicals including pesticides, carbon tetrachloride from industry; petroleum hydrocarbons such as ethylene dibromide and ethylbenzene; and radionuclides, e.g., uranium and radium isotopes. Setting MCLs is an ongoing process – materials can be added to the lists as needed. There is a list of secondary contaminants that influence taste and smell of water, or may cause discolorations, but are not health risks. These are not enforced but recommended for better water quality.

11.7 Drinking Water Issues

Some water contaminants are natural and have persisted throughout time. Biological contaminants fit this description; with the caveat that the problem has grown with increasing populations of humans and confined animal feeding operations. Inorganic materials also fit this description and, again, environmental problems have increased with anthropogenic changes – such as mining, mineral pulverizing for road fill, reservoir building, and many industrial uses. There is a wide range of natural fertilizers, but water quality problems became an issue with the widespread use of manufactured fertilizers. Organic chemicals of concern are created by humans. Radioactive materials became a concern when humans learned to manipulate such materials. Pharmaceuticals, discussed in Chapter 4, other micropollutants, and microplastics are more recent contaminates of concern.

11.7.1 Microbiological Contaminants

The pathogenic organisms (parasites, bacteria, viruses, and other microorganisms) were discussed in Chapter 4. The modern era sees more outbreaks of disease from these organisms because they enter the water supply primarily from human sewage and confined animal units. The more crowded conditions become, the more likely that more-developed countries will have wastewater treatment plant failures. It's also likely that less-developed countries will have no sewers and no wastewater treatment. People in more-developed countries are likely to have pets that leave waste on lawns and park grounds. Ponds in parks attract geese and other waterfowl that leave significant mounds of waste; so much that walking becomes an obstacle course.

The two most prevalent pathogenic organisms in more-developed countries are *Cryptosporidium parvum* and *Escherichia coli*. Current outbreaks include **Cryptosporidium**, a microscopic, single-celled parasite that is 20 times smaller than the width of a human hair. It can survive for months in water and is resistant to chlorination. *Cryptosporidium* is widespread in nature, and has been found in humans, cattle, sheep, dogs, cats, deer, raccoons, foxes, beaver, rabbits, squirrels, and other animals.

Water source contamination can occur anywhere in a watershed – especially after a heavy rain or snowmelt. The largest reported waterborne disease outbreak in the US occurred in Milwaukee, Wisconsin, in 1993. One hundred people died and 400,000 became ill from a *Cryptosporidium parvum* outbreak when one of the city's water treatment filtration systems was contaminated by raw sewage. The cost to upgrade Milwaukee's treatment plants was about US$100 million, and no doubt, many other cities will need similar improvements [26]. In 1996, 2,000 people became ill during a similar epidemic in Cranbrook, British Columbia, and a few weeks later over 10,000 became ill in Kelowna, British Columbia.

Escherichia coli is a type of bacteria commonly found in the intestines of animals, including humans. According to the USEPA, the presence of *E. coli* in water is a strong indicator of the presence of sewage or animal waste contamination. However, not all *E. coli* is harmful to humans. *E. coli* O157: H7 is one of hundreds of strains of *E. coli*, and produces a powerful toxin that can cause serious human illness. In 2000, seven people died and 2,300 become ill in Walkerton, Ontario, from drinking water contaminated with *E. coli* O157:H7. Heavy rains appear to have caused manure from a confined feedlot to drain into a poorly protected well. The feedlot practiced modern techniques in waste management, but contamination still occurred [27]. The connection between protecting the well and managing the feedlot was not made, illustrating the importance of seeing the whole picture when trying to protect a watershed.

Figure 11.11 was developed to help explain how the cause of the 2020 pandemic, COVID-19 virus, can be transported from homes to wastewater treatment and back into the environment [28]. It can also apply to other biological contaminants not completely treated during waste management. The Flint, Michigan, story (later in this chapter) is a good example of how a treatment system can fail tragically.

Wastewater treatment in more-developed countries can be very highly effective in removal of pathogens. However, these facilities can vary in the level and quality of treatment. Rural areas and small-town systems may provide minimal treatment. Older treatment systems often lack the more efficient newer technologies. The situation in many countries goes from one extreme to the other depending on location in that country. People living in countries that still lack basic sanitation are highly susceptible waterborne pathogens. One major reason is obvious, technologically advanced systems require money to design, build, for operation and maintenance, and for the highly trained personnel necessary to run the plants. Figure 11.12 shows some tertiary treatments beyond chlorine disinfection shown in Figure 11.9.

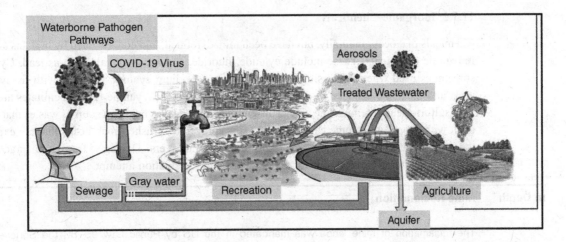

Fig 11.11 Possible pathways for waterborne pathogens such as the COVID-19 virus that caused the 2020 pandemic. These pathways could also apply to bacterial pathogens not properly treated during wastewater treatment. Notice that industrial processes may cause aerosols and possibly spreading pathogens in the air. Agriculture use of post treatment wastewater (PTWW) may deposit pathogens on crops or into surface water through runoff. PTWW may also be added directly into water used for recreation or surface water streams. The disinfectants and soaps in gray water make it less likely to have pathogens. Large groups of infected people, such as in a pandemic, increase the risks of finding pathogens in PTWW (Image modified from Bogler, A., Packman, A., Furman, A. et al., 2020, "Rethinking wastewater risks and monitoring in light of the COVID-19 pandemic," *Nat Sustain*, https://doi.org/10.1038/s41893–020-00605-2 (Creative Commons Attribution v4.0 International license, CC BY)

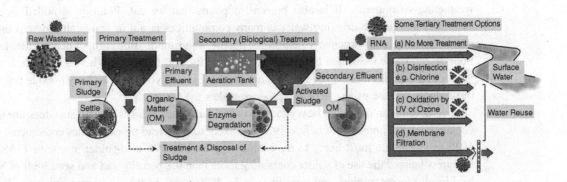

Fig 11.12 Wastewater treatment diagram with treatment levels, primary settling, secondary biological, and tertiary disinfection. In secondary biological treatment, enzymes degrade organic materials (OM), including viruses and other pathogens. Viruses not removed by settling will likely combine with suspended organic material and be transported in primary and secondary effluent. (a–b) Secondary effluent can be directly discharged to surface waters (a) or disinfected prior to discharge (b). (c–d) Further treatment can be accomplished with advanced oxidation processes (c) and/or filtration via different membrane systems (d). Ultrafiltration, such as in a membrane bioreactor, can eliminate viruses [29].

Some other tertiary options are sand filtration, managed aquifer recharge, removal of nutrients (primarily excess nitrogen (microbial treatment) and phosphorus (flocculation treatment). Nitrogen is harmful to aquatic organisms, and excess phosphorus can cause eutrophication in surface waters (Image modified from Bogler, A., Packman, A., Furman, A. et al., 2020, "Rethinking wastewater risks and monitoring in light of the COVID-19 pandemic," *Nat Sustain*, https://doi.org/10.1038/s41893–020-00605-2) (https://creativecommons.org/licenses/by-sa/4.0/)

11.7.2 Inorganic Chemicals

Chemicals that occur naturally, but have been mined, refined, and concentrated by humans are called **inorganic chemicals**. These include cyanide, fluoride, and heavy metals such as lead. Cyanide is sometimes used when processing gold or silver ore. Sodium cyanide is mixed with finely ground rock, and then zinc is added to precipitate zinc, gold, or silver cyanide. The precipitates are treated with sulfuric acid to remove the zinc. This entire process leaves behind a soup of wastes that must be dealt with properly. In some cases, cyanide was sprayed on crushed rock heaps that are exposed to the weather and allowed its transfer into streams and rivers. Damages from such practices are widespread and can continue for years if there is no reclamation attempt.

In Depth   **Mine Reclamation**

The reclamation of mine lands was mandated in the US by Public Law 95–87, the Surface Mining Control and Reclamation Act of 1977. Still, in the little town of Zortman, Montana, a 52,000-gallon spill of cyanide mine waste contaminated the drinking water supply in the year 2000. Interestingly, the spill was reportedly discovered when a mineworker recognized a cyanide odor coming from a faucet in their house [30].

The mine owners had been sued in the early 1990s by the USEPA, Fort Belknap tribes, and the State of Montana for multiple violations of water quality. In 1997, the mine owners agreed to build another water treatment plant, among other things, but the real result was a bankruptcy filing. This left a US$33 million bill for restoration. The State of Montana estimates that water treatment will have to be done in perpetuity because the damage is so extreme.

Another inorganic chemical of interest is fluoride. It occurs naturally in water from the weathering of minerals. It is also present in plants that we eat. Fluoride is added in carefully controlled amounts to drinking water in many communities to aid in oral health to help fight tooth decay. Globally, it is particularly helpful for people who cannot afford proper dental care. In excess, fluoride can cause weakened bones, impaired mobility, and fractures. Whether fluoride is good for you or bad depends on how much is in your drinking water, your diet, and any outside supplements or products you are using. Here, again, dose makes the poison.

Lead is an important heavy metal to consider. Water in lakes and streams does not generally contain lead, but homes built before 1930 are likely to have lead pipes. Homes constructed between 1930 and 1986 are most likely to have copper pipes, but with lead-solder joints. In 1986, the US Congress banned the use of solder containing more than 0.2 percent lead and set a limit of 8 percent lead in all faucets, piping, and pipe fittings. In 2020, the USEPA, in keeping with the Reduction of Lead in Drinking Water Act (RLDWA), passed by Congress, set a weighted average limit of 0.25 percent lead in the system. It also provided the calculations to make certain that manufacturers and installers understood and were using the same standard methods of compliance [31].

In Depth   **The Case of Flint Michigan's Water Quality Crisis**

One of our goals in writing this text was to emphasize the importance of understanding, gaining knowledge, and examining problems from as many viewpoints as necessary before taking actions that can have significant impacts on ecosystems and humans. From a scientific standpoint, this is much like following the scientific method discussed in Chapter 1. What do we want to accomplish? What research do we need to do to make certain what we are trying to do works well for the entire system? What do we know and what do we need to know? You get the point. What follows is a brief history, then a necessarily abbreviated version, of some of the decision failures causing what became the Flint, Michigan, water quality crisis.

History

By 2014, the city leaders in Flint, Michigan, needed to reduce water costs rates that were among the highest in the nation. The city's economic structure was failing. It was built on industries that had closed and a population that had declined from 200,000 to about 99,000. It was still a large city, but 42 percent of the remaining residents lived below the federal poverty level. Many citizens could not afford to pay their ever-increasing water bills and the city was losing money. Detroit Water and Sewer Department (DWSD) had supplied good quality treated water from Lake Huron for 50 years, but ever-increasing cost was a problem. How to fix this problem was the question.

The City of Flint, which was now under the direction of state officials, decided to stop using water from the DWSD system and instead take water from the Flint River. This required reopening and upgrading a closed water treatment plant, but the decision was made that this was feasible and would be cheaper than Detroit water. When officials wanted to begin using the new water source, Flint's utility administrator, Michael Glasgow did not believe the plant was sufficiently updated and wanted more time to train his staff, and to solidify their monitoring plans to meet legal and safety standards. He notified officials at the Michigan Department of Environmental Quality (MDEQ) of his concerns, but the plant was opened despite his legitimate and soon confirmed concerns [32].

- **What was the failure in the decision to open the water treatment plant?**

Water Treatment

Within a month after activating the new water system, residents were seeing brown, cloudy, foamy, foul-smelling water coming from their taps. Understandably, it was not water anyone wanted for any domestic use – let alone drinking. What happened? MDEQ had told water treatment plant operators that corrosion control, to stop corrosion in distribution lines, was not necessary. Without the corrosion control, the water lines deteriorated and released unpleasant and dangerous chemicals into the city's water.

Failure to properly treat the river water broke a critical health and safety federal law, the Safe Drinking Water Act (SDWA) [33], that requires proper treatment of all source water to meet designated use standards. The highest standards are for drinking water [34]. Why MDEQ chose not to require corrosion control seems to still be an unanswered question in 2021. The proper steps – making certain the pH was properly buffered to reduce corrosion and adding a chemical agent, such as orthophosphate which forms insoluble lead carbonates that inhibit the movement of lead into the water – were simply not taken because operators were told not to follow these critical procedures.

- **This was the big failure that caused this crisis and the basis for lawsuits that followed. Imagine yourself as a knowledgeable technician – can you think of ways this could have been avoided?**

Bacterial Contamination

Subsequent water quality testing in Flint, Michigan, showed bacteriological levels of contamination, from fecal coliforms and legionella. This should not happen because chlorine treatments that were being used should have been sufficient to control bacteria. Why weren't they effective? Chlorine easily binds to metals such as iron and lead and decreases its effectiveness. These new biological contamination levels violated SDWA standards and required that water users be notified. The management decision was made to increase the level of chlorine treatment. The elevated levels of chlorine treatment caused the formation of trihalomethanes, discussed under chlorine treatment earlier in this chapter. These are byproducts of the chlorine disinfection process which over time can have serious negative health effects. This was another violation of the SDWA.

- **What were the failures in this scenario? Were there red flags here that went unheeded? Is there a connection between the negative physical features and their probable source that went unnoticed?**

Citizen Action

Let's look at just three of the many citizen actions that finally brought the entire Flint problem to light in the national press.

Bringing another water quality issue, lead contamination, into focus required the efforts of LeeAnne Walters, a concerned mother of four. She first contacted the Midwest Regional EPA water division and spoke to Miguel Del Toral, Region 5 Ground Water Drinking Water Branch Regulations Manager. Ms. Walters had concerns that lead in her drinking water was making her children sick. She showed Region 5 USEPA two water quality tests, conducted at her request, by the city utility manager, Mike Glasgow. The results indicated lead concentrations of 104 and 397 ppb. Mr. Del Toral was genuinely concerned that these lead levels were 27 times higher than the legal drinking water maximum standard. Mr. Del Toral notified MDEQ.

The State of Michigan, and the USEPA, in an effort to not upset the citizens who were consistently told that their water was safe, decided to publicly disregard his concerns. Some even labelled Miguel Del Toral as a rogue USEPA agent. While there is no safe level of lead in drinking water, the USEPA's maximum concentration allowed by law in drinking water is 15 ppb. The WHO uses a drinking water standard of 10 ppb [35]. One other note, before the Walters moved into their home someone had stolen the plumbing system. Ms. Walters replaced the entire plumbing with plastic pipe. Organized citizen groups that were already protesting the fact that their water was not fit to drink became even more angry as the national news picked up Ms. Walter's story [36].

When Dr. Mona Hanna-Attisha, the pediatrician at Hurley Medical Center heard about LeeAnne Walter's story, she became worried about her patients. Lead is a neurotoxin that has some of its worse effects on infants and young children. Using medical records, she began researching the blood lead levels of Flint's youngest children, before and after the change of water supply, and compared them to children living elsewhere in Genesee County (where Flint is located). She found that the rate of elevated lead concentrations, in the blood of children younger than five, had doubled, and in some areas, tripled, following the switch to Flint River water.

At the same time, Virginia Tech professor Dr. Marc Edwards, a water distribution system corrosion expert, gathered a field team to sample the water delivered to homes in Flint for lead. The team also looked for water quality data from the city and state. His team found that one in six Flint homes had lead water levels exceeding the USEPA's safety threshold. Dr. Mona Hanna-Attisha, Dr. Marc Edwards, and LeeAnne Walters all bravely spoke publicly, facing backlash from city, state, and federal officials, about their findings. Shortly after these findings were made public, officials could no longer declare the water safe, and the entire Flint water disaster could no longer be denied.

- **There are many types of failures in these last scenarios: name some. If the problem did not originate in Ms. Walter's home, where did it start? Would the City of Flint still be fighting poor water quality without citizen action? What would you consider the worse failure in these scenarios? Explain. What did city, state, and federal officials really lose in Flint, Michigan?**

As you read through these examples, consider that it took 18 months for officials to admit wrongdoing and to start fixing the many problems – all created by one broken federal law for no apparent reason. Bottomline, despite this series of failures and evidence to the contrary, citizens were assured by local, state, and federal officials that the water was safe, and that the scientists and doctors were wrong. During this time, serious harm was done to the citizens of Flint, many of them children. It took determined citizens to organize protests and gather support. It took people who would not stop asking why and explored what could be done (in coordination with scientists willing to find and present facts) to force a solution. This was a serious failure in community health planning that did damage to humans. As a result, funding was finally provided to rebuild infrastructure and to remove lead pipes from homes in the city. In 2020, the State of Michigan proposed to provide financial assistance to citizens, with 80 percent going to families with children. It was estimated that the cost for infrastructure improvements and assistance to individuals could exceed US$1 billion. Final note: The anti-corrosion control chemicals, that should have been originally used in the Flint drinking water treatment system, cost approximately US$150 per day. By comparison, the US$1 billion-cost of mitigating the problem would have provided proper water quality treatment for a period of 18,265 years. (The federal assistance of US$1 billion divided by US$150/day for proper water treatment equals 6.7 million days, or 18,265 years of proper water quality treatment for the citizens of Flint, Michigan.)

11.7.3 Fertilizers

Nitrogen from agricultural, lawn, and golf course applications, and as waste from sewage treatment processes, can be a serious concern for drinking water treatment. Nitrates – an oxidized form of nitrogen (NO_3) – can leach into groundwater or wash into surface water from confined animal feedlots, agricultural fields, from septic systems, or basically anywhere nitrogen fertilizers are applied (see Figure 11.13). High nitrate concentrations (USEPA MCL 10 ppm) have been shown to contribute to spontaneous abortion and infant mortality from *methemoglobinemia* (blue baby syndrome). The iron in normal hemoglobin exists in a reduced state (ferrous) and can pick up oxygen and carry it through the body. Nitrates in the bloodstream can oxidize the iron (ferric), preventing it from picking up oxygen. Infants can turn blue due to a lack of oxygen and will die if not treated. This is a problem in rural areas with pervious soils, and also in developing communities with septic systems. High nitrates in groundwater can become a problem anywhere that nitrogen is used as fertilizers, and where human and animal wastes are processed in a way that provides a path for nitrates to the groundwater.

11.7.4 Organic Chemicals

Chemicals that contain carbons, and have been created by humans in a laboratory or chemical process, are called **organic chemicals**. Examples include industrial solvents such as TCE (trichloro-ethane), pesticides like DDT (dichlorodiphenyl trichloroethane), PCBs (polychlorinated biphenyls), and dioxins. In 1962, the US Public Health Service Drinking Water Standards set limits, for the first time, for all organics – including synthetic and natural, toxic, and non-toxic. However, it was not until 1976 that the USEPA published specific synthetic organic limits to prevent contamination from these compounds. The USEPA maintains a list of about 30 Synthetic Organic Chemicals (SOC) that can be found in drinking water [37]. Most of these are pesticides, defoliants, fuel additives or

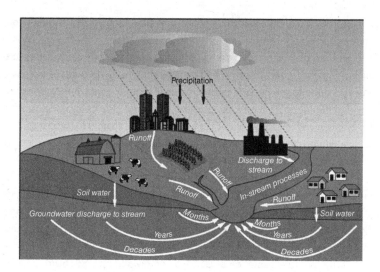

Fig 11.13 Nitrates can enter water systems, including groundwater, by several paths. Notice that human land use provides pathways and sources of nitrates into streams and groundwater. Also notice that the time it takes for groundwater to move back to the surface can vary depending on the geologic makeup of the aquifer.

Since oxygen comes from the atmosphere, younger groundwater in porous aquifers is more likely to be **oxic** (with oxygen) than older or more consolidated aquifers. Since microbes convert nitrate to nitrogen gas only under **anoxic** (without oxygen) conditions, aquifer characteristics are important in determining the fate of nitrates in groundwater (Image courtesy of USGS)

ingredients in other organic compounds. Some of these have been banned from use, e.g., toxaphene and chlordane, but are persistent in the environment and continue to be in the testing panel.

Human response to these excess organic chemicals varies with the specific chemical. Responses range from an increased risk of cancer to damage of major organ systems, including the nervous system, circulatory system, and endocrine system, as well as specific injury to kidneys, liver, and stomach. Not all people develop disease symptoms from these contaminants, but for those that do the consequences can be dire. These potential threats are more hazardous to those who are infants to under age five, adults with weakened immune systems, or who are pregnant.

11.7.5 Radioactive Elements

Some radioactive elements, such as radium and uranium, occur naturally at low levels in nature. Others can be present as byproducts of power plant operations and military installations. These byproducts are called **radionuclides**. (Nuclides are simply all forms of an element. Forms of an element vary in the number of protons and neutrons in the nucleus.) Currently, the USEPA has set standards for some radioactive elements, called *elements of concern* [38].

In 1997, the US Geological Survey studied the elements of concern in the USEPA list in the US in 1997; and found they did not occur in amounts that would violate USEPA standards (5 picocuries per liter for radium), or proposed standards (20 ppb for uranium) [39]. In the year 2000, the uranium standard was changed to 30 ppb because the USEPA determined that 20 ppb was an unreachable standard, and 30 ppb still provided adequate protection. Locations where increased radiation could occur, e.g., radioactive waste disposal sites, areas with high natural levels, and mining and some manufacturing sites were identified and required to follow testing guidelines and report their findings. Mining is only one area of concern for radionuclide contamination.

Globally, the unsafe use of radionuclides can cause contamination. Nuclear-weapons testing is known to cause contamination because many countries experimented with these weapons in the previous century. In addition, accidents in nuclear power plants have dramatically shown the consequences. In 1986, the Chernobyl accident in the Ukraine, for example, is still producing terrible effects. The most powerful earthquake to hit Japan (a magnitude of 9.1) occurred in March 2011, and generated tsunami waves up to 40.5 meters (133 feet) high. Three nuclear reactors at Fukushima Dai-ichi plant suffered meltdowns, categorized as a Level 7 nuclear emergency, the highest level on the scale created by the International Atomic Energy Agency. This was the same category as the Chernobyl accident. Japan chose to close its nuclear power plants for inspection and maintenance after the devastation of this event. Most were still closed in 2019. As of 2020, contaminated water at the site was still being treated to remove radioactive elements. No end is in sight for clean up of this disaster.

Think About It: Energy Needs

Increasing energy needs are a real problem worldwide. Japan's closure of nuclear power plants has led to the planned building of 22 new coal-fired plants estimated to generate almost as much CO_2 annually as all the passenger cars sold each year in the United States [40]. All our energy production methods have some level of environmental damage to aquatic systems during production, through wastewater, or contaminates from the air. Should the discussion of alternatives be limited to environmentally "safe" alternatives? Define "environmentally safe" to yourself, then think about how to maximize energy production while decreasing environmental damage. Are there other ways to handle an ever-increasing demand for energy? Has the world scientific community focused adequately on developing safe energy alternatives? Are there problems with alternative energies? Justify your answer.

11.7.6 Microplastics

Concern about the presence of microplastic particles in soil, air, and water and their effect on biota and human health is increasing among scientists, policymakers and the public. This is due to steadily improving knowledge of the scale and impacts of pollution by plastic, in general, and by microplastics, either intentionally produced, or formed by the degradation of larger plastic items. Heightened media attention to marine and land-based plastic pollution - with images of floating garbage patches, littered beaches, entangled and suffocated animals, and zooplankton ingesting plastic particles - is also contributing significantly to public awareness.

Starting Consideration of the Statement by the European Commission Group of
Chief Scientific Advisors [41]

Microplastics in drinking water are a relatively recent environmental and health concern. Their presence in groundwater, surface water, and wastewater has caused concern about their potential presence in drinking water. There are few studies that address this issue (approximately 50 in 2019), but they have reported microplastics in drinking water (tap) and bottled water. A recent review of the 50 studies reporting microplastics in freshwater sources, surface, groundwater, and wastewater was completed and looked at commonly used quality assurance methods in research of this type. These quality criteria were sampling methods, sample size, sample processing and storage, laboratory preparation and clean air conditions, negative controls, positive controls, sample treatment, and polymer identification. Each aspect was assigned a score to quantitatively determine the reliability of the data [42]. Only 4 of the 50 studies met all criteria. The study authors concluded that more high-quality data are needed to better understand potential exposure and human health risks.

The WHO report on microplastics in drinking water [43] emphasizes the need for more information. How much risk comes from a potential toxic substance is related to exposure and toxic levels (dose). Microplastic risks come from the physical damage from the particles, toxic chemicals that make up the particles, and potentially pathogenic microbial colonization on the particles as biofilms. Limited evidence shows a low level of concern from these risks and there is no evidence that microplastics have human health effects. WHO's concern is that limited resources should focus on known deadly microbes, and identified chemicals, because these have profound effects on human health discussed throughout this book. Wastewater treatment plants in developed countries can remove 90 percent of the microplastics using current techniques (see Figure 11.12 for review). Drinking water treatment plants, using methods to produce low turbidity water, can remove particles smaller than a micrometer (μm) through processes of coagulation, flocculation, sedimentation/flotation, and filtration. Microplastic size is set at 5 millimeters (mm), which is 5,000 micrometers (μm).

The WHO report reiterates the fact that not all countries can provide these levels of protection, and in these countries water treatment for disease protection must remain paramount. For review, United Nations Children's Fund (UNICEF)/WHO reports that approximately 67 percent of the population in low and middle-income countries lack access to sewage connections, and about 20 percent of household wastewater collected in sewers does not undergo at least secondary treatment [44].

The fact that we do not have enough high-quality data, concerning the possible effects of drinking water microplastics on human health, does not mean that they are safe. It only means, we don't know. Researchers will continue using high-quality studies to find actual data, to establish levels of microplastics in drinking water, and to determine if they harmful to human health.

However, regardless of drinking water, issues with plastics of all quantities and sizes should be controlled. They are pollutants in terms of air quality, sanitation, and quality of life, for both humans and wildlife. Those images of entangled sea life, huge floating plastic masses in the oceans, littered roadsides, parks, or your favorite hiking and fishing areas are real, and anything we can do to help should be done. It might surprise you to know that the most common single-use plastics (in order of use) are cigarette butts, plastic drinking bottles, bottle caps, food wrappers, plastic grocery bags, plastic lids, straws and stirrers, other types of plastic bags, and foam take-away containers. These are throwaways that individuals can control.

Industries also need to reduce the overuse and waste of plastics. Wrappings on products use large amounts of single-use plastics. In 2020, the Lego toy company, built on plastics, switched from plastic bags in their toy sets to recyclable paper bags. Their CEO, Niels Christiansen, said that ideas from children inspired the Danish toymaker to change. Citizen action by children made a difference. Over 60 countries have banned or taxed single-use plastics with varying degrees of success, And most voluntary citizen actions are only successful if the public is properly educated. Some single-use plastic items, e.g., syringes and surgical gloves, are essential; however, their proper disposal must be part of their planned use. Erik Solheim, Head of UN Environment, begins their report on single-use plastics with the statement: "Plastic isn't the problem. It's what we do with it. And that means the onus is on us to be far smarter in how we use this miracle material" [45].

11.8 Source Water Protection

Source water is the untreated water located in rivers, lakes, and groundwater that will be used as drinking water for public water supply systems. Generally, some level of treatment is required to ensure the safety of drinking water. Protecting source water from contamination can greatly reduce treatment costs, and ultimately provide safe drinking water supplies. In the US, the Safe Drinking Water Act of 1996 requires that states develop USEPA-approved plans that:

1. Define the land area contributing water to the public drinking water supply system.
2. Identify major potential sources of contamination to the drinking water supply.
3. Determine the susceptibility of the drinking water supply from these potential contamination sources.
4. Develop actions to reduce potential sources of contamination to protect the community drinking water system, such as zoning laws and the acquisition of conservation easements.

The Drinking Water State Revolving Fund **(DWSRF)** was created in 1996 with enactment of the Clean Water/Clean Air Bond Act, as well as the passage of the 1996 amendments to the Safe Drinking Water Act by the US Congress. The DWSRF provides a financial incentive for municipally and privately owned public water systems to undertake needed drinking water infrastructure improvements. This is accomplished by providing market-rate and below market-rate financing, for the construction of eligible public water system projects, for the protection of public health. Qualifying communities, with demonstrated financial hardship, can be given interest rates of zero percent and grants may also be provided. One of the key factors in determining projects to support is the value of the resource that will be improved, restored, or protected based on the best use classification of the receiving waterbody [46].

In Australia, the 2011 (updated version 2019) Australian Drinking Water Guidelines (ADWG) [47] provide extensive guidelines for all source water protection issues within a **watershed** (catchment). They state that, "For reliability, there is no substitute for understanding a water supply system from source to consumer, how it works, and its vulnerabilities to failure."

Wellhead protection, for example, is an area of concern for people using groundwater as their drinking water source. This could be thought of as watershed protection, where the receiving body of water is a well. The term **wellhead** refers to any land contributing water to a well, since a well is a direct conduit into the groundwater system. Suppose someone carelessly poured motor oil, from a car oil change, into the wellhead area and a heavy rain occurred soon after. This oil could move into the well and reach the groundwater, potentially contaminating millions of cubic meters of water. Common wellhead source problems involve poor planning especially when wells were constructed before land uses changed. For instance, placement of animal feedlots and septic tanks too close to a well can result in microbial contamination. Similarly, chemical contamination can occur from petroleum tanks, improper placement of industrial wastes, and local fertilizer and pesticide use and storage.

To prevent these problems, whether municipal or rural, managers need to identify and inventory wells and water sources, plan, and then manage a wellhead area or management zone. The planning process is broken down into two parts. First, delineation of the wellhead protection area, drinking water supply management area, and an evaluation of the vulnerability of the well(s) is conducted. Second is the creation of the wellhead protection plan – including goals, objectives, plan of action, evaluation program, and contingency plan. This type of regulation is designed to prevent problems. As almost always, prevention is usually easier, cheaper, and better than clean up.

11.9 Desalination

Humans cannot drink saline water, but salts can be removed to provide freshwater. The salt concentration in water can be expressed several ways. It is commonly given as the weight of salt per weight of water. A milligram of salt in a kilogram of water is called a "part per million" (ppm). That is $(0.001 \text{ g}/1,000\text{g}) = 0.000,001 \times 1,000,000 = 1$, or 1 part in a million parts. The following figures provide parameters for saline water:

Freshwater – less than 1,000 ppm
Slightly saline water – from 1,000 ppm to 3,000 ppm
Moderately saline water – from 3,000 ppm to 10,000 ppm
Highly saline water – from 10,000 ppm to 35,000 ppm

The processes of **desalination** remove salts and other chemicals to amounts less than 1,000 ppm – freshwater levels. Desalination is not a new process – distillation has been used for centuries in heating water to produce steam. In ancient times, many civilizations used distillation technology on ocean-going ships to create drinking water from seawater. Steam has little salt remaining in it and is essentially freshwater. The hydrologic cycle also uses this basic principle – energy from the Sun causes water to evaporate, where it condenses and later falls as precipitation [48].

Seawater is the primary source of water used in desalination and is growing in use in arid regions. The International Desalination Association (IDA) notes there are over 17,000 desalinization plants contracted for construction with 19,000 already online worldwide in 174 countries. In 2019, desalinization plants had a cumulative capacity of 107 million m^3/day. There are more than 300 million people around the world who rely on desalinated water for some or all their daily needs [49]. Cost of desalinization remains high. In a California study, it was the most expensive water supply option examined, with a median cost of US$2,100 per acre-foot (325,851 gallons) for large projects and US$2,800 per acre-foot for small projects [50]. One method of desalination is **reverse osmosis**. This process uses membrane filters to separate dissolved salts from a drinking water supply. The process is often used to produce freshwater from salt or brackish (saltier than freshwater, but not as salty as seawater) water. Reverse osmosis requires water to be placed under relatively high pressure, on the inlet side of the reverse osmosis system, to force it through a series of confined, semipermeable membranes (filters within pipes). It has become more widely used throughout the US, Middle East, and other regions of the world in the past few decades.

Desalination impacts vary widely and can include significant environmental impacts. For example, installation of seawater intake pipes, along ocean shorelines, can disturb seafloor and sand dune ecology. A greater issue is the disposal of the salt brine generated during the desalination process. Salt brine effluent is a byproduct and can be twice as salty as ambient seawater. If the brine is discharged back to the seawater environment, the heavier effluent is denser and tends to sink to the bottom. Sensitive organisms near the outfall could be harmed, although marine environments can vary greatly. Brine from inland locations is generally disposed in sanitary landfills, deep groundwater injection wells, or percolation galleries (groundwater recharge sites).

Think About It: Technology and Population

Another significant environmental impact of desalination can be the urban growth induced by increasing freshwater supplies in a water-short region. Water supply management, as a population growth management issue, is becoming a greater issue in some regions, and provides challenges to local jurisdictions and its citizens. A central debate issue is – should drinking water, or the lack of it, be used to slow down population growth and migration? What are some water quality and quantity issues associated with internal migration and population increases?

11.10 Emerging Drinking Water Health Issues

Trace chemical contaminants, particularly carcinogens, became a concern after World War I. Scientists discovered that freshwater contained contaminants from synthetic organic chemicals created during the conflict. Three items have become recent problems in more-developed countries – methyl tert-butyl ether (MTBE), perchlorate, and arsenic.

MTBE causes health problems when inhaled and has been detected in many water sources. It was widely used as a gasoline additive, but its use has greatly decreased in recent years. Limited research data exist on the health effects of MTBE if ingested (swallowed); however, MTBE is a potential human carcinogen at high doses. It can enter the environment wherever gasoline is stored or transferred since it dissolves easily and is not bound to the soil. This allows pollution to migrate faster and farther than other gasoline components. In addition, MTBE does not degrade (break down) easily, and is difficult to clean up from groundwater. In 1996, MTBE caused serious contamination in water supply systems in Santa Monica, California, and around Lake Tahoe, California, in 1997. Leaking underground storage tanks, pipelines, and automobile accident sites can be sources of contamination.

11.10.1 Arsenic

Arsenic is a naturally occurring element that is a widely distributed in the Earth's crust. It can be found in soils and many types of rock, particularly in ores that contain copper or lead. Arsenic is no longer produced in the US but is imported for commercial applications, such as wood preservatives (voluntarily discontinued after 2003), pesticides, as additives in animal feed, and in metal production of alloys. However, the greatest current uses of arsenic are in lead-acid batteries for automobiles, computer semiconductors, and light-emitting diodes.

Most arsenic compounds dissolve in water, and can migrate into lakes, rivers, and groundwater. In addition, arsenic contaminants can bind with sediments on the bottom of rivers and lakes. If fish and shellfish ingest contaminated sediments, arsenic can build up in their tissue, called arsenobetaine, an organoarsenic ("fish arsenic"). The contaminated fish, if consumed by humans, pass the arsenic through the food chain. Arsenic in other forms can be highly toxic. In 2001, the USEPA lowered the limit for arsenic in drinking water from 50 to 10 parts per billion (ppb) [51].

11.10.2 Perchlorate

Perchlorate salts are a naturally occurring and human-made chemical. Most perchlorate in the US was manufactured as solid rocket propellant for missiles, fireworks, and other applications. Unfortunately, improper disposal has contaminated water supplies, which can persist for decades in groundwater and surface water sources. Perchlorate from runoff contaminates the drinking water of as many as 16 million Americans. A USGS study in 2004, published 2010, collected samples from 171 sites on rivers in 19 states and 146 wells in five states. Perchlorate was detected in samples collected in 15 states, in 34 of 182 samples from rivers and streams, and in 64 of 148 groundwater samples at concentrations equal to or greater than 0.4 micrograms per liter [52].

> In 2011, USEPA announced its decision to regulate perchlorate under the Safe Drinking Water Act (SDWA). Specifically, USEPA determined that perchlorate meets SDWA's criteria for regulating a contaminant; that is, perchlorate may have an adverse effect on the health of persons; perchlorate is known to occur or there is a substantial likelihood that perchlorate will occur in public water systems with a frequency and at levels of public health concern; and in the sole judgment of the Administrator, regulation of perchlorate in drinking water systems presents a meaningful opportunity for health risk reduction for person served by public water systems [53].

Perchlorate causes serious health problems. It impairs the proper function of the thyroid gland to produce metabolic hormones, affecting normal growth and cognitive development. Researchers from Vanderbilt University found that perchlorate inhibits iodine uptake necessary for normal thyroid development [54]. Because this has the most harmful influences on fetal development, pregnant women, and infants, the American Academy of Pediatrics urged for the "strongest possible" federal limits [55].

Think About It: Controversy and Science

In 2020 the USEPA completely reversed its position and decided to not regulate perchlorate in drinking water [56]. Erik D. Olson, former general counsel for the US Senate Committee on Environment and Public Works, with 30 years of experience in public health, was shocked. His work with the Food and Drug Administration (FDA), to protect children from eating or drinking food additives (such as perchlorate that can enter our food supply through contaminated water or soil) has greatly improved food safety. Olson commented:

> In an extraordinary decision ... Environmental Protection Agency (USEPA) administrator ... has decided to defy a court-ordered consent decree requiring the agency to issue a drinking water standard for the widespread contaminant perchlorate. Studies show this chemical poses threats to the brain development of fetuses and young infants and has been found in millions of Americans' tap water. The decision, which not only ignores the science but violates a court order and the law, is expected to be announced publicly in coming weeks [57].

A US senator on the Environment and Public Works Committee said in a statement that the USEPA "abdicated its responsibility to set federal drinking water standards for a chemical long known to be unsafe, instead leaving it up to states to decide whether or not to protect people from it."

What do you think? Does the US and every other country need federal guidance on laws such as the Safe drinking Water Act that affect every US citizen? Who is responsible if citizens are harmed by chemical contaminates in their drinking water, industries, governments? How difficult should it be to alter drinking water protections, and what should be the basis for any such action?

11.10.3 Safe Drinking Water

Another critical modern health issue is occurring right now in less-developed countries across the globe. Over 1 billion of the world's population still has unsafe drinking water sources, causing the waterborne diseases that have plagued civilization for centuries. Unsafe drinking water leads to the death of nearly 2 million children annually from diarrhea and other waterborne illnesses – over 5,000 children each day. An infant born in sub-Saharan Africa is 500 times more likely to die from diarrheal disease than a baby born in a more-developed country of the world. Sudan, Zimbabwe, and Syria are particularly affected. Caring for sick children adds to the heavy workload of many women in poverty. In some societies, children (particularly girls) are denied the right to education because they are required to fetch the daily water needs of their family from distant sources [58].

An article in the *New Yorker* magazine describes a line of Indian women that began at dawn in a slum on the southern edge of New Delhi. They hold bright-blue plastic jugs, many of the silent women with babies strapped to their backs. It may be an hour or more before the water tanker truck arrives; on a bad day, the trucks may not arrive at all. Even in the prosperous areas of the city, water is available for only a few hours each day; this forces millions to get out of bed in the middle of the night to collect water for the coming day [59]. Children are particularly susceptible to polluted drinking water because their immune system is less developed. Additionally, in proportion to their weight, children drink more water than adults, and take in larger doses of contaminants present in their drinking water.

The availability of safe drinking water is worse today in many less-developed countries than it was for residents of ancient Rome and Athens, as described earlier. Women and children bear the significant burden of collecting water and dealing with illnesses caused by polluted water. A typical situation occurs in the Angolan village of Mabuia, where women and girls spend up to four hours a day collecting water from a nearby river; the community's source of freshwater. The water is polluted, and the drinking water spreads disease – particularly to the young. In other villages of Angola, drinking water might be collected from a hole dug along a ditch or riverbank. During drought, these water sources may go dry, and new, even more polluted sites are used. In villages such as Mabuia, mothers and daughters spend many hours caring for sick children. It is not unusual for infants and youngsters to be sick multiple times each year from the bacteria and parasites present in their family's water supply [60].

Agencies of the United Nations and the WHO have estimated that up to one-third of all global diseases are caused by environmental factors such as polluted water. Each year, diarrheal diseases (caused by dirty water and poor sanitation) claims the lives of over half a million children. It is the leading cause of malnutrition in children under five years old, and globally there are nearly 1.7 million case of childhood diarrheal disease every year [61]. In some locations, clean water might be available from private water companies, but the cost is often unaffordable for the poor. If tap water is provided, too often it is not clean, tastes bad, or may have a strange color.

Another aspect of water and poverty is education. The lives of many young girls are dictated by collecting water. Fetching water dominates life in Mabuia – for some, three trips a day are required – back and forth to the river to fill jugs with water. If families are sick, women and young girls generally are the caregivers. School attendance suffers in these situations and continues the cycle of limited education and poverty. Too often, a poor nation is also an ill nation.

Think About It: Achieving Safe Drinking Water for All

At this point, you may be asking yourself "How can this continue to happen?" Simple filters and boiling techniques would prevent much of this suffering. Why doesn't someone help these people? The truth is, organizations such as UNICEF and WHO, people for example the Carter Foundation, the Bill and Melinda Gates Foundation, Doctors Without Borders, and many community volunteers are helping, and progress is being made. The answers are just seldom simple. Solutions are often complicated, not only by water issues but also by inefficient governments, local elders, taboos, war, transportation, lack of funding, and sometimes the inability of convincing someone to care. These problems continue to be an insult to human dignity and a condemnation of our civilization's effectiveness at helping one another. What do you think can be done? Can an individual make a difference? Is there more that science can accomplish?

Guest Essay By James B. Chimphamba

James B. Chimphamba is a lecturer in physical geography at Chancellor College, University of Malawi where he teaches hydrology, geomorphology, soil science, and biogeography. He has worked for over 30 years with the Ministry of Agriculture and Livestock Development of the Republic Government of Malawi, in the Department of Land Resources Conservation as a land use planner and water and soil conservationist, before retiring as Chief Land Management Training Officer. He was the Principal of Land Husbandry Training College from 1992 to 2000 and managed a Southern African Development Community (SADC) Program on Integrating Soil Conservation into the Farming Systems. He was also the regional resource person for the land husbandry course at Sokoine University, Tanzania. He is the University of Malawi Coordinator of the SADC Integrated Water Resources Management Masters' Program, Chair of the Malawi Environmental Observatory Network, and Coordinator of the United Nations Office for the Outer Space Affairs Pilot Project on mapping of deforestation, forest fires and mudslides for sub-Saharan Africa.

Stream Water Dependency for Rural Communities in Southern Africa

Domestic water supplies are one of the fundamental requirements for human life. The quantity of water withdrawn by households is an important aspect of domestic supplies that influences hygiene and, therefore, public health [62]. However, the quantity of water that can be drawn in part depends on the availability of water from the source and the delivery services. In the southern African region, about 1,000 cubic meters (264,000 gal) of water per capita is annually renewable as stream runoff in the wet season, of which about 300 cubic meters (80,000 gal) is available in the dry season [63]. This drop in availability of water between the wet and dry seasons translates into extreme water scarcity generally experienced over Southern Africa during the dry period, and poses a major challenge for accessing stream water sources by rural communities.

Figure 11.14 shows the proportion of the population of states in the southern African region that directly depend on rivers as the main source of their water supply. Malawi, Tanzania, Burundi, and Rwanda – with an urbanization level of less than 15 percent – have over 10 percent of their rural communities depending on stream water sources. Zambia, Zimbabwe, and Swaziland have between 5 and 10 percent of their rural population depending on river water sources. South Africa, Namibia, and Botswana have less than 5 percent of their population depending on river water sources, but for quite different reasons.

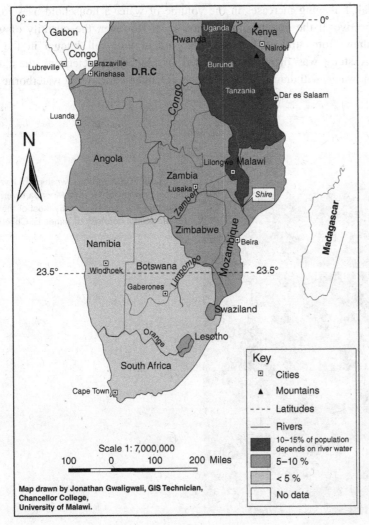

Fig 11.14 Map of southern Africa, showing the proportion of the population in each country dependent on rivers for water supply

Urbanization and improved rural water delivery systems have contributed to a low dependence on river water sources in South Africa, while desertification has led to drilling of boreholes (wells) as a major source of freshwater for rural communities in Botswana and Namibia. In some countries of Southern Africa, the number of people depending on river water might be rising. In our study, several households were observed to bypass boreholes and to draw water from nearby streams because they preferred the taste of stream water to that from boreholes. In the Chikanda Township of the Municipality of Zomba, Malawi, low-income households preferred drawing water from Mponda River, instead of community piped water, because they could not afford the cost of communal water (see Figure 11.15). In addition, community taps remained dry for most of the time. This clearly indicates that dependency on river water for some southern African states will remain a reality for some time to come.

However, the quantity of water that a household withdraws has a direct impact on the incidence of waterborne diseases, and therefore that household's health and well-being [64]. This is because waterborne diseases are caused by lack of water for hygiene or, alternatively, lack of knowledge of the need of hygiene. Waterborne diseases are those whose incidence of occurrence will fall following increases in the volume of water a household uses for hygiene purposes – irrespective of the quality of the water [65]. However, the quantity of water that households withdraw from streams for domestic uses is not well known in the southern African region. A study was therefore conducted in Malawi to determine the quantity of household stream water withdrawal and establish its interaction with waterborne diseases (Figure 11.16).

Fig 11.15 (a) Chikanda Township; (b) low-income people drawing water from Mponda River; (c) communal taps remain dry most of the time (Photographs courtesy of James B. Chimphamba)

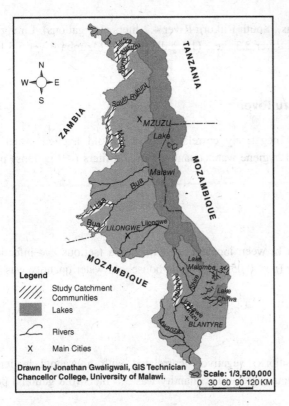

Fig 11.16 River study sites in Malawi (Map drawn by Jonathan Gwaligwali, GIS Technician, Chancellor College, University of Malawi)

Lessons from Malawi

A study was conducted to determine household water withdrawal from streams. Nine rivers were purposely selected from the north to the south of Malawi (see Figure 11.14). These rivers were Lufira, North Rukuru, South Rukuru, Mzimba, Lusa, Bua, Lirangwe, Lunzu, and Lisungwe. For each river, two communities that depended on the river for the supply of water were selected and 40 households were sampled for collection of data in each of the villages. The study aimed at quantifying the household hygiene water quantity and its interaction with waterborne diseases. Three most prevalent waterborne diseases in Malawi were investigated. These were conjunctivitis, trachoma, and scabies [66]. A questionnaire was designed and administered once a month for the months of May and June. Data were collected on the following:

- household size,
- household stream water quantity (this was physically measured with calibrated pails),
- household water quantity used for hygiene (this was also physically measured),
- incidence of waterborne diseases such as conjunctivitis, scabies, and trachoma contracted by any member of the household during the study month, and
- mean household total water quantity and hygiene water quantity of the sampled 40 households for the study month were also quantified.

There was no correlation between household hygiene water quantities with any of waterborne diseases in the communities that drew water from these rivers, even though they also contracted waterborne diseases. The mean household hygiene water quantity per capita per day was:

Rufira River 4.5 liters (1.2 gallons), South Rukuru River 4.2 liters (1.1 gallons), Lirangwe River 4.6 liters (1.2 gallons), Lisungwe River 5.3 liters (1.4 gallons), and Mzimba River 5.5 liters (1.5 gallons).

North Rukuru River and Lunzu River

Conjunctivitis ("pink eye") was negatively correlated to household hygiene water quantity ($p < 0.05$). The mean household hygiene water quantity was 3.4 liters (0.9 gallons) per capita per day.

Lusa River

There was a negative correlation between incidences of trachoma (serious eye infection) with household hygiene water quantity ($p < 0.05$). The mean household water quantity was 2.7 liters (0.7 gallons) per capita per day.

Bua River

Incidence of scabies (skin infections) negatively correlated with household hygiene water quantity ($p < 0.05$). The mean household water quantity was 3.0 liters (0.8 gallons) per capita per day.

Discussion

There is strong evidence that household hygiene water quantity is a determinant of the transmission of waterborne diseases. Households that used hygiene water quantities of below 3.5 liters (0.9 gallons) per capita per day contracted waterborne diseases, and the incidence of such infections were significantly correlated to household hygiene water quantity. For instance, households of Lusa River had a mean hygiene water quantity of 2.7 liters (0.7 gallons) per capita per day and contracted trachoma – indicating that this mean hygiene water quantity could be responsible for the transmission of trachoma. Similarly, households of Bua River used a mean hygiene water quantity of 3.0 liters (0.8 gallons) per capita per day and contracted scabies, while those of North Rukuru River used a mean of 3.4 liters (0.9 gallons) per capita per day and contracted conjunctivitis.

On the other hand, households that used a hygiene water quantity of over 4.2 liters (1.1 gallons) per capita per day also contracted waterborne diseases, but the incidence of these was not correlated with household hygiene water quantity. This appears to suggest that use of a volume of 4.2 liters (1.1 gallons) of hygiene water per person per day is adequate to protect the household from contracting waterborne diseases. However, the incidence of waterborne diseases for households that use 4.2 liters (1.1 gallons) of hygiene water quantity per capita per day could be attributed to poor hygiene behavior such as not washing hands before eating, or not washing hands after defecation. Our results therefore confirm the observation that waterborne diseases are caused by lack of water for hygiene as is the case for households that used less than 3.5 liters (0.9 gallons) per capita per day or, alternatively, lack of knowledge of the need of hygiene as is the case for those households that used at least 4.2 liters (1.1 gallons) of hygiene water per person per day [67].

The significance of our results in the promotion of household health protection resides in the fact that, at present, even the WHO of the United Nations has not given guidelines as to how much water should be used for basic health protection [68]. In the southern African region including Malawi, the ministries of health and other health-related organizations are promoting hygiene behavior – such as washing hands before eating, or after defecation, and washing cooking utensils – but do not give a recommendation as to how much water a household should be keeping for such services. This is hampering the effective implementation of such hygiene behavior since households are not able to plan for the right quantity of water from rivers, wells, boreholes, and other water reservoirs.

Our study suggests that 4.2 liters (1.1 gallons) of hygiene water per person per day appears to be adequate to protect a household from the transmission of waterborne disease, while hygiene water quantities of less than 3.5 liters (0.9 gallons) per person per day are not safe. However, caution should be taken in using our results since they were only carried out for two months (May and June) and therefore did not take into account the seasonal variation of river flows, which might also have affected how much river water households draw.

Conclusion

Our study has confirmed that household hygiene water quantity is a determinant of transmission of waterborne diseases. Hygiene water volumes of less than 3.5 liters (0.9 gallons) per capita per day appear to be responsible for such transmission while those of 4.2 liters (1.1 gallons) per capita per day are not. Our study also suggests that household hygiene behavior could also be a determinant of transmission of waterborne diseases even when households use adequate water for personal hygiene purposes. Lack of guidelines of the volume of water for hygiene is hampering efforts for promotion of protection health in southern African region, where a large proportion of people still depends on rivers located outside their homes.

11.11 Early Wastewater Treatment

Pollution has affected humans for thousands of years. As civilization developed, settlements grew in area and population. This created a new health issue – human waste disposal. For example, as far back as 2500 BCE, cities along the Indus River and its tributaries, such as the Mohenjo-Daro and Harappa, contained as many as 40,000 residents. Remarkably, these communities had sanitation systems that included bathrooms connected to sewers. In ancient Greece, the city of Athens had ceramic tile pipelines, placed beneath city streets, to remove sewerage from the city.

Extensive aqueducts provided water to ancient Rome. These elaborate systems provided water to flush latrines through small channels of water that continuously flowed beneath these facilities. The first wastewater drainage system in Rome, called *Cloaca maxima* (biggest sewer), still exists today. Wastewater was discharged directly into the Tiber River. Most Romans lived in three–six-story apartment buildings called *insulae*, and generally used a chamber pot in their homes. It was not unusual to dump the contents, or other waste, out a convenient window or doorway and onto the street. Pedestrian life was risky – people walking beneath an open window at the wrong time could experience unexpected and disgusting consequences. Under Roman law, the victim could file an objection to reclaim loss caused by injury or loss of wages [69].

Around 1200 CE, Philippe II Auguste (1165–1223), the King of France, ordered the streets of Paris paved. The King also incorporated drainage ways (open troughs) in the middle of the cobblestone streets to remove wastewater from the city. The River Seine provided drinking water, often in the same location where wastewater was returned to the river. At the time, Paris was a city of more than 70,000 people, but only had 10 kilometers (6 miles) of rudimentary sewers [70]. This system of waste removal was not highly effective, and its failure contributed to the spread of diseases, such as the Black Death.

As described earlier, the European Middle Ages were a time of little or no advancement in drinking water treatment, and the same was true regarding human waste disposal. The Black Death, from 1347 to 1350 CE, swept throughout Europe and Asia. Perhaps 40 percent of the population of Europe perished during this time, one of the worst pandemics in human history. Worldwide, as many as 75 million people died from this disease, which may have been spread by fleas and rats. People knew nothing of contagious diseases, and the filth and human waste that littered streets and homes contributed to infestations of vermin and fleas.

In Depth   Rivers of Waste

Centuries ago, waste disposal from animal butchering was a serious problem in British cities. Butchery byproducts were regularly dumped onto streets and highways and created great piles of filth. In other locations, waste was dumped in adjacent ditches, where it stunk and rotted. People were afraid of these locations – particularly during the outbreaks of the Black Death during the Middle Ages. London experienced plagues during 1349, 1361, 1369, 1379, 1382, 1390, 1391, and 1407. The plague of 1349 – the worst of all – killed at least one-third of the population. In 1354, King Edward III of England (1312–77) became concerned about sanitation and waste disposal. Ultimately, at the urging of the King, City of London officials ordered butchering waste to be disposed into the Thames River so that the tides would carry it away from the city and out to sea. In addition, the King ordered that all animal butchering would occur outside the city. However, these efforts did not always work effectively, particularly when incoming tides washed sewage back beneath bridges and along shorelines [71].

After the plague outbreak of 1391, King Richard II (1367–1400) sent a writ to the Mayor and Sheriff of London, ordering a penalty of £100 for anyone not complying with his previous order. The following year, butchers were granted a license to dispose offal and other byproducts into the Thames River at ebb tide (when the water level was falling due to an outgoing tide). Thus, at the end of the fourteenth century, a system was in place, under statute, to provide some waste control for the citizens of London. In 1409, King Henry IV (1367–1413) took the issue one step further and ordered the Mayor of London to forbid waste disposal on city streets.

These royal edicts and municipal orders were not successful, and many butchers simply disposed dead animals and other waste alongside highways outside London. The dumping practice was widespread. In 1472, it became so terrible that King Richard III (1452–85) sent a writ to the Mayor and Sheriff of London ordering them to prohibit the disposal of any filth or garbage into any ditches, rivers, or other waters about the city or suburbs. An exception was provided to the butchers of London. They were licensed to dispose of waste into the Thames, at mid-river, during ebb tide. Carts were used to transport the animal waste to bridges across the river, but the waste often fell onto city streets. Imagine, if you dare, the sights and smells of London during this era.

Sanitation was a matter of life and death. In Paris, the first underground sewer was constructed in 1370 CE beneath the rue Montmartre and drained into the River Seine. In 1388 CE, the city of Cambridge, England, became the first in Great Britain to adopt an urban sanitation law. The Statute of Cambridge prohibited the disposal of animal waste or offal (animal internal organs) from slaughterhouses into rivers or ditches. Toward the end of the Middle Ages, below-ground privy vaults and cesspools were commonly used. Sanitation workers carried off the waste, at the owner's expense, and spread it onto fields as fertilizer. Sometimes the waste was simply dumped into waterways or onto vacant land. Sadly, most people did not make the connection between human health and waste disposal. People still thought that bad air (the miasma problem discussed earlier) was the cause of disease.

A few hundred years later in France, Napoleon Bonaparte I (1769–1821) ordered the construction of the first vaulted sewer network beneath the streets of Paris. Pierre Bruneseau, the city's municipal works director, oversaw the legendary project, which extended for 30 kilometers (19 miles). The complete rework of the subterranean sewer system of Paris occupied seven years, from 1805 to 1812. In 1808, Bruneseau extended the sewer lines and disinfected and purified the whole network. Victor Hugo, French playwright of *Les Misérables*, and friend of Bruneseau, commented in 1862: "Paris has another Paris under herself; a Paris of sewers; which has its streets, its crossings, its squares, its blind alleys, its arteries, and its circulation, which is slime, minus the human form" [72] [73]. Since the mid-1800s, tours have been given of the sewers of Paris. The Musée des égouts de Paris (Museum of the Sewers of Paris) is located at Pont de l'Alma, facing 93 quai d'Orsay near the Eiffel Tower. The museum even has a souvenir shop, and tours remain quite popular today.

The summer of 1842 in London was miserable, "marked by perhaps the greatest incidence of unemployment, destitution, and social protest than any other in the nineteenth century." This was according to Edwin Chadwick's *Report on the Sanitary Condition of the Laboring Population of Great Britain*, published privately and at his own expense, that same year. The terrible conditions of the summer of 1842 – highlighted by the smell and filth of the Thames River polluted with human waste – began a "great sanitary awakening." The Public Health Act of 1848 in Great Britain was ultimately passed, in part due to the cholera epidemic of 1847–8. For the first time, a government was charged with the responsibility of safeguarding the health of its citizens [74]. The General Board of Health of England was also created by the Act. Local boards of health were established in areas where the death rate exceeded 23 per 1,000 residents. Unfortunately, London, Scotland, and Ireland were excluded from the Act.

That same year, the "Great Stink" of 1858 created havoc throughout London and caused the House of Commons to recess. The odor of raw sewage from 3 million people in the Thames River was overwhelming in the summer heat.

In Depth   **Father of Public Sanitation**

Edwin Chadwick (1800–1890) was trained as a lawyer and worked as a journalist in London. Chadwick was considered a "fanatic" and a "bore," because he could become obsessed with a cause, particularly sanitation – which he pursued through much of his career. In 1848, Chadwick became the Sanitary Commissioner of London and campaigned successfully for the Public Health Act of 1848. He believed that public health should be administered locally so that people participated in their own protection. He also believed that sewage and filth in rivers were less dangerous than the sanitary condition of sewers. Dr. John Snow disagreed with Chadwick about what caused epidemics. Snow believed in the *germ theory* of disease, while Chadwick was a strong believer that dirt and dust caused epidemics. Chadwick was unpopular because he regularly upset many vested interests. Ultimately, he was forced to resign from his public career in 1858, the same year the Central Board of Health was terminated. Chadwick is considered by many to be the Father of Public Sanitation in England – albeit a disgraced one during his own lifetime.

An Irish Newspaper, *The Nation*, published the following

It is no exaggeration to say that in 1845 … there is hardly an unpolluted river in the whole of England. Between the sewage of towns and the outscourings of manufactories, distilleries, breweries, and the like, every stream and river in the country is poisoned and rendered unfit for domestic use. Sparkling brooks that not many years ago were frequented by speckled trout and silvery salmon are now transformed into gigantic cesspools, which a clean-living toad would be ashamed to haunt. No wise man or woman will touch a drop of London water until it has been boiled and filtered, and even then, they will use as little of it as they can [75].

It took nearly 20 more years before the British River Pollution Control Act of 1876 was finally approved to make it illegal to dump sewage into a stream. By 1892, Europe's last great outbreak of cholera occurred, in Hamburg, Germany. The neighboring city of Altona did not lose many to the disease since it had a water purification system. In 1895, sewage clean up of the Thames River allowed the return of some fish species.

A year earlier, in 1894, New York City appointed Colonel George Waring, Jr. (1833–1898) to head the Department of Street Cleaning. Waring was a veteran of the Civil War, and helped the city of Memphis, Tennessee, develop a sanitary sewer system after devastating epidemics in 1878 and 1879. Waring had studied the British systems closely, and later helped other cities improve their systems – in both the US and Cuba. In New York City, it was estimated the city collected 3 to 4 million pounds (over 1 million kilograms) of horse manure from streets and city stables every day from the city's 150,000 horses. During wet weather, manure was tracked onto sidewalks, into stores, and carried into carriages across the city. Street sweepers attacked the problem with white coats, brooms, and wheeled dustbins.

In 1899, the Rivers and Harbors Act (also called the Refuse Act) was passed by the US Congress and made it illegal to throw garbage or other refuse into navigable waters without a US Army Corps of Engineers permit. Liquid sewage from streets or sewers was exempted from this law.

11.12 Emerging Wastewater Treatment Innovations

In 1969, the Cuyahoga River caught fire in Cleveland, Ohio, capturing national attention. Floating debris, oil, sludge, industrial wastes, and sewage provided the fuel, and created enormous public awareness of the need to reduce wastewater dumped into the nation's rivers. Less than 50 years ago, blood waste and offal from large cattle- and hog-packing plants was routinely drained directly into the Missouri River at Omaha, Nebraska. Large grease balls clogged water intakes downstream along the river, and thousands of fish died. In 1972, the US Congress adopted the Clean Water Act to restore the chemical, biological, and physical nature of the nation's waterways. The Clean Water Act of 1972 required that every city in the US with a population over 100,000 to install a wastewater treatment plant, and clean water became a national goal. Industrial facilities were the main target of the law since these were viewed as the main culprit of contaminated water. The US government spent billions of dollars in grants to clean up industrial plants and to construct municipal wastewater treatment plants.

Today, in Europe and North America, 30–60 percent of sewage sludge is applied to fields as fertilizer; a practice used for millennia in China. A downside of this process is that some disease-causing microorganisms can survive in the soil and can be transferred back to humans by the consumption of crops grown on those lands. Heavy metals and industrial contaminants are also contained in some of these sludge materials [76].

As a country becomes more-industrialized, pollution and economic inequality tend to increase. However, as the economic output of a nation increases and efficiency improves, more attention is given to human welfare, safe drinking water and proper waste management – and the environment. These trends are known as the *Kuznets Curve*, named after Simon Kuznets, a professor of economics at Harvard University, and the Nobel Prize winner for economics in 1971 [77].

Unfortunately, ~26 percent of the world's population (more than 2 billion people) lacks basic sanitation facilities. In some regions, communal waste (defecation) fields create enormous health issues. In the Medak district of Andhra Pradesh in central India, for example, restroom facilities were not available in some schools until recently. Students were forced to find locations outside, and privacy was not always available. This caused some students, particularly girls, to avoid school because of no toilet facilities in the school compound. In Vietnam, overall sanitation facilities in rural areas are quite low. UNICEF estimates that less than 30 percent of rural households have access to basic latrine facilities in that country [78].

Summary Points

- **Drinking water treatment** was developed before science provided reasons to treat water. For example,
  - As early as 2000 BCE, the ancient Greeks filtered water through charcoal, sand, and gravel to improve its drinkability.

  - Ancient Hindu culture required the boiling, sunlight exposure, or dipping of a hot copper rod, and filtering of water before drinking.
  - Tomb paintings show that the Egyptians used a settling device to remove sediments as early as 1447 BCE.
  - The island of Samos, Greece, in 500 BCE had both water delivery and sewage removal systems.
  - Hippocrates, 460 BCE, recommended filtering water through cloth, and choosing clear water to drink.

- **Water filtration** methods have evolved through the centuries, although some methods have not changed significantly during that same time.

- Invention of the **microscope** by Hans Janssen had a profound effect on human understanding of waterborne contaminants.
  - **Robert Hooke** found living organisms in a drop of water and helped move forward the discovery of bacteria.

- **Anthony van Leeuwenhoek** documented observations and experiments with tiny animalcules.
- **Louis Pasteur** proposed the germ theory of disease.
- **Dr. John Snow** followed the route of a cholera outbreak to prove that germs in the water caused the outbreak.

- The **germ theory of disease** was verified in 1876 by **Robert Koch** with his discovery of the cholera bacterium *Vibrio cholerae*.

- Many scientists worked on methods to purify water on a large scale. Sand filtration was particularly successful.

- The first municipal water treatment plant was developed in Paisley, Scotland, in 1804.

- **Chlorine treatment** was first used in the 1850s. In 1997, *Life* magazine named water chlorination as perhaps the most significant public health advance in millennia.

- Federal water quality standards in the United States were created to protect the drinking water supply beginning in 1914. The **Safe Drinking Water Act of 1974** made water treatment the law.

- Water quality can be degraded in a wide variety of ways, both by nature and by humans. Human contamination is avoidable.
  - Biological contaminants include pathogenic organisms.
  - Inorganic chemicals include cyanide, fluoride, and heavy metals such as lead.
  - Fertilizers can cause disease in humans and other mammals.
  - Organic chemicals, industrial solvents, some pesticides, can also be harmful.
  - Radionuclides from nuclear testing, radioactive wastes, and mining can be deadly.
  - Microplastics in drinking water and their effects on humans are an area of recent concern that needs more high-quality research to accurately define the issues.

- **Desalination** of seawater for drinking is technologically possible but still very costly.
  - Desalination also produces large amounts of very saline wastes that must be handled safely. Desalination is also very energy intensive.

- **Inadequate wastewater treatment** causes widespread human health issues.
  - As early as 2500 BCE cities had bathrooms connected to sewers, but wastes continued to contaminate water in cities, towns, and villages.
  - Perhaps as much as 40 percent of the population of Europe died during the era of Black Death from 1347 to 1350 CE.
  - Children are more susceptible to disease than adults and today, worldwide over half a million children per year die from preventable diarrhea and waterborne diseases.

- Poverty and a lack of education aid the spread of disease.

- **Sanitation** was and still is a matter of life and death.

- **The Clean Water Act of 1972, US,** was passed to restore the chemical, biological, and physical nature of the nation's waterways.

- **Worldwide, ~26 percent of the population – more than 2 billion people – still lack basic sanitation facilities.**

Questions for Analysis

1. What were some ancient methods of purifying drinking water?
2. Why is Dr. John Snow considered the Father of Epidemiology?
3. How did the process of sand filtration improve over time?
4. What problems can biological contaminants create in a drinking water supply system?
5. What problems were created by inadequate wastewater treatment during previous centuries? What are some current issues with inadequate wastewater treatment?
6. What are some current wastewater issues around the world?
7. Can good water quality be guaranteed without adequate sanitation? Explain your thoughts.

Further Reading

1. Baker, M. N., 1981, *The Quest for Pure Water*, 2nd ed, vol. 1, Denver, CO.: American Water Works Association.
2. Carcopino, Jerome, 1947, *Daily Life in Ancient Rome*, New Haven, CT.: Yale University Press.
3. Specter, Michael, 2006, "The last drop," *New Yorker*, October 23, 2006.

References

1 World Health Organization (WHO), November 2002, "General Comment No. 15 on the implementation of Articles 11 and 12 of the 1966 International Covenant on Economic, Social and Cultural Rights."

2 National Driller, "Building from the past," www.nationaldriller.com/CDA/Archives/ c9275bc6c6197010VgnVCM 1000000f932, March 2007.

3 M. N. Baker and M. J. Taras, 1981, *The Quest for Pure Water: The History of Water Purification From the Earliest Records to the Twentieth Century*, New York: American Water Works Association.

4 Water History.org, "Water and wastewater systems in Imperial Rome," www.waterhistory.org/histories/rome/, March 2007.

5 M. Cartwright, 2013, "Slavery in the Roman World." *Ancient History Encyclopedia*, November 1, www .ancient.eu/article/629/ (accessed January 2021).

6 National Driller, "Building from the past," www.nationaldriller.com/CDA/Archives/ c9275bc6c6197010VgnVCM 1000000f932, March 2007.

7 Kirk D. French and Christopher J. Duffy, 2010, "Prehispanic water pressure: A new world first," *Journal of Archaeological Science* 37, 5), 1027–1032.

8 National Driller, "Building from the past," www.nationaldriller.com/CDA/Archives/ c9275bc6c6197010VgnVCM 1000000f932, March 2007.

9 N. A. Darmani, 1995, "Avicenna: the prince of physicians and a giant in pharmacology," *Journal of the Islamic Medical Association of North America* 26, 78–81; also "Avicenna: the prince of physicians and a giant in pharmacology," hwww.afghan-network.net/Culture/avicenna.html, *March* 2007.

10 Brian J. Ford, 1992, "From dilettante to diligent experimenter: a reappraisal of Leeuwenhoek as microscopist and investigator," *Biology History* 5, 3.

11 Letter from Antony van Leeuwenhoek to the Royal Society of London, September 17, 1683.

12 M. Bentivoglio and P. Pacini, 1995, "Filippo Pacini: a determined observer," *Brain Research Bulletin* 38, 161–165.

13 See Steven Johnson, 2006, *Ghost Map*, London: Riverhead, for an excellent discussion of this London epidemic and the work of Dr. John Snow. Additional information can be found at the University of California at Los Angeles (UCLA) Department of Epidemiology, School of Public Health website, "John Snow," www.ph.ucla.edu/epi/snow.html

14 Edward Stitt, 1922, *The Diagnosis and Treatment of Tropical Diseases*, Philadelphia, Penn.: P. Blakiston's Son.

15 Jennings, George Henry, 1881, *An Anecdotal History of the British Parliament, from the Earliest Periods to Present Times*, London: D. Appleton and Company, p. 276.

16 M. Madigan and J. Martinko, eds., 2005, *Brock Biology of Microorganisms*, 11th ed, New York: Prentice Hall.

17 University of California Los Angeles (UCLA) Department of Epidemiology, School of Public Health, "John Snow," www.ph.ucla. edu/epi/snow.html, March 2007.

18 National Driller, "Building from the past," www.nationaldriller.com/CDA/Archives/ c9275bc6c6197010VgnVCM 1000000f932, March 2007.

19 Daniel A. Okun, 1996, "From cholera to cancer to cryptosporidiosis," *Journal of Environmental Engineering* 122, 453–458.

20 American Chemistry, "Chlorhexidine: controlling infection in humans and animals," www .americanchemistry.com/s_chlorine/sec_content.asp?CID-1269&DID=4749, March 2007.

21 Water Quality and Health Council, "The history of chlorine," www.waterandhealth.org/drinkingwater/ history.html, March 2007.

22 American Chemistry Council, 2020, "Chlorine and drinking water," https://chlorine.americanchemistry .com/Chlorine/DrinkingWaterFAQ/ (accessed 2020).

23 World Health Organization, 2017, *Guidelines for Drinking-Water Quality: Fourth Edition Incorporating the First Addendum*, Geneva: License: CC BY-NC-SA 3.0 IGO (download PDF at www.who.int/water_ sanitation_health/water-quality/guidelines downloaded 2020).

24 WaterWorld Magazine, 2017, "Drinking water plant is world's largest ozone installation," www.waterworld .com/drinking-water/treatment/article/16191858/drinking-water-plant-is-worlds-largest-ozone- installation#:~:text=Drinking%20Water%20Plant%20Is%20World%27s%20Largest%20Ozone% 20Installation,Texas%20Municipal%20Water%20District%20%28NTMWD%29.%20May%201st%2C% 202017

25 USEPA, www.epa.gov/safewater/

26 Kathleen Blair, Epidemiologist, City of Milwaukee Health Department, 1995, *"Cryptosporidium* and public health," *Drinking Water and Health Newsletter*, http://waterandhealth.org/newsletter/old/03-01-1995.html, March 2007.

27 S. E. Hrudey et al., 2003, "A fatal water-borne disease epidemic in Walkerton, Ontario: comparison with other waterborne outbreaks in the developed world," *Water Science and Technology* 47, 7–14.

28 A. Bogler et al., 2020, "Rethinking wastewater risks and monitoring in light of the COVID-19 pandemic," *Nat Sustain*, https://doi.org/10.1038/s41893–020-00605–2

29 Ibid.

30 Earthworks, 2020, Fort Belknap Reservation, "Cyanide spill contaminates drinking water," www .earthworks.org/stories/fort_belknap_reservation/

31 USEPA, 2020, "Use of lead free pipes, fittings, fixtures, solder, and flux for drinking water," www.epa.gov/ sdwa/use-lead-free-pipes-fittings-fixtures-solder-and-flux-drinking-water#:~:text=At%20the%20time% 20%22lead%20free%E2%80%9D%20was%20defined%20as,be%20in%20compliance%20with% 20voluntary%20lead%20leaching%20standards

32 Bridge, 2020, "Disaster day by day: A detailed Flint crisis timeline, A Truth Squad Companion," www.bridgemi.com/truth-squad-companion/disaster-day-day-detailed-flint-crisis-timeline (accessed 2020).

33 USEPA, 2020, "Safe Drinking Water Act," www.epa.gov/sdwa (accessed 2020).

34 USEPA, 2020, "Drinking water requirements for states and public water systems," www.epa.gov/dwreginfo/surface-water-treatment-rules

35 World Health Organization (WHO) 2011, Guidelines for drinking water quality, 4th ed, www.who.int/water_sanitation_health/publications/dwq-guidelines-4/en/ (accessed and downloaded 2020).

36 Lurie, Julia, 2016, *Meet the Mom Who Helped Expose Flint's Toxic Water Nightmare*, Mother Jones News, January 21, www.motherjones.com/politics/2016/01/mother-exposed-flint-lead-contamination-water-crisis/

37 USEPA, 2020, Chemical Contaminant Rules, "Drinking water requirements for states and public water systems," www.epa.gov/dwreginfo/chemical-contaminant-rules (accessed 2020).

38 USEPA, 2020, "Drinking water requirements for states and public water systems," Radionuclides Rule, www.epa.gov/dwreginfo/radionuclides-rule (accessed 2020).

39 US Geological Survey (USGS), 1997, *Radioactive Elements in Coal and Fly Ash: Abundance, Forms, and Environmental Significance, Fact Sheet FS-163–97*, Washington, DC: US Government Printing Office.

40 Hiroko Tabuchi, 2020, "Japan to build up to 22 new coal power plants despite climate emergency," *New York Times* climate and environment," February 4, www.nytimes.com/2020/02/03/climate/japan-coal-fukushima.html (accessed 2020).

41 GCSA, 2018, *Initial Statement by the Group of Chief Scientific Advisors (European Commission). A Scientific Perspective on Microplastic Pollution and its Impacts*, doi:10.2777/087124. https://ec.europa.eu/research/sam/pdf/topics/mp_statement_july-2018.pdf

42 Albert A. Koelmans, Nur Hazimah Mohamed Nor, Enya Hermsen et al., 2019, "Microplastics in freshwaters and drinking water: critical review and assessment of data quality," *Water Research* 155, 410–422, doi:10.1016/j.watres.2019.02.054 (www.sciencedirect.com/science/article/pii/S0043135419301794 (accessed 2020).

43 World Health Organization, 2019, *Microplastics in Drinking-Water*, World Health Organization, https://apps.who.int/iris/handle/10665/326499. License: CC BY-NC-SA 3.0 IGO (PDF downloaded 2020).

44 UNICEF/WHO, 2019, *Progress on Household Drinking Water Sanitation and Hygiene 2000–2017: Special Focus on Inequalities,* New York: United Nations Children's Fund and World Health Organization, www.who.int/water_sanitation_health/ (accessed 2020).

45 UNEP, 2018, *Single-use plastics: a roadmap for sustainability*, Nairobi: United Nations Environment Programme, www.unenvironment.org/resources/report/single-use-plastics-roadmap-sustainability (accessed 2020).

46 New York State Environmental Facilities Corporation (EFC), 2020, "Drinking water state revolving fund," www.efc.ny.gov/drinkingwater

47 Australian Drinking Water Guidelines, Water and Human Health, 2019, www.nhmrc.gov.au/about-us/publications/australian-drinking-water-guidelines#block-views-block-file-attachments-content-block-1 (accessed 2020).

48 USGS, 2020, Desalination, www.usgs.gov/special-topic/water-science-school/science/desalination?qt-science_center_objects=0#qt-science_center_objects (accessed 2020).

49 International Desalination Association, 2020, "Desalination and water reuse by the numbers," https://idadesal.org/

50 Heather Cooley and Rapichan Phurisamban, 2016, "The cost of alternative water supply and efficiency options in California," https://voiceofoc.org/wp-content/uploads/2016/10/Pacific-Institute.pdf (downloaded 2020).

51 USEPA, 2020, "Drinking water requirements for states and public water systems, chemical contaminant rules," www.epa.gov/dwreginfo/chemical-contaminant-rules20standard%20by%20January%2023%2C% 202006 (accessed 2020).

52 S. J. Kalkhoff, S. J. Stetson, K. D. Lund, R. B. Wanty, and G. L. Linder, 2010, "Perchlorate data for streams and groundwater in selected areas of the United States, 2004: U.S. Geological Survey Data Series 495," https://pubs.usgs.gov/ds/495/pdf/ds495.pdf (downloaded 2020)

53 USEPA, 2020, "Perchlorate in drinking water frequent questions," www.epa.gov/sdwa/perchlorate-drinking-water-frequent-questions#why-regulate

54 A. Llorente-Esteban, R. W. Manville, A. Reyna-Neyra et al., 2020, "Allosteric regulation of mammalian Na^+/I^- symporter activity by perchlorate," *Nat Struct Mol Biol* 27, 533–539, https://doi.org/10.1038/s41594-020-0417-5

55 Erick D. Olson, 2020, *EPA Refuses to Protect Children from Perchlorate-Contaminated Tap Water*, National Resources Defense Council, www.nrdc.org/experts/erik-d-olson/epa-refuses-protect-children-perchlorate-contaminated-tap-water

56 USEPA, 2020, EPA Issues Final Action for Perchlorate in Drinking Water, www.epa.gov/newsreleases/epa-issues-final-action-perchlorate-drinking-water

57 Erick D. Olson, 2020, *EPA Refuses to Protect Children from Perchlorate-Contaminated Tap Water*, National Resources Defense Council, www.nrdc.org/experts/erik-d-olson/epa-refuses-protect-children-perchloratecontaminated-tap-water

58 UNICEF, *Water, Sanitation and Hygiene*, www.unicef.org/wash/ (accessed 2020).

59 Michael Specter, 2006, "The last drop," *New Yorker*, October 23, 61–63.

60 UNICEF, n.d., "Voices of youth: explore," www.voicesofyouth.org/, March 2007.

61 WHO, 2017, "Diarrhoeal disease," www.who.int/news-room/fact-sheets/detail/diarrhoeal-disease (accessed 2020).

62 G. Howard and J. Bartram, 2003, *Domestic Water Quality, Service-Level and Health*, Geneva, Switzerland: World Health Organization.

63 S. Cairncross, 1993, "Control of enteric pathogens in developing countries," in *Environmental Microbiology*, R. Mitchell, ed., New York: Wiley-Liss, pp. 157–189; Malawi Government, 1999, *Water Resources Management Policies and Strategies*, Lilongwe, Malawi: Ministry of Water Development.

64 D. Bradley, 1977, "Health aspects of water supplies in tropical countries," in *Water, Wastes, and Health in Hot Climates*, R. Feacham, M. McGarry, and D. Mora, eds., Chichester, UK: John Wiley, pp. 3–17; M. Falkenmark, J. Lundqvist, and C. Widstrand, 1989, "Macro-scale water scarcity requires micro approaches: aspects of vulnerability in semi-arid development," *Natural Resources Forum* 13, 258–267; E. A. Petersen, L. Robert, M. S. Toole, and D. E. Peterson, 1998, "The effect of soap distribution on diarrhea: Nyamithuthu refugee camp," *International Journal of Epidemiology* 27, 520–524.

65 Ibid.

66 R. G. Feachem, 1986, "Water supply in low-income communities in developing countries," *Journal of Environmental Engineering* 101, 687–700.

67 J. Wirima and P. A. Reeve, 1990, *Common Medical Problems in Malawi*, Lilongwe, Malawi: Ministry of Health.

68 D. Bradley, 1977, "Health aspects of water supplies in tropical countries," in *Water, Wastes, and Health in Hot Climates*, R. Feacham, M. McGarry, and D. Mora, eds., Chichester, UK: John Wiley, pp. 3–17.

69 Jerome Carcopino, 1947, *Daily Life in Ancient Rome*, New Haven, Conn.: Yale University Press.

70 Paris Kiosque, "The remarkable sewer of Paris," www.paris.org/Kiosque/mar97/egouts.html, March 2007.

71 Ernest L. Sabine, 1933, "Butchering in Mediaeval London," *Speculum* 8, 335–353.

72 Atlas Obscura, Paris Sewer Museum Paris, France, "The curious underground history of keeping Paris clean," www.atlasobscura.com/places/paris-sewer-museum (accessed 2020).

73 Victor Hugo, 1862, *Les Misérables*.

74 *The Nation*, 1875, "The coming measures," March 4, p. 11, https://thisdayinwaterhistory.wordpress.com/2018/03/04/march-4-1877-birth-of-garrett-a-morgan-1875-british-public-health-act-debated/ (accessed 2020).

75 Edwin Chadwick, 1842, "Report on the sanitary conditions of the labouring population of Great Britain," self published, a summary is available at https://victorianweb.org/history/chadwick2.html (accessed 2021).

76 UNICEF, "Water, sanitation and hygiene," www.unicef.org/wash/

77 The Nobel Prize, "Simon Kuznets facts," www.nobelprize.org/prizes/economic-sciences/1971/kuznets/facts/ (accessed 2020).

78 UNICEF, "Water, sanitation and hygiene," www.unicef.org/wash/

12 Water Allocation Law

Whiskey is for drinking, and water is for fighting over.

Often attributed to Mark Twain, American writer and humorist (1835–1910)

Chapter Outline

12.1 Introduction

Who owns the water in a region? Who controls the water? How often has the subject come up in your daily life? Most of us don't spend much time thinking about water as property. We turn on a shower, a hose, a faucet, and water appears; this is our water. We may purchase it from the city or a water district. However, where do cities get their water? Is there a well, a reservoir, a lake? Who determines how much water a city can divert from a stream, lake, or aquifer? What happens if the city needs water and the local factory is using more than their share, and causes the city to have a shortage?

If a creek runs through your land, is the water yours to use as you wish? Can you build a dam and stop the creek from going downstream to your neighbors? What happens if they decide to sue you in court for drying up the creek? On a wider scale, what about international borders with rivers running through many countries or aquifers that extend across several states? Global climate change is intensifying water conflict by increasing extreme drought. This can create water scarcity and make water laws even more complicated (see Figure 12.1).

Water has always been a contentious subject, and the fact that all these questions need answers brings us to the subject of water allocation law, or: "Who gets the water – and how much?"

Water allocation law deals with the ownership, control, and use of water resources for a wide range of purposes. Water allocation law can be closely related to real estate law – particularly in arid regions of the world. On the other hand, wetter locations may have relatively unrestrictive water regulations. In recent decades, water laws have been adopted to protect water quality and the environment. Water law can have a variety of economic and environmental impacts, both

Fig 12.1 Water scarcity is an issue of global importance and a source of conflict (Getty credit: tre)

positive and negative. However, our need or desire for economic growth and development, and the need to protect human and environmental health, has led to our current system of water laws around the world.

Water can be exceedingly difficult to allocate and regulate because of its unique features. Unlike land, water can be moved from one location to another, and the quantity of supply changes every year, and sometimes even daily. In some circumstances, water can be used multiple times by multiple users. In other locations, water is used once and then returned to a river, lake, stream, aquifer, or ocean. In some situations, water users may have differing needs for the water supply but are able to "share" that use – such as river rafters in the springtime and irrigators later in the summer. Canoeing in many locations is regulated by a public agency; the Bureau of Land Management (BLM), for example, controls much of the public lands used for recreation in the American West. A consideration of recreation management in these locations is the timing of water releases from reservoirs to enhance canoeing and other boating experiences (Figure 12.2).

Water allocation law is undergoing profound changes in some regions of the world. In South Africa, Chile, and Australia, for example, surface water is being redistributed toward more public and environmental purposes. In other locations, groundwater allocation law is being developed – such as in Louisiana and West Virginia. In Nebraska and Idaho, groundwater law is being combined with existing surface water allocation statutes. The art and science of water allocation law is changing as water resource conflicts and needs change.

As you read this chapter, keep in mind the climate of a region as we discuss the development of water allocation law around the world. Dry climates and scarce water resources require strict allocation and conservation of water supplies. Wetter climates provide more ample supplies of water, although shortages can still occur seasonally or during periods of prolonged drought. In these more humid (wetter) regions, water allocation law most likely developed around the riparian (Latin *ripa* for "bank" or "shore") needs of landowners adjacent to a stream – navigation, waterpower for mills, and other low-consumptive water uses. The ancient Romans, residents of Western Europe, some regions of Africa, New Zealand, Southeast Asia, South America, and the eastern portions of Canada and the US developed laws more suited to these riparian issues

Fig 12.2 Army Major Anthony L. Smith, a veteran amputee wounded in the Iraq conflict, helps steer his raft through a rapid on the Salmon River in Idaho during a four-day river rafting trip August 14, 2006, in Salmon River, Idaho. This recreational opportunity is provided by Higher Ground, a program run by Sun Valley Adaptive sports in Ketchum, Idaho. They are a non-governmental organization (NGO) that provides sports-based rehabilitation experiences for disabled veterans (Photograph by Brent Stirton/ Exclusive by Getty Images)

12.2 Historical Development of Water Allocation Laws

12.2.1 Mesopotamia

Throughout this book, you've seen that history shows the important connection between the development of an adequate water supply and the social and economic advancement of a civilization. In Chapter 2, we discussed the early growth of agriculture in Mesopotamia, the land between the Tigris and Euphrates Rivers in the Middle East. Recall from that earlier discussion the harsh climate of the region; an annual rainfall of only 15–20 centimeters (6–8 inches) seared by summer temperatures of 50 °C (120 °F) in the shade. As early as 7000 BCE, native vegetation was cleared by settlers to create tracts of land to raise crops and domesticated animals for human consumption. Even though the climate was very dry, the lush soils of the flood-prone valleys provided significant opportunities for irrigated agriculture to flourish. Ancient irrigators constructed breaches (cuts) through riverbanks to allow water to flow onto the fertile floodplains.

These crude first attempts at irrigation in the desert climate of modern-day Syria, Iraq, and Turkey were later replaced and improved with the construction of elaborate networks of irrigation canals, dikes, and reservoirs. By 3000 BCE, every major city in the region was the center of an irrigation canal system. These large irrigation projects required much more cooperation than the smaller, earlier efforts, and created the need for regulations and institutions previously unknown. Civilization evolved as government and laws were formed and refined to accommodate their more elaborate social and economic systems.

Ancient rulers regulated water use and waterworks under their control. Irrigation provided food, security, and allowed societies to evolve beyond hunting and gathering. King Hammurabi reigned over Babylonia from 1795–1750 BCE) and was known as the "King of Justice." He developed extensive rules of law, called the **Code of Hammurabi** which included some of the first rules of water use in the world – primarily to eliminate wasteful use of water.

Think About It: Water Conservation

As we discuss water allocation law in this chapter, pay close attention to the historic importance placed on water conservation. Waste of water was not allowed in early laws, because the production of food – and life itself – depended upon using limited water supplies efficiently. Survival was of primary concern. Where do water and survival continue to be a daily issue today? Are laws sufficient to prevent disputes over how and by whom water is used?

12.2.2 Egypt

Unlike Babylonian rulers, the Pharaohs of ancient Egypt did not develop written laws pertaining to water (see Figure 12.3). Granted, nilometers were installed by Egyptian rulers to measure the flow of the 6,695-km (4,160-mile)-long Nile River. This helped measure and forecast the extent of monsoon floodwaters that would flow from the highlands of Ethiopia to the south. However, irrigation from the Nile was left to the devices and ingenuity of ancient irrigators.

Part of the reason for a lack of strict water allocation laws in Egypt may have been that Nile River floodwaters were more predictable and reliable than the sporadic torrents of water in the Tigris and Euphrates Rivers. It also didn't hurt that water supplies along the Nile were greater and more dependable than the river systems of Babylonia.

No written ancient Egyptian water laws have been found, although the importance of irrigation to the Pharaohs and Egyptian society was tremendous. Irrigation was carried out by local farmer groups that used the water. Maintenance issues were not great because Nile River floods were expected, did not occur rapidly, and water did not have to be transported great distances away from the river.

At its peak during seasonal flooding, the entire floodplain of the Nile was covered to a depth of 1.5 meters (5 feet). With great predictability, flows in the Nile River began to rise in southern Egypt in early July, and reached the northern end of the valley near Cairo about four to six weeks later. Floodwaters began to recede in early October, and by late November most of the irrigated river valley was dry. Irrigators planted crops around December and harvested them in the

Fig 12.3 Ancient Egyptian painting showing the pouring of water. Water is a common theme found on Egyptian wall paintings

spring. Then, magically to the ancient Egyptians, the entire process repeated itself. Even in modern times, June 17 is celebrated as the "Night of the Drop," meaning that tears from the Goddess Isis begin to fall, and the Nile River would flood again.

Changes in Egyptian government did not affect the local administration of agricultural and irrigation systems. Egyptians did not require the strong involvement of the state to operate its irrigation systems, and supplies were generally stable. This was a big difference from the system in Mesopotamia described earlier. No other location in the world has remained in continuous cultivation for so long as has the Nile River Valley of Egypt [1].

However, their centuries-old system of irrigation is now threatened by a newly constructed dam upstream in Ethiopia. The US$4.5 billion Grand Ethiopian Renaissance Dam, completed in 2020, is Africa's largest. Downstream Egyptian farmers are rightfully concerned about over-taxed Nile River flows, the country's growing population – 100 million in 2020 – climate change, and the new upstream dam that will divert water to other users. Negotiators are working on a water allocation scheme that will be acceptable to both countries (Figure 12.4) [2].

12.2.3 China

Yu the Great was the first manager of Chinese waters, and later became Emperor of China. Around 2280 BCE, Yu was instrumental in dredging and constructing dams and dikes along the Huang He (Yellow River) to reclaim agricultural lands and control flooding. Government-sponsored projects for irrigation were developed later. Around 560 BCE, the Cheng State Irrigation Canal was completed, and around 300 BCE, the Changshui and Cheng Kuo irrigation canals were completed. Chinese water management was intimately tied to the State. An ancient water code was enacted in China about 500 BCE.

12.2.4 Roman Empire

Roman society established an elaborate system of laws for its citizens, and the property they controlled, including water. In 528 CE, the Roman Emperor Justinian I (483–565 CE) ordered that all Roman laws in existence be compiled into a single set of law, called a code of law. It became known as the **Justinian Code** (also known as the *Corpus juris civilis*, or Body of Civil Law), and was the most comprehensive and elaborate compilation of law in the world at that time.

Early Roman law recognized three classes of water rights: (1) private, which allowed the owner unrestricted and unlimited use of water, subject to the sale of land on which the waters were located; (2) common, meaning that others were entitled to use of water for any purpose, without limit or needing permission from others or the state; and (3) public, which meant that the use of that water was subject to state (Roman) control.

It's important to note that Roman law has greatly influenced legislation in most modern nations. These laws were written to provide social and economic stability, and to provide rules for all citizens to follow – including aspects of surface water use. Environmental protection, and the allocation and use of groundwater, were not yet regulated or legislated.

12.2.5 Great Britain

English law, including aspects related to water use, developed over a long period of time. Custom and conventions were commonly developed, used, and refined throughout England for the past

Fig 12.4 The seasons of the Egyptians corresponded with the cycles of the Nile, and were known as Inundation (pronounced *akhet*, which lasted from June 21 to October 21), Emergence (growing season; pronounced *proyet*, which lasted from October 21 to February 21), and Summer (Harvest season; pronounced *shomu*, which lasted from February 21 to June 21)

2,000 years. Queen Martia, wife of a king in a small English kingdom, wrote one of the first compendiums of English law. It was later incorporated into the English common law by William the Conqueror around 1066 CE. During this same time, tin mines in Wales had extensive bodies of law, some of which pertained to the use and transportation of water in canals (called *leats*) to the mines. In 1769, Sir William Blackstone completed the *Commentaries on the Laws of England*, which became the bible of most American lawyers in the eighteenth century.

12.2.6 Spain

The King of Castile (a powerful state on the Iberian Peninsula of today's northern Spain), known as Alfonso the Wise (1212–1284), followed the lead of the Roman Emperor Justinian I and compiled all Spanish laws into *Las siete partidas* (seven divisions of law). It is one of the outstanding landmarks in Spanish and world law. A primary feature was that all water, land, and minerals belonged to the Royal Crown (the King or Queen), and only the King or Queen of Spain could grant private ownership. The *partidas* allowed natural rainfall and diffuse water flow (water not in a stream or captured in a lake) on the land surface to be used without permission. In addition, water use for domestic purposes – drinking, washing, and cleaning – was unlimited.

Note the tremendous contrast in the control of water allocation and use between the customs of the Egyptian Pharaohs, and the King and Queen of Spain thousands of years later. In Egypt, individuals and groups of farmers retained control of their lands and irrigation water. Taxes were imposed on these ancient Egyptian irrigators, but local water management decisions were vested with the farmers. This was not so during the time of Alfonso the Wise in Spain, or with his predecessors. Water supplies in Spain were limited, and more irregular, than the flow of the massive Nile River. This unreliable nature of water supplies required more strict governmental control. Water law evolved as a function of climate, culture, and perhaps through the personalities of rulers of the day.

12.2.7 Mexico

Spanish settlers in Mexico used the Arabic *acequia* system to distribute water to irrigate arid lands in North America. A local person was appointed by the governor as the *mayordomo*, or superintendent, to maintain and supervise the delivery of water in an irrigation canal. Irrigators received a certain proportion of water in the ditch for their fields; the amount could change daily if the flow of water in the river or stream fluctuated. Mexico was a great distance from Madrid, and communication could take months or even years. Water decisions were at times needed quickly, and so local control was given to the *ayuntamiento* (town council); today called the board of directors of an irrigation company, or a water board of a town, city, or regional water agency.

The governor of the state, appointed by the King or Queen of Spain, still had ultimate authority over landownership and water allocation issues in Mexico. This extended to the use of groundwater since it was also in limited supply in some areas. The *ayuntamiento* allocated water between communities that diverted water from the same river. In addition, they could allocate limited water supplies between competing uses, such as domestic and irrigation purposes. The town council could even limit population densities in drier regions; however, drinking water was never restricted. Some of the water management concepts utilized by the members of the town council were borrowed from the native peoples of Mexico. They had been irrigating for centuries in arid regions of the country, and had devised methods to best utilize scarce water resources. The Spanish settlers were wise to adapt their long-held practices to the management methods of local irrigators.

12.2.8 France

After the French Revolution, Napoléon I (1769–1821) followed the ideas of King Hammurabi of Babylonia, Emperor Justinian I of Rome, and King Alfonso the Wise of Spain, and ordered the codification (collection and organization) of all French laws – including those pertaining to water. In 1804, the **Code Napoléon** (also called the *Code Civil des Français*, or the French Civil Code; see

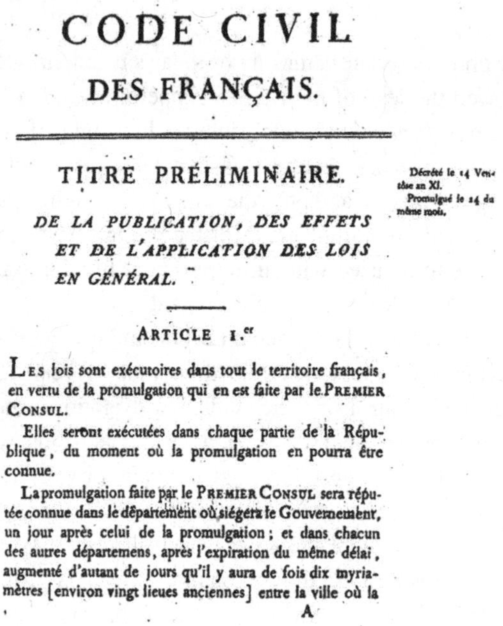

CODE CIVIL

DES FRANÇAIS.

Fig 12.5 This is a faithful reproduction of the first page of the 1804 Napoleonic Code
(Image courtesy of David Monniaux at http://en .wikipedia.org/wiki/File: Code_Civil_1804.png)

TITRE PRÉLIMINAIRE.

DE LA PUBLICATION, DES EFFETS ET DE L'APPLICATION DES LOIS EN GÉNÉRAL.

Décrété le 14 Ven- tôse an XI.
Promulgué le 24 du même mois.

ARTICLE 1.er

Les lois sont exécutoires dans tout le territoire français, en vertu de la promulgation qui en est faite par le Premier Consul.

Elles seront exécutées dans chaque partie de la République, du moment où la promulgation en pourra être connue.

La promulgation faite par le Premier Consul sera réputée connue dans le département où siégera le Gouvernement, un jour après celui de la promulgation ; et dans chacun des autres départemens, après l'expiration du même délai, augmenté d'autant de jours qu'il y aura de fois dix myriamètres [environ vingt lieues anciennes] entre la ville où la

A

Figure 12.5) was completed, probably one of his greatest achievements. The Code defined riparian water rights, navigation, and the ownership of streambeds [3]. These laws were important aspects of water use and management that would further establish the rights and responsibilities of riparian landowners.

12.2.9 US and Canada

As the Spanish developed settlements in Mexico and the desert southwest region of the US in the 1500s–1700s, France and England were settling the eastern portions of Canada and the United States. English common law became the method of allocating water in the more humid eastern regions of the US, except for the state of Louisiana – which was greatly influenced by French settlers and the Code Napoléon. Legal doctrines that were developed in the dry regions of Spain, Portugal, and Mexico were introduced to locations known today as California, Arizona, and New Mexico. The system of *acequias* and *mayordomos*, introduced by the Spanish, worked well along the small rivers and streams of this region where irrigation was necessary to provide food. When the US became an independent country in 1776, the foundation was laid for individual states to determine their method of water management – both for surface water and groundwater resources.

In Canada, industrial activity of the 1800s was located along its rivers, primarily for waterpower and as waterways for the timber industry. As such, the riparian doctrine of water rights was generally adopted initially. In 1898, the Northwest Irrigation Act vested water ownership with the Crown of England, and established western Canada water law on more of a priority system of allocation, particularly in Alberta and Saskatchewan. These were constitutionally protected in 1982. More recently, recognition of First Nations water rights has dominated water law challenges and changes in Canada.

12.3 Development of the Riparian Doctrine

Landowners have utilized water from streams for centuries. In wetter regions of the ancient world, rivers and streams could be used by anyone for navigation, fishing, swimming, or drinking water. As human populations increased, however, some adjacent landowners to a stream tried to limit access as water passed across their property. Tolls (charges) for the passage of boats, and other methods of limiting transportation on rivers, became more widespread. These landowners, called "riparian," used their ownership of land along a river to argue that they had unique legal rights to the use of water in the stream. The Roman Justinian Code, of the sixth century CE, may have been the first written law regarding the rights of riparian landowners, and their influence on the use of an adjacent stream for navigation.

> The public use of the banks of a river is part of the law of nations, just as is that of the river itself. All persons, therefore, are as much at liberty to bring their vessels to the bank, to fasten ropes to the trees growing there, and to place any part of their cargo there, as to navigate the river itself. However, the banks of a river are the property of those whose land adjoins; and consequently, the trees growing on them are the property of the same persons.

12.3.1 Justinian Code

The Justinian Code (completed 533 CE) also defined that water in a stream could be used by fishermen as well as for navigation, and could not be controlled by private individuals. The riparian landowner, however, was allowed to make use of water from a river or stream for domestic, agriculture, and milling purposes as long as it was *de minimis* (Latin "as little as possible") i.e., the use was reasonable and had a negligible effect on the stream. Water used by the riparian landowner had to be returned to the stream relatively unchanged.

Thirteen centuries later, in 1827, a water dispute over "who owned the river" was settled in the Rhode Island Supreme Court. In 1827, Joseph Story ruled in *Tyler v. Wilkinson* that the water in a stream was owned "in perfect equality of right" by riparian landowners. He stated that a riparian landowner had the right to use water from a stream but could not sell that right apart from the land. Wilkinson wanted to dig a canal to provide waterpower to a new mill along the Blackstone River. Tyler had been there since the 1700s and was concerned that his mill would suffer if more water were diverted out of the stream. This ruling was handed down from the Supreme Court of Rhode Island; a region of relatively high precipitation. Just because Tyler diverted water first, the court held that all riparian landowners had equal rights. The ruling might have gone in Tyler's favor if it had been litigated in a more arid region of the US, as we'll see later.

Think About It: Riparian Doctrine Today

The Riparian Doctrine is defined in the Justinian Code of the sixth century, and its precepts survive to this day. It's important to note that this Doctrine is primarily used in humid (wetter) regions of the world, and did not work well in drier regions, as we'll see later in this chapter. The concepts of the Riparian Doctrine generally worked well for navigation protection, and to provide water to power mills of the 1800s and early 1900s. What effect, if any, would the Riparian Doctrine have on water quality, fish populations, or wildlife habitat (see Figure 12.6)?

Fig 12.6 Riparian zone on Pole Creek in the Stanley Basin area of Idaho US, where water can be pumped for irrigation with a permit from the State. The USFWS diverts water for fisheries habitat maintenance. This area is considered critical to the Salmon River salmon runs
(Photograph by Dean Pennington)

12.4 Development of the Doctrine of Prior Appropriation

Now, let's look at another set of water allocation issues. In arid regions of the world, sharing water does not work. The idea of allowing only riparian landowners access to water from a stream quickly failed; too little land could utilize the scare resource. Instead, the concept of "water rights" was developed. This allowed a person or company to divert water for uses on lands great distances from a river. The three main components of obtaining a water right, under the Doctrine of Prior Appropriation, are evidence of intent to divert water, construction of the diversion structure, and placing water to a beneficial use. Additionally, who should receive a better priority for a water right – the person who arrived first and developed the water right claim, sometimes referred to as "first possession" (or "senior appropriator"); or someone that came later (called a "junior appropriator")?

The Doctrine of Prior Appropriation for water rights is often thought to have developed in the western US during the gold-mining era of the 1800s. However, the concepts of first possession and prior appropriation were not new to the gold miners of California or Colorado during the mid-1800s. Prior appropriation, and the diversion of irrigation water from a stream, had some basis in earlier colonial Spanish water law and English common law. Many miners involved in the California Gold Rush of 1848, and the Colorado Gold Rush of 1859, traveled from other parts of the world – such as Spain, Portugal, Mexico, and several countries in South America. In these regions, water allocation methods were used – such as the *acequias* systems in Spain, Portugal, Mexico, and the southwestern regions of the future US (discussed earlier). Other miners traveled from Great Britain (Cornish and Welsh tin miners) and brought a wealth of knowledge to the gold mines of the New World regarding the transportation and allocation of water.

Gold was discovered in California in 1848. Imagine the chaos and excitement that surrounded the rivers and streams of that region. Miners wanted to divert water away from a stream for use on mining claims, so they could strike it rich. Time was of the essence, but the riparian rules of western Europe and the eastern US just didn't work between gold miners trying to take water from small mountain streams.

Ditches were constructed to carry water around the contours of hills, at not too steep an incline, to the mining claims. Water was used in sluice boxes to remove sediments from gold ore, and later for hydraulic mining to remove overburden. However, few legal institutions existed to secure a claim; California was still a territory and governed by a military governor of the US Army. In addition, Mexico retained title to all land in the region. The gold miners needed to create a "water right" that would give them certainty and recognition by other miners. The concept of "first possession" was used to allocate limited water from a stream and was used a decade later by many of the same miners in the goldfields of Colorado.

The **Doctrine of Prior Appropriation**, also known as the *Colorado Doctrine* of water law, uses the concept of "first possession" and **first in time, first in right**. The Doctrine of Prior Appropriation allows water to be transported to land great distances from a stream. This doctrine is generally administered under conditions of scarcity, since either the flow of a river is inadequate to meet demands, or too many water rights have been claimed on the river for its limited water supplies. *Coffin v. Left Hand Ditch Company* [4] is often cited as the landmark ruling because the Colorado Supreme Court formally established the Doctrine of Prior Appropriation and rejected the older common law Doctrine of Riparian Rights.

Prior Appropriation effectively severed water rights from the land and allowed for the sale of a water right separate from the land – the direct opposite of the riparian doctrine. Property rights for water were necessary to give certainty and incentives to people to allow them to maintain and invest in resources for economic growth of a region. However, we need to note that it did little to protect riparian habitat or environmental needs of a riverine system.

Although the state retains ownership of water under the Doctrine of Prior Appropriation, individuals, corporations, and municipalities can obtain the right to use water (called a **usufructory right**) for a beneficial use. The first person in line to use water (called a **senior appropriator**) acquires the water right (called a **priority**) for a future use ahead of later users (called **junior appropriators**). To acquire a water right, one must first make an appropriation by diverting water and applying it to a **beneficial use**. A diversion is made by removing a quantity of water from its natural course. The beneficial use is made by irrigation, mining, industrial, or municipal use, or another non-wasteful activity. Figure 12.7 shows irrigated agriculture in southern Idaho where irrigation is a beneficial use of water.

Fig 12.7 A siphon-irrigated field in the sagebrush grassland of southern Idaho, US, where agriculture cannot exist without irrigation
(Photograph by Karrie Pennington)

Think About It: Define Beneficial Use

In recent years, the definition of *beneficial use* has been expanded by some US legislatures to include such things as snowmaking for ski resorts, dust control for environmental or health purposes, instream flow rights for recreation, fish and other aquatic species, and other applications unheard of a few decades ago. How would you decide which beneficial uses are appropriate to establish and enforce?

Unlike riparian water rights, an appropriator under the Doctrine of Prior Appropriation may have no geographic limitations on the use of the water right. However, the use of a water right may be limited to historical practices, e.g., irrigation use. Riparian landowners obtain no right to use water, even though their land may be adjacent to a stream. Water rights are considered property rights, but with strict rules of use and selling transactions. A lack of use of an appropriation, for an extended period of time, may result in the abandonment of the right to divert and use water.

A direct flow water right entitles the appropriator to divert a given rate of flow, not a total volume of water. For example, the holder of a 0.06 cubic meter per second (2.0 cubic feet per second) water right means that they are entitled to divert that flow rate from a stream or groundwater well. They can continue to divert this quantity of water if they are "in priority," meaning that the water is physically available in the stream and can be put to beneficial use. By contrast, a storage water right in a lake may be measured by volume. An appropriator could store up to a given amount, such as 49,000 cubic meters (13 million gallons, or 40 acre-feet). This water could be stored in a reservoir for later beneficial use. Storage rights are usually for only one filling per year of a particular storage vessel.

12.5 Evolution of the Doctrine of Prior Appropriation

The prior appropriation doctrine was refined in the 1800s to meet the needs of miners, irrigators, and people who were developing cities in the western US. No regard was given to the protection of wildlife or the environment when it was first created. Over the past century, millions of people have moved to these same desert regions which utilize this method of water allocation; consequently environmental pressures have also increased dramatically. The current challenge of the Doctrine of Prior Appropriation is to protect historical senior water rights while providing for new water needs, increased demands, and environmental protection.

12.5.1 Threatened and Endangered Species

We've just discussed numerous situations where water law was developed to protect property rights and to encourage economic development. However, we've said almost nothing regarding environmental protection – because, basically, it just was not much of a factor, historically. Fortunately, this situation changed in the US about 50 years ago. The National Environmental Policy Act (NEPA) was adopted by the US Congress in 1969, and the Endangered Species Act (ESA) was approved in 1973. This was nearly 100 years after development of the Doctrine of Prior Appropriation in the US. Water rights ownership and administration were long established in the western states and made it difficult to establish new water uses – such as the need to protect the ecology of rivers and streams. The ESA is intended to protect threatened and endangered plant and animal species, and their habitats. The strict application of prior appropriation for water allocation is sometimes directly in conflict with the water needs of the environment. Resolving these conflicts is an ongoing challenge and creates tremendous pressures on both purposes.

12.5.2 Instream Flows

Fish and wildlife depend on adequate flows in rivers for their habitats. However, the Doctrine of Prior Appropriation does not generally take into account the impacts of water diversions on the ecology of a region. In some areas, government agencies can purchase water rights from willing sellers, and allow the water to remain in a stream. Some states have established minimum stream-flows to protect the baseflow of a stream for ecological purposes.

12.5.3 Public Trust Doctrine

A second issue evolving in states using the Doctrine of Prior Appropriation is the Public Trust Doctrine. This principle is based upon the Justinian Code and English common law. It states that

Table 12.1 Comparison of surface water allocation laws

Issue	Riparian Doctrine	Prior Appropriation Doctrine	Hybrid Systems
Climate	Humid/Wet	Dry/Arid	Humid & Arid
Location of Water Use	Riparian lands	Limited by land terrain and water availability	Elements of both Riparian and Prior Appropriation
Type of Water Use	Reasonable use (cannot interfere with the reasonable use of other riparian uses downstream)	All decreed beneficial uses (generally domestic, municipal, agricultural, industrial, and recreational)	Elements of both Riparian and Prior Appropriation
Transferability of Water Right	Remains with the property	Private sale can transfer water right to another location and use	Elements of both Riparian and Prior Appropriation
Restrictions of Use	Pro-rata reduction of water use may be required during periods of limited streamflow	Junior water rights can be curtailed during periods of limited streamflow	Elements of both Riparian and Prior Appropriation
State Regulations	Permits required in some states; no unreasonable use allowed	Strict – based on priority (first in time, first in right)	Elements of both Riparian and Prior Appropriation

there are three things common to all humans – air, running water, and the sea and its shore. Title to these remains with the State, as sovereign (supreme, permanent authority), in trust for its citizens. Under this rule of law, the State must administer these resources (air, running water, and the sea) to maintain public rights for all citizens. This is a very controversial concept that butts directly with the Doctrine of Prior Appropriation.

12.5.4 Water Quality Protection

A third issue regarding the Doctrine of Prior Appropriation is water quality protection. Water allocation law does not account for the volumes of water needed to dilute pollutants in a waterbody. Less water in a watercourse or reservoir means that any chemicals, sediments, wastewater, or organics coming into the waterbody will be more concentrated. This could increase pollutant levels – sometimes drastically affecting fish and wildlife. Water quality and water allocation laws are often not administered in unison. The need to consider both is becoming more apparent as the multiple demands for water increase (Table 12.1).

12.6 Groundwater Allocation Laws

The need to regulate groundwater pumping is a recent phenomenon around the world, and students of water resources should pay close attention to this topic in coming years. Historically, groundwater pumping was limited by technology. In ancient times, a well was dug by hand, sometimes hundreds

Fig 12.8 Old windmills can be found in pastures throughout the American West. This one sustains a stock pond with groundwater for cattle drinking water
(Photograph by Karrie Pennington)

Fig 12.9 This scene from West Texas farmland shows the characteristic circular landscape of center pivot irrigation as seen from the air
(Photograph courtesy of Google Earth)

of feet (or meters) deep, to access supplies of water for human consumption or livestock watering. The amounts withdrawn were so small that restrictions on use were unnecessary. The same was true for in more recent times, when windmills pumped groundwater into livestock ponds, or to irrigation pastures, and did not generally use large volumes of water (see Figure 12.8).

Today, growing pressures on surface water supplies worldwide are causing, and, in too many cases, forcing water users to acquire more and more groundwater supplies. Technology has greatly improved in the past 50 years, particularly around pumps and application methods for irrigation. The **center pivot** (a revolving mechanical irrigation device) revolutionized agricultural water use, and created enormous demand for groundwater around the world (see Figure 12.9). Arid regions – such as Egypt, Saudi Arabia, South Africa, Brazil, and portions of the US and Canada – have experienced explosive growth in the numbers and impacts of center pivot systems pumping groundwater.

Many regions of the world are engaged in lawsuits, and other conflicts, over the use of groundwater, even in more humid regions. New Zealand, for example, is considering implementing groundwater regulations to protect aquifers from over-pumping, and to protect groundwater quality. Conflict is increasing between the neighboring states of Tennessee and Mississippi in the US over groundwater pumping by the City of Memphis. It's one of the largest cities in the world that relies on groundwater and is allegedly depleting the Memphis Sand Aquifer in neighboring Mississippi. In 2005, a lawsuit was filed by the State of Mississippi to stop Memphis from tapping into "their" groundwater resources, and to require the city to instead treat and use surface water from the Mississippi River. The US Supreme Court appointed a Special Master, and he provided recommendations to the court in 2019. In other areas of the world, groundwater-dependent ecosystems are being negatively impacted by unregulated groundwater withdrawals. Regulations are increasingly being developed to protect these sensitive environmental zones.

Groundwater allocation laws vary greatly – from no restrictions on pumping, to annual pumping limits – measured by meters on wells and verified by local or state government agencies. In the US and Canada, groundwater laws vary by state and province since water law is not a federal responsibility. In the eastern, more humid regions of Canada and the US, groundwater pumping has historically not been limited, although this is rapidly changing as unlimited well pumping has reduced flows in nearby streams.

12.6.1 Reasonable Use Rule

The **Reasonable Use Rule** limits a landowner's use of groundwater to that amount which is reasonable for the use on the overlying land. It allows the landowner to pump as much groundwater as desired, if none is wasted or used off-site from the property. Twenty-one states in the US have adopted or prefer the Reasonable Use Rule these are as follows: Alabama, Arizona, Arkansas, Delaware, Florida, Georgia, Illinois, Kentucky, Maryland, Missouri, Nebraska, New Hampshire, New York, North Carolina, Oklahoma, Pennsylvania, South Carolina, Tennessee, Virginia, West Virginia, and Wyoming.

12.6.2 Correlative Rights Rule

The **Correlative Rights Rule** maintains that the authority to allocate groundwater is held by the courts in that state. Overlying landowners are still required to make reasonable use of the groundwater, and cannot waste water, but it may be used on adjoining lands. However, a permit is required from the court to protect the public's interest and the interests of private parties. Six US states use this method or have indicated a preference for the Correlative Rights Rule: California, Hawaii, Iowa, Minnesota, New Jersey, and Vermont.

12.6.3 Rule of Capture

> *To whomever the soil belongs, he owns also to the sky and to the depths.*
>
> (English Law, 152 Eng. Rep. 1235 [1843])

The **Rule of Capture** concept (also called the Absolute Dominion Rule) was developed from Spanish and English laws. It allows a landowner to dig a well and use as much groundwater as desired, even if it interferes with neighboring landowners' wells. It has been adopted or modified for use by eight US states: Connecticut, Indiana, Louisiana, Maine, Massachusetts, Mississippi, Rhode Island, and Texas.

12.6.4 Prior Appropriation Rule

This strict rule of groundwater allocation follows the concept of the Doctrine of Prior Appropriation for surface water: "first in time, first in right." It maintains that the first landowner to divert and beneficially use surface water is granted a priority of right to that water. Some states use a permitting system to allocate groundwater under this rule, while Colorado has incorporated this procedure to include shallow groundwater in alluvial aquifers. Twelve western US states have adopted or indicate a preference for the Prior Appropriation Rule. These are: Alaska, Colorado, Idaho, Kansas, Montana, Nevada, New Mexico, North Dakota, Oregon, South Dakota, Utah, and Washington. Several states have adopted combinations of the above rules. Four have adopted Reasonable Use Rules along with the Prior Appropriation Doctrine described earlier: Arkansas, Delaware, Missouri, and Wyoming. Nebraska uses both the Reasonable Use Rule and the Correlative Rights Rule.

It's important to note that local groundwater management agencies – such as groundwater districts, conservancy districts, and natural resources districts – are also actively involved in a wide variety of groundwater management issues. In some states, efforts other than water allocation are largely voluntary, but in other regions, state legislatures have given powers to local groundwater management agencies to manage groundwater resources for water quality and environmental protection.

In other regions of the world, such as in Mexico and Saudi Arabia, the national governments are in control of groundwater allocation. By contrast, in Australia, the federal and state governments are cooperatively trying to resolve growing issues between groundwater exploitation and the ecological needs of the environment. The restriction of groundwater pumping, the purchase of groundwater rights, and the shutdown of wells without compensation will all become more prevalent around the world as groundwater supplies are overused, contaminated, or lost for environmental protection or enhancement (Table 12.2).

Table 12.2 Comparison of groundwater allocation laws

Issue	Reasonable Use Rule	Correlative Rights Rule	Rule of Capture	Prior Appropriation Rule
Climate	Humid or Arid	Humid or Arid	Humid or Arid	Dry/Arid
Location of Water Use	Overlying property	Overlying property	Overlying property	Overlying property
Type of Water Use	Generally unrestricted	Generally unrestricted	Generally unrestricted	Restricted to decreed beneficial uses
Transferability of Water Right	Water use stays with the property	Water use stays with the property	Water use stays with the property	Water use stays with the property
Restrictions of Use	Few restrictions	Limited to reasonable use based on land ownership	Few restrictions	Junior water rights are curtailed when not in priority
State Regulations	Permits may be required	Water use may be limited based on aquifer-life parameters	Strict – based on priority (first in time, first in right including surface water rights)	Strict – based on priority (first in time, first in right including surface water rights)

12.7 Interstate Compacts

Have you ever heard of an interstate compact? Article 1, Section 10, of the US Constitution authorizes states to enter into compact agreements with other states, with the consent of Congress. During the early years of the American Republic, interstate compacts were limited to boundary disputes, navigation, and fishing issues. One of the earliest compacts was adopted in 1783 between the states of New Jersey and Pennsylvania over a navigation issue on the Delaware River.

The Colorado River Compact of 1922 was the first interstate compact for the allocation of surface water within a river basin. Arizona, California, Colorado, Nevada, New Mexico, Utah, and Wyoming signed the agreement, which allocates the flow of the Colorado River between these seven states, and later with Mexico in 1944. Later, the Delaware River Basin Compact (1961) was unique in that the federal government was a participating member, as well as the four states directly affected – New York, New Jersey, Pennsylvania, and Delaware – to conduct river basin planning and to improve water quality.

Interstate compacts are now used extensively in the western US to allocate scarce surface water resources between states. Numerous rivers contain compact requirements for water deliveries to downstream states, and designate the amount of surface water that can be diverted in upstream states. Each compact is unique, but all have been ratified by the US Congress. However, most interstate compacts on rivers only pertain to water quantity, and not to water quality or ecosystem needs. What problems does this create on rivers like the Colorado? Water quality and environmental needs may be addressed by interstate compact, but to date that use has been limited. Interstate compacts can be important tools for water management – particularly for water quality and environmental protection.

Think About It: Environmental Protection

The Colorado River Compact of 1922 did not have any provisions to protect the environment, but that is not surprising. Model T cars of that era did not have anti-smog devices or seat belts, and were powered by leaded gas; environmental protection was not yet at the forefront of people's concerns. Jobs, economic development, and settlement of the American West were the focus of attention. If you live in the US, is there an interstate compact in your region? If so, is it used to manage water quantity, water quality, or other features?

The results of the 1922 Colorado River Compact were tremendous – cheap hydropower, water for irrigation, municipal and industrial growth, and massive recreation opportunities. The impacts to the environment were also great. Impacts included increased salinity in downstream river water, inundation of ancient canyons, and the diversion and consumption of so much water from the Colorado River that it ceased to flow the length of its course through Mexico. Some would label it a miracle of modern technology, but others might call the development of Colorado River water an environmental disaster.

The Endangered Species Act of 1973 greatly complicated management of the Colorado River. One of the threatened and endangered species on the Colorado River is the Colorado pikeminnow (see Figure 12.10). Instead of only turning turbines and irrigating crops, the river was now used to recover an ecosystem drastically altered and destroyed in some areas. Who would

Fig 12.10 Colorado River endangered species, Colorado pikeminnow

bear the cost of reallocation of water supplies, and who would compensate the hydropower and agricultural industries for their losses to protect endangered species? Should the federal government bear the burden of these "new" uses along the Colorado?

Consider the impacts to Mexico. Before the Colorado River Compact was implemented and dams constructed along its course, the Colorado River delta of Sonora and Baja California in northern Mexico was lush with vegetation and wildlife. Over the course of 80 years and the construction of 29 dams, the delta in Mexico is almost gone, limited to wetlands in the Cienega de Santa Clara estuary. The limited water supplies in this region come from agricultural drainage off fields, or from limited groundwater seepage, and is often laden with fertilizers and pesticides.

The Colorado River Compact was a tremendous accomplishment in its era – the era of large dam construction projects in the US. However, that time is past, new issues have emerged, and new solutions are needed.

12.8 Emerging Water Allocation Laws

Numerous countries are currently developing or revising their water allocation laws. In South Africa, for example, the National Water Act of 1998 changed the regulatory regime in the country. It has abolished the old system of private water rights and introduced public rights into the system. The South African national government is the public trustee of the nation's water resources, and has the duty to provide equitable access to water, to protect the environment, and to pursue justifiable social and economic development plans. Catchment (river basin) management agencies have been created, as well as water user associations and advisory committees to help with this complex and aggressive new program of water management [5].

In 2002, the Republic of China adopted a new water law system to emphasize the need for unified management of water resources [6]. Highlights include more efficient water consumption and the importance of balancing water resources between the needs of population growth, economic development, and environmental values and protection. The revised law is intended to build a more water efficient and pollution-free society, create a more sustainable use of water resources, and promote sustainable social and economic development – remarkable goals.

In Australia, the national government implemented an AUS$10 billion, 10-point plan to improve water efficiency and address the over-allocation of water in rural Australia, particularly in the Murray–Darling River Basin. The plan has the following specific elements:

1. Improve Australia's irrigation infrastructure by lining or piping major delivery channels to reduce water losses.
2. Improve on-farm irrigation technology and metering.
3. Share water savings on a 50/50 basis between irrigators and the Commonwealth to increase water supplies for irrigators and to increase environmental flows in rivers.

4. Address water over-allocation in the Murray–Darling Basin.
5. Create a new set of governance arrangements for the Murray–Darling Basin.
6. Place a sustainable cap on surface and groundwater use in the Murray–Darling Basin.
7. Create major water engineering works in the Murray–Darling Basin.
8. Expand the role of the Bureau of Meteorology to provide the water data necessary for good decision-making by governments and industry.
9. Create a taskforce to explore future land and water development in northern Australia.
10. Complete the restoration of the Great Artesian Basin.

In 2007, Prime Minister John Howard described the plan in these words:

Tackling Australia's water security is an immense challenge. It requires a comprehensive, bold plan. It requires a commitment of resources and above all requires people to think as Australians above any other parochial identification or consideration. I commend this plan not only to you and to my fellow Australians, but I commend this plan to all of those around the nation who have responsibility in government. This is our great opportunity to fix a great national problem. It can only be solved if we surmount our parochial differences, it can only be realized if, above all, we think as we should on the eve of Australia Day, overwhelmingly as Australians [7].

Many water allocation water laws across the world are outdated, unable to protect water quality and environmental values, and inadequate to meet the needs of growing populations. In addition, many water delivery and treatment systems are antiquated and in dire need of improvements. Financial and political challenges make improvements difficult, particularly during times of economic stress. Emerging water allocation laws will need to respect the rights of prior water users while at the same time finding ways to protect the rights of all humans and the environment. It will take a generation of new leaders to explore and find reasonable and effective solutions to mitigate the impacts of growing populations and climate change.

Think About It: Know Your Watershed Laws

Find out what type of water allocation law or laws are used, for both surface water and groundwater, in your watershed. Is there a local water agency that oversees and administers the use of water in your region? What agency is responsible for providing your drinking water, and what, if any, restrictions are placed on water diversions? Who controls the use of the groundwater residing beneath your feet? How is the environment protected by those water laws?

Summary Points

- **Water allocation law** deals with the ownership, control, and use of water resources for a wide range of purposes.

- Water can be very difficult to allocate and regulate because of its unique features. For example:
  - It can be moved from place to place.

- The amount available varies annually, seasonally, and sometimes exceptionally as in flood or drought.
- Water can be used more than once and by more than one party.

- Water resources have been regulated in some parts of the world for centuries and totally unregulated in others.

- Interestingly, some of the first water laws involved the elimination of wasted water; this was the *Code of Hammurabi* from Babylonia.

- Nile River flows in **Egypt** were predictable and provided enough water for irrigators. There were no written water laws.

- **Yu the Great**, around 2280 BCE, was the first **Chinese** water manager trying to harness the Yellow River. Chinese water law grew from there.

- **Roman law** was highly organized. Emperor Justinian I (483–565 CE) had all laws compiled into a code of law called the *Body of Civil Law.* Roman water law has greatly influenced the laws of modern nations.

- **Great Britain** refined their laws over many years. **Sir William Blackstone** completed the *Commentaries on the Laws of England* in 1769. These laws have guided American law.

- **Spain**, under **Alfonso the Wise** (1212–1284 CE), declared all water, land, and minerals belonged to the Monarchy.

- **Napoléon I** (1769–1821 CE) had French law compiled into the *Code Napoléon.* The laws defined water rights.

- **Water allocation law** is undergoing profound changes in some regions of the world.

- The **Riparian Doctrine** and the **Doctrine of Prior Appropriation** are the two most common forms of water allocation used in modern times.
  - **Riparian Doctrine** states that all riparian landowners have equal right to the water. They can use the water but not sell it apart from the land.
  - The **Doctrine of Prior Appropriation** allows a person to develop a water claim. The person who developed their claim first retains the right to the water over all those who come later. This is called **first possession** and **"first in time, first in right."**

- Threatened and Endangered Species are protected in the US by the **Endangered Species Act** (ESA), 1973.
  - Water allocation law generally does not protect the environment. It protects water users. The ESA has caused conflict when water is diverted to protect an endangered species.

- Water law also does generally not protect water quality. This also can cause conflicts with water users and people trying to protect the resource.

- Groundwater laws are evolving around the world today. Managing groundwater has its own challenges due to its unique features discussed in our chapter on groundwater.
  - **Reasonable Use Rule** says that a groundwater user can pump the amount of water necessary to irrigate the land to which it is applied.

- **Correlative Rights Rule** says that the right to use groundwater is held by the state courts.
- **Rule of Capture** says that a well can be dug and used regardless of how much it harms a neighboring water user.
- **Restatement of Torts Rule** maintains that a well may be used as long as it does not cause unreasonable harm to a neighbor.
- **Prior Appropriation Rule** applies "first in time, first in right" to groundwater.

- States may enter into agreements concerning the allocation of water from a river that runs through multiple states. These are called **Interstate Compacts**.

- **Australia's** recent efforts to revise their water law to provide multiple users with water are commendable. They are trying to protect both the resource and the users.

- **South Africa** passed a public right to water in their 1998 **National Water Act**.

Questions for Analysis

1. Who controls the allocation of water where you live?
2. Why was it necessary for ancient rulers to manage water resources wisely in their regions?
3. What are some fundamental aspects of the Riparian Doctrine?
4. What are some fundamental aspects of the Doctrine of Prior Appropriation?
5. What method of water allocation is used in your watershed, and is it the most efficient to protect local environmental values? Is it the most efficient method to promote regional economic development?
6. Is groundwater regulated in your watershed? How?
7. Should the law protect the water needs of endangered species over economic development or other human interests? Explain your reasoning.

Further Reading

Matthews, Olen Paul, 1984, *Water Resources: Geography and Law*, Washington, DC: Association of American Geographers.

Postel, Sandra, 1999, *Pillar of Sand: Can the Irrigation Miracle Last?*, New York: W.W. Norton.

Reisberg, Marc, 1993, *Cadillac Desert*, New York: Penguin Books.

Worster, Donald, 1985, *Rivers of Empire: Water, Aridity, and the Growth of the American West*, New York: Pantheon Books.

References

1 Sandra Postel, 1999, *Pillar of Sand: Can the Irrigation Miracle Last?* New York: W. W. Norton.

2 Declan Walsh, and Somini Sengupta, 2020, "For Thousands of Years, Egypt Controlled the Nile. A New Dam Threatens That," *New York Times*, February 9, www.nytimes.com/interactive/2020/02/09/world/africa/nile-river-dam.html (accessed February 20, 2020).

3 Olen Paul Matthews, 1984, *Water Resources: Geography and Law*, Washington, DC: Association of American Geographers, p. 25.

4 *Coffin v. Left Hand Ditch Co.*, 6 Colo. 443 (1882).

5 *National Water Act* (Act 36 of 1998), South Africa, www.info.gov.za/gazette/acts/1998/

6 *Water Law of the People's Republic of China* (Order of the President No. 74), October 1, 2002.

7 John Howard, 2007, "A National Plan for water security," Address to the National Press Club, Parliament House, Canberra, January 25.

13   Roles of Federal, Regional, State, and Local Water Management Agencies

From every conceivable angle – economic, social, cultural, public health, or national defense – conservation of natural resources is an objective on which all should agree.

Hugh Hammond Bennett, father of soil conservation (1881–1960) [1]

Chapter Outline

13.1 Introduction

This chapter will center on the United States. Other countries have different names for various levels of government, but like the US, range from a central body to the local water users. As you can imagine, local, state, regional, and federal governments get involved in the management of water resources in any country. Consider the resource: water is vital to many aspects of our environment, economic growth, culture, and to life itself. Governmental intervention is crucial to protect water users and the environment. However, government agencies have different goals and objectives. A federal agency in charge of wildlife protection may be in direct mission conflict with a local or state agency that constructs water projects. Personalities, budget constraints, and political agendas also factor into the work of government agencies – a reality that must be noted.

Think About It: Agency Perspectives

So, as you read this chapter, think about the wide range of competing goals, missions, and values held by the many different organizations involved with water regulation. Use these examples to discuss the pros and cons of different agency perspectives. You may want to research the current status of these issues in your area to map their progress. Why is consensus so difficult to reach in many water management situations?

13.2 US Federal Water Agencies

Numerous federal agencies play a role in the protection and management of water resources and the environment in the United States. At times, competition and cooperation between agencies ebbs and flows; sometimes due to budget constraints and other times due to conflicting agency roles values, or politics on a particular project or issue. Through the years, various measures have been implemented, such as the President's Council on Environmental Quality (CEQ), to coordinate the efforts of federal agencies. This chapter will discuss the background and duties of various federal, regional, state, local, and nonprofit agencies. We will consider numerous case studies and look at how an issue was mitigated, and in some situations negatively impacted, by the actions of the various agencies.

13.2.1 US Environmental Protection Agency

The US Environmental Protection Agency (USEPA) was created by Congress and signed into law by President Richard Nixon in 1970 to protect the nation's human health and environment, including water resources. Approximately 14,000 people work for the agency in the Washington, DC, headquarters office, 10 regional offices, and more than a dozen laboratories. The Administrator of the USEPA is head of the agency and is appointed by the President of the United States. The agency develops and enforces regulations, provides financial assistance, conducts environmental research, sponsors voluntary programs and partnerships, provides environmental education opportunities, and publishes information.

The USEPA, with the help of states, tribes, and other partners, works on these water resources goals:

(1) to ensure that Americans have clean air, land, and water;
(2) to reduce environmental risks through the use of best available scientific information;
(3) to provide fair and effective enforcement of federal laws to protect human health and the environment.

In 1969, the **National Environmental Policy Act (NEPA)** was adopted by the US Congress. It requires that all federal agencies integrate environmental values into their decision-making processes. This is accomplished by considering the environmental impacts of proposed federal actions and developing reasonable alternatives to those actions. All federal agencies are required to prepare an Environmental Impact Statement (EIS) to meet this requirement. The USEPA is the lead agency to review and comment on any EIS prepared by other federal agencies to comply with NEPA.

In 1972, the Federal Water Pollution Control Act Amendments were enacted by Congress, and later amended in 1977. This law is commonly known as the **Clean Water Act of 1972**. It establishes criteria for regulating discharges of pollutants into waters of the United States. It also gives the USEPA authority to set pollution standards, and to implement pollution control programs for industries. The Clean Water Act made it unlawful for anyone to discharge a pollutant from a point source into navigable waters unless a permit was obtained. In addition, wastewater treatment plants were funded, and non-point source pollution programs were implemented. It is the most important water quality protection law in the history of the US.

Enforcing the Clean Water Act has enabled American rivers, streams, and lakes that were seriously impaired literally to come back to life. It provides the rules, but also demands funding and support for programs to make them work. The Clean Water Act is a law with teeth. It has worked extremely well to control point source pollution. Its success with non-point sources will be more difficult. We will discuss this in more detail later.

13.2.2 US Fish and Wildlife Service

The mission of the US Fish and Wildlife Service (USFWS) is "working with others, to conserve, protect, and enhance fish, wildlife, and plants and their habitats for the continuing benefit of the American people." The agency is located within the Department of Interior. It was authorized by Congress and approved by President Franklin D. Roosevelt in 1939 by combining the Bureau of Fisheries and the Bureau of Biological Survey. The USFWS employs approximately 9,000 people in facilities across the country and at its headquarters in Washington, DC. This includes seven regional offices and nearly 700 field units; such as national wildlife refuges, national fish hatcheries and management assistance offices, law enforcement, and ecological services field stations.

Some of the USFWS conservation programs are among the oldest in the world. Its origins are in the US Commission on Fish and Fisheries (created in 1871) in the US Department of Commerce, and the Division of Economic Ornithology and Mammalogy (formed in 1885) in the US Department of Agriculture. It was later named the Bureau of Biological Survey. In 1900, the Lacey Act was passed by Congress to prohibit the interstate shipment of illegally taken wildlife. President Theodore Roosevelt established the first federal bird reservation on Pelican Island, Florida, in 1903, and placed it under the jurisdiction of the Bureau of Biological Survey. Bird reservations were later redesignated as National Wildlife Refuges in 1942.

The Migratory Bird Treaty Act was passed between the US and Great Britain (for Canada) in 1918, to regulate migratory bird hunting. Some of the most well-known American birds are the eagles which are protected from hunting; Figure 13.1 shows a golden eagle. In 1946, the USFWS established a River Basins Study Program to help reduce impacts to fish and wildlife from federal water projects.

Perhaps the most controversial of the environmental acts is the **Endangered Species Act (ESA)**. It was passed by Congress in 1973 to protect threatened and endangered species – both domestically and internationally. The controversy stems from the protection of animal life, in some situations, above human wishes or needs.

It is fairly easy to champion a majestic bird, but it is harder to evoke public support for a spotted frog (Figure 13.2). The Columbia spotted frogs (*Rana luteiventris*) in the Great Basin of Nevada have been a candidate for ESA protection since 1993, and some landowners have agreed to

Fig 13.1 This beautiful golden eagle is protected from being hunted in the US by the US Fish and Wildlife Service
(Photograph courtesy of USFWS)

Fig 13.2 The threatened Columbia spotted frog in a newly constructed habitat in Nevada, US (Photograph courtesy of USFWS)

Fig 13.3 Newly created pond habitats can be seen in this valley on Warners Ranch, Nevada, US (Photograph courtesy of USFWS)

build frog habitat to save the frogs (Figure 13.3). Another battle, this one for salmon, has sparked controversy in several western states. "Water for salmon" has become a battle cry in parts of the American West as salmon compete with irrigators, developers, and industry for water. The ESA provides a voice for the fish, which both angers and frustrates many water users. The USFWS and the US National Oceanic and Atmospheric Administration (NOAA) National Marine Fisheries Service assumed responsibility to administer the ESA for salmon.

13.2.3 US Army Corps of Engineers

In 1802, the US Army Corps of Engineers (USACE or Corps) was created by Congress and approved by President Thomas Jefferson, and is the nation's oldest water resource agency. It is located within the Department of Defense and is made up of approximately 37,000 civilian and military members. Their mission is to provide quality, responsive engineering services to the nation including:

- Planning, designing, building, and operating water resources and other civil works projects (navigation, flood control, environmental protection, disaster response, etc.);
- Designing and managing the construction of military facilities for the Army and Air Force (military construction); and
- Providing design and construction management support for other Defense and federal agencies (interagency and international services).

The Chief of Engineers and Commander of the USACE is located in the Pentagon in Washington, DC, and serves as the Army's topographer and head of the world's largest public engineering, design, and construction management agency. The headquarters is also located in Washington, DC, with eight divisions and 41 districts located throughout the US (based on watershed regions), Asia, and Europe. In 2004, a provisional district office was created to oversee operations in Afghanistan and Iraq.

The USACE operates the Institute of Water Resources (IWR) to develop forward-looking programs and research for the government's civil works program. It was created in 1969, and has offices in Alexandria, Virginia, Davis, California, and New Orleans, Louisiana.

The Corps has been involved with regulating activities in navigable waterways through the granting of permits since passage of the Rivers and Harbors Act of 1899. At first, this program was meant to prevent obstructions to navigation, although an early twentieth-century law provided regulatory authority over dumping of trash and sewage into rivers. Passage of the Clean Water Act in 1972 greatly broadened this role by giving the Corps authority over dredging and filling in "waters of the United States," including many wetlands. When determining if a permit for dredging or filling should be issued, the Corps is required, by federal law, to consider both economic development and environmental protection.

The Corps is also involved in drinking water delivery by way of its construction of the Washington, DC, water supply system in the 1850s. It continues to operate that system today. The Corps allows many cities and industries across the country to use water stored in flood control reservoirs. Currently, 10 million people obtain water from a USACE flood control project in 115 cities. In addition, some arid regions receive water supplies for agricultural irrigation (Figure 13.4).

13.2.4 US Bureau of Reclamation

The US Bureau of Reclamation (USBR or BuRec) has its roots in the **Reclamation Act of 1902**, approved by Congress and signed into law by President Theodore Roosevelt, which created the Reclamation Service. The purpose of the law was to encourage settlement on the drier lands of the 17 western states through construction of irrigation projects. Today, 338 reservoirs have been constructed with a total storage capacity of 173 billion cubic meters (45.6 trillion gallons, 140 million acre-feet). Approximately 4 million hectares (10 million acres) of land are irrigated, and 53 hydropower plants are operated. Approximately 60 percent of the nation's vegetables are grown

Fig 13.4 This is an aerial view of Navarro Mills Lake and Dam on Richland Creek in Navarro County, Texas, US. The US Army Corps of Engineers constructed this dam in 1963 for flood control and water supply for Navarro County (Photograph courtesy of US Army Corps of Engineers Digital Visual Library)

Fig 13.5 The Grand Diversion Dam was one of the first reclamation projects on the Colorado River. Irrigation of the Grand Junction, Colorado, area from the Highline Canal began in 1915 (Photograph courtesy of US Bureau of Reclamation)

under a USBR irrigation system (Figure 13.5). The western states economic development would have been much slower without these water projects.

No new USBR dams or reservoir systems have been constructed for several decades. Instead, the agency has focused on improved water management, water quality, and water conservation issues. Some of the best-known projects include Hoover Dam in Nevada/Arizona, Grand Coulee Dam in Washington, Shasta Dam in California, and the Colorado–Big Thompson Project in Colorado.

Think About It: Development and Water

Dams and large water projects have come under a great deal of criticism for many reasons, some of which we have discussed in this book. Can you imagine the desire to open up the American West at the beginning of the twentieth century to attract settlers to a better life? Was there a better way to provide jobs and homes? Is it the business of government to provide opportunities for employment? If so, how can it best be done in a developing area with little water?

13.2.5 US Geological Survey

The mission of the US Geological Survey (USGS) is to provide "reliable scientific information to describe and understand the Earth; minimize loss of life and property from natural disasters; manage water, biological, energy, and mineral resources; and enhance and protect our quality of life." The USGS was approved in 1879 by Congress, and confirmed by President Rutherford Hayes, to classify public lands held by the federal government. Ten years later, the agency began measuring streamflow, and today has ongoing water resource data collection efforts in all 50 states. Many data collection sites can be viewed online, some in real time (as it is collected) [2], and others as databases of years of accumulated measurements. The USGS and its 8,600 employees are located within the Department of Interior and have more than 400 locations around the country. The headquarters for the USGS are in Reston, Virginia, and its Central and Western Region offices are in Denver, Colorado, and Menlo Park, California, respectively.

The USGS is actively involved in stream gage monitoring and has over 7,200 gaging stations (Figure 13.6). This information is used to predict floods, forecast water supplies, monitor drought, determine contaminant loads, and provide information for international treaties, management of aquatic habitat, water rights administration, and basic hydrologic research. The USGS also conducts extensive monitoring of groundwater and provides technical assistance for remediation (cleanup) of contaminated groundwater. The agency continues its original mission of cartography with production

(a)

Fig 13.6 (a) Panorama of Crater Lake from near Victor Rock, Crater Lake National Park, Klamath County, Oregon, ca.1901. Professor J.S. Diller piloted the first USGS extensive geologic study of the Crater Lake caldera starting in 1883. The public had little knowledge of these wonders of nature in the late nineteenth and early twentieth centuries. These early geologic studies help lead to the formation of Crater Lake National Park. It is hard to appreciate the true beauty of the deep multiple shades of blue waters of Crater Lake in a black and white photo. It is breathtakingly lovely (Photograph courtesy US Geological Survey *(Data Owner)*, Joseph Silas Diller, Photographer)

(b)

Fig 13.6 (b) During this study, researchers measured lake depth, from soundings and evaporation rates, using an evaporation pan located on the square raft. These studies help scientists determine important characteristics of the caldera and the lake. The USGS has been monitoring water in lakes, streams, and rivers since 1889. Improvements in technology have made it possible to evaluate the many water quality and physical parameters discussed in this text. Students with a desire for field work could find it in many areas of study like these
(Photograph courtesy US Geological Survey *(Data Owner)*, Joseph Silas Diller, Photographer)

of USGS topographical maps that show land and water features, elevations, and other valuable land-use information. The original USGS surveyors were explorers going to places few people had seen to make maps. Today, their exploration includes aerial photography and satellite data.

13.2.6 Natural Resources Conservation Service

Since 1935, the Natural Resources Conservation Service (NRCS) (originally called the Soil Conservation Service) has provided leadership to help America's private landowners and managers conserve soil, water, and other natural resources. Hugh Hammond Bennett, the "father of soil conservation," was said to be part showman, part scientist. He was not opposed to using the dramatic to make his point. Bennett's speeches inspired action for soil conservation around the country, whether at farm-field demonstrations, scholarly gatherings, or in the Congress. When a dust storm from the Great Plains moved over Washington, DC, in the spring of 1935, during the height of the Dust Bowl, Bennett was testifying before a Congressional committee on the bill that would create the Soil Conservation Service. He knew the storm was coming and used it dramatically to demonstrate the need for soil conservation.

All NRCS programs are voluntary, which makes the decision to use conservation practices a personal decision. The following is a list of NRCS responsibilities today:

- The Conservation Technical Assistance (CTA) program provides voluntary conservation technical assistance to land-users, communities, units of state and local government, and other federal agencies in planning and implementing conservation systems.
- NRCS manages natural resource conservation programs that provide environmental, societal, financial, and technical benefits.
- Science and technology activities provide technical expertise in such areas as animal husbandry and clean water, ecological sciences, engineering, resource economics, and social sciences.
- NRCS deploys expertise in soil science and leadership for soil surveys and for the National Resources Inventory, which assesses natural resource conditions and trends in the United States.
- NRCS provides technical assistance to foreign governments and participates in international scientific and technical exchanges.

NRCS has six mission goals: high-quality, productive soils; clean and abundant water; healthy plant and animal communities; clean air; an adequate energy supply; and working with farm and ranch lands (Figure 13.7). To achieve these goals, NRCS has programs with benefits specifically intended to address the issues, with water quality and quantity issues being high in priority. The Environmental Quality Incentive Program and the Wetland Reserve Program provide financial and technical support for many of the management practices we have discussed in this text.

NRCS is part of the US Department of Agriculture, with its main offices in Washington, DC. State offices are in all 50 states, and local offices are located in counties based on the state office's evaluation of where they are needed.

13.2.7 Centers for Disease Control and Prevention

The Centers for Disease Control and Prevention (CDC) is part of the Department of Health and Human Services. It was founded in 1946 to help control the spread of malaria, and was first called the

Fig 13.7 Classic gully erosion from surface water runoff can be very destructive. The Natural Resources Conservation Service helps landowners prevent erosion before it destroys the landscape (Photograph courtesy of NRCS)

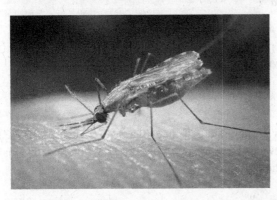

Fig 13.8 Fighting this villain, the *Anopheles* mosquito, the vector for spreading malaria, was the first job of the Center for Disease Control (Photograph courtesy of CDC)

Communicable Disease Center. Its initial mission focused primarily on killing mosquitoes (Figure 13.8). Since then, the CDC has developed into the premier health agency in the world with public health efforts to prevent and control the spread of infectious and chronic diseases, including environmental health threats and bioterrorism. CDC's mission is "to promote health and quality of life by preventing and controlling disease, injury, and disability." Its main office is in Atlanta, Georgia.

The CDC conducts research and investigations to improve people's daily lives. It also responds to health emergencies, such as the COVID-19 pandemic. It works with states and other partners to provide a system of health surveillance to monitor and prevent disease outbreaks by implementing disease-prevention strategies. It also maintains national health statistics and strives to guard against international disease transmission, with personnel stationed in more than 25 foreign countries. Waterborne diseases are a major interest of the CDC. It investigates drinking water issues, but also is concerned with swimming pools, lakes, streams, and other natural waters, especially in recreational areas.

13.3 Selected US Federal Agency Water Case Studies

The following brief case studies are presented to provide examples of water resource issues involving various US federal water agencies. The examples focus on federal water agency activities, although stakeholders include local, state, and regional water resources agencies. This diversity of government agencies – and their sometimes-competing missions – adds to the complexity and slow progress that too often ensues when trying to mitigate water resource issues.

13.3.1 Case Study: Water Quantity

Overview

The Colorado River was divided by an interstate compact in 1922 and faces severe demands on its water supply – dwindling due to drought, warming temperatures, and explosive urban growth. Some 25 million people in seven western states – Arizona, California, Colorado, New Mexico, Nevada, Utah, and Wyoming – rely on the Colorado River for water and power. The river also supports a diverse riparian system that has suffered as water flows have decreased (Figure 13.9).

Primary Federal Agencies Involved

US Bureau of Reclamation (lead agency, water supply, compliance with NEPA), US Fish and Wildlife Service (fish and wildlife protection), US Environmental Protection Agency (water quality), and the US National Park Service (environmental issues).

Other Agency Involvement

International Boundary and Water Commission, United States and Mexico (IBWC); states of Arizona, California, Colorado, Nevada, New Mexico, Utah, and Wyoming; regional and local governments, and private conservation groups.

Issues

The 1922 Colorado River Compact provided for the division of water from the Colorado River among the seven western states – 9.3 billion cubic meters (2.4 trillion gallons or 7.5 million acre-feet) for Arizona, California, and Nevada, and the remaining flows to the four Upper Basin States. Unfortunately, for the Upper Basin, the period of record used in 1922 to allocate river flows was during a very wet period for the Colorado River Basin. Subsequently, the 7.5 million acre-feet figure for the three Lower Basin states was too high. Currently, flows in the Colorado River cannot support the long-term needs of all potential users.

   The Secretary of the Department of Interior (the Secretary) is responsible for managing the mainstream waters of the lower Colorado River pursuant to the 1922 compact and federal law. The BuRec is responsible to act on behalf of the Secretary and is the lead federal agency to ensure

Fig 13.9 The Colorado River at Horseshoe Bend near Page, Arizona, US, remains beautiful despite the current demands on its limited water supply
(Photograph courtesy of Paul Hermans – Own work, CC BY-SA 3.0, https://commons.wikimedia.org/w/index.php?curid=22298121, converted to black and white)

compliance with the NEPA. The BuRec, under federal law, is required to consult with the USFWS regarding potential impacts to fish and wildlife, and the USEPA regarding water-quality issues – particularly salinity. In 1997, US Department of Interior Secretary Bruce Babbitt announced rules to allow the interstate transfer of Colorado River water from agricultural users to urban users; the first ever allowed under federal law.

Water quality, particularly salinity, has become a profoundly serious issue in the lower reaches of the river, particularly in Mexico. The Salinity Control Act of 1974 was adopted to require the US to construct facilities to control salinity levels in the Colorado River at the Mexican border. This program must also meet the criteria of the Clean Water Act of 1972, and the Mexican Water Treaty of 1944. Although the BuRec is the lead federal agency, the USEPA and US Department of Agriculture have been very involved in these water quality activities.

13.3.2 Case Study: Water Quality

Overview

Nitrogen and phosphorus in waters of the Mississippi River are flowing into the Gulf of Mexico and killing marine life off the Louisiana coast. Phytoplankton blooms are increasing, consuming dissolved oxygen from Gulf waters, and robbing it from other biological needs. A "dead zone" of aquatic life has been observed since the late 1980s, growing seasonally in size into an area larger

Fig 13.10 The Mississippi River deliveries sediments and nutrients into the Gulf of Mexico (NASA image by Robert Simmon, based on Landsat data provided by the UMD Global Land Cover Facility)

than New Jersey (Figure 13.10). Water mixing during the hurricane season eliminates the dead zone, so its extent varies seasonally.

Primary Federal Agencies Involved

US Environmental Protection Agency, US Department of Agriculture, US Army Corps of Engineers, US Fish and Wildlife Service, and the National Oceanic and Atmospheric Administration.

Other Agency Involvement

Various upstream states.

Issues

Excess nitrogen and phosphorus loading has created a "dead zone" in the north-central Gulf of Mexico. Federal and state government agencies are working to improve erosion control and fertilizer practices to reduce runoff into the Mississippi River and, ultimately, to the Gulf. The goal is to reduce the size of the hypoxia zone – the area lacking enough oxygen to sustain aquatic life. Nitrogen and phosphorus loading, coastal water circulation patterns, discharge from the Mississippi River and salinity levels all contribute to the seasonal variation of this problem. In addition to fertilizer practices, river channelization, loss of wetlands, loss of floodplain area, and removal of vegetation along the Mississippi River have also contributed to increase loading of nutrients to the Gulf. market mechanisms – such as nutrient trading between agricultural producers and industry – may provide positive incentives for environmental improvement. However, the need for conservation starts at the landowner level. For example, 1.6 million hectares (four million acres) of the Mississippi Delta drain into the Mississippi River. Almost 75 percent of that land is privately owned and cultivated for agriculture production. That is why agencies such as the US Department of Agriculture Natural Resources Conservation Service work with individual landowners to put conservation best management practices (BMPs) into practice. This is true for all the states draining into the Mississippi River.

13.3.3 Case Study: Flood Control

Overview

The Los Angeles River – encased in concrete through the city bearing its name – is in crisis (Figure 13.11). During the twentieth century, freeways have been developed with little or no mitigation for the riparian habitat of the river system in Los Angeles. It has been channelized to control infrequent but devastating flooding compounded by urban development – impervious rooftops, parking lots, and roads. Unfortunately, flood control in the lower reaches of the river has been inadequate, and improvements are needed – exacerbating the magnitude of concrete channels containing the urban river.

Primary Federal Agencies Involved

US Army Corps of Engineers.

Other Agency Involvement

City and County of Los Angeles, Friends of the Los Angeles River, and Unpave L.A.

Issues

The Los Angeles River was the birthplace of California's largest city. Unfortunately, the river has been denuded by concrete to contain the tremendous floodwaters generated by the watershed. New plans by the USACE, to expand the length and height of the concrete drainage channels and to provide additional safety from floods, are being met with resistance by those that would like to see the channel function again as a river – complete with riparian features for wildlife. This could be accomplished with riverside restoration projects to restore the river's natural ecosystem. Concerns

Fig 13.11 The Los Angeles River – running through the city with its urban riparian zone
(Photograph by Ed Fitzgerald at http://en.wikipedia.org/wiki/File:274851934_bfe9d6728c_b.jpg)

over flood protection are a reality, and incorporation of riparian features present challenges. Talks continue.

13.3.4 Case Study: Water Security

Overview

A computer hacker penetrated the security system of a water filtering plant near Fort Collins, Colorado in 2019. The Federal Bureau of Investigation (FBI) immediately launched a probe into one of numerous cyber incidents in the past two decades.

Federal Agencies Involved

US Environmental Protection Agency, Department of Homeland Security, and the FBI.

Other Agency Involvement

Federal, state, and local governments.

Issues

The cyber security of drinking and wastewater treatment facilities has become a high priority of local, state, and federal agencies in the past 20 years. The US Congress has provided hundreds of millions of dollars to improve the safety of water systems across the country and has adopted legislation requiring security vulnerability assessments of dams, lakes, and water distribution systems; in addition to water treatment facilities.

Vulnerability assessments have been required to protect water supplies from potential terrorist attacks. Response plans have also been developed and practiced for emergencies and other incidents. New security measures have been adopted by state and local governments, including health agencies, emergency responders, environmental professionals, and law enforcement agencies. Milwaukee, Wisconsin, for example, fenced its water treatment plants, improved lighting and security, and installed video surveillance of its properties. In addition, local health officials track illnesses in their communities, to detect the early outbreak of a waterborne epidemic, as part of bioterrorism preparedness.

13.3.5 Case Study: Threatened and Endangered Species

Overview

The world's only wild flock of whooping cranes (*Grus americana*) is continuing its comeback, and now numbers a record 825 birds (Figure 13.12). The endangered cranes winter at the Texas Coastal Bend area along the Gulf of Mexico. Spring nesting grounds are located in the boreal forest of Wood Buffalo National Park – Canada's largest national park (established in 1922) – in northern Alberta and the Northwest Territories. The surviving cranes are all descended from 15 that survived habitat loss and hunters since 1941.

Primary Federal Agencies Involved

US Fish and Wildlife Service, Parks Canada, and Environment Canada.

Fig 13.12 Whooping crane (*Grus americana*)
(Photograph courtesy of USFWS)

Other Agency Involvement

Local, state, and provincial governments, and conservation groups.

Issues

The ESA was passed by the US Congress in 1973 to conserve and restore species at risk of extinction. The ESA is successfully returning endangered species to a recovery status, primarily because of recovery plans, protection of critical habitat, and the federal listing of the species itself. However, critics argue that recovery of only 46 species in 32 years indicates failure. Regardless, the ESA is one of the nation's most powerful environmental laws.

13.3.6 Case Study: Ecosystem Restoration

Overview

In 1948, the USACE was authorized by Congress to construct an elaborate water management and drainage system in the Everglades (Figure 13.13) of central Florida to protect the area from flooding and other high-water problems. The project was very effective, but negatively affected the ecosystem of the region.

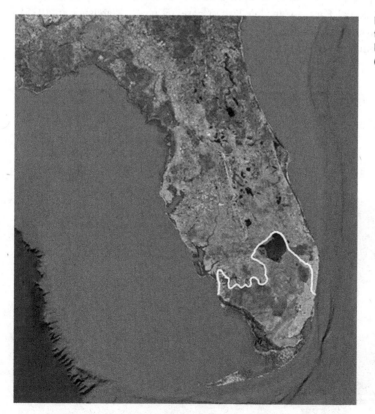

Fig 13.13 The white line delineates the northern boundary of the Florida Everglades ecoregion
(Photograph courtesy of NASA)

Primary Federal Agencies Involved

US Army Corps of Engineers, US Fish and Wildlife Service, US National Park Service, and the US Environmental Protection Agency.

Other Agency Involvement

State of Florida, Southwest Florida Water Management District, and Tribal Councils.

Issues

In the mid-1800s, the US Congress authorized numerous "Swampland Acts" to drain wetlands and other marshy areas to improve human health and to promote economic development. In the early 1900s, high water from devastating hurricanes, and seasonal rains, prevented settlement in much of the interior of Florida. In the mid-1900s, extensive drainage systems were installed by the USACE to improve development potential. However, the Florida Everglades ecosystem was devastated.

In 2000, the Comprehensive Everglades Restoration Plan (CERP) was approved to restore, protect, and preserve the water resources and environment of central and south Florida. Cost was estimated at US$7.8 billion over the next 30 years. Freshwater flows that now leave the area will be captured and redirected to areas of greatest environmental need, reviving dying ecosystems. In addition, some water will be used to benefit cities and agricultural producers in the region. The project is the most comprehensive and ambitious ecosystem restoration project ever attempted in the US.

13.3.7 Case Study: Climate Change

Overview

The world has warmed 1.1 °C since the Industrial Revolution of the late 1800s. Scientists predict that if we surpass 2 °C, the risk escalates to destroy coral reefs, thaw Arctic permafrost, and collapse the West Antarctic ice sheet. In 2015, the Paris Climate Agreement was adopted by over 160 countries to reduce global emissions. It was a landmark international agreement which the US rejoined under President Joe Biden on February 19, 2021.

Primary Federal Agencies Involved

US Environmental Protection Agency, US Fish and Wildlife Service, US Bureau of Reclamation, US Department of Agriculture, US Army Corps of Engineers, and the US Geological Survey (Figure 13.14).

Other Agency Involvement

Numerous local, state, and regional governments and private organizations.

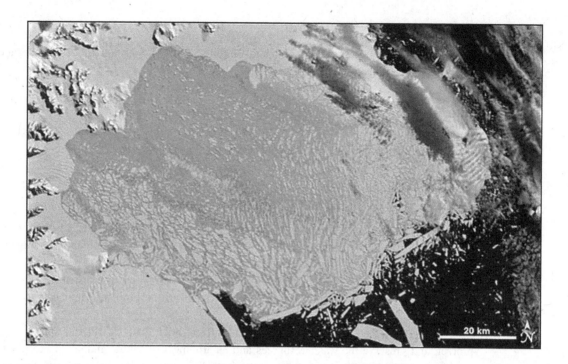

Fig 13.14 The Larsen-B Ice Shelf on the Antarctic Peninsula collapsed over 35 days in early 2002, prompted by 3 °C of warming since the 1940s (NASA image by Jesse Allen, based on MODIS data)

Issues

Future effects of global climate change could have a significant effect on water resources, ecosystems, and biodiversity in the US and around the world. Climatic change introduces uncertainty for future water supplies. It could lead to a global redistribution of annual and long-term precipitation patterns. In addition, changes in precipitation timing, intensity, and duration could affect water quality in rivers, lakes, and aquifers. These changes could lead toward less reliable water supplies in some regions, and wetter conditions in others. Water conservation and other management issues may be required to adapt to these conditions. Government agencies at all levels will be required to develop new programs and practices to meet these future changes.

Ecosystems and organisms attempt to adapt to regional climatic conditions over time. Humans rely on these systems for natural, cultural, recreational, and economic stability. Global warming could alter the number and type of biological diversity present in an ecosystem; due to changes in the start and end of breeding seasons, shifts in migration patterns, distribution, body size, and population numbers. Conditions could also change that have a beneficial effect on some species, although certain invasive species (such as weeds or mosquitoes) could be the beneficiaries.

13.3.8 Case Study: Navigation

Overview

In 2003, a federal judge held that the USACE was in contempt of court for refusing to lower water levels in the Missouri River to protect endangered fish and wildlife (Figure 13.15). The Secretary

Fig 13.15 Fort Randall Dam on the Missouri River, Pickstown, South Dakota, US, is one of the USACE dams capable of regulated water releases
(Photograph courtesy of US Army Corp of Engineers Digital Visual Library)

of the Army was ordered to comply or pay a US$500,000 fine each day the order was not obeyed. The USACE argued the ruling conflicted with a federal court order requiring adequate river flows to allow for barge traffic and power generation. The Missouri River has numerous federal on-channel reservoirs that are used to regulate flows of the river.

Primary Federal Agencies Involved

US Army Corps of Engineers (for navigation and flood protection), and the US Fish and Wildlife Service (for fish and wildlife protection, endangered species).

Other Agency Involvement

States of Iowa, Missouri, Montana, Nebraska, North Dakota, and South Dakota, and private conservation groups.

Issues

Recent drought has led to increasing conflict between the use of the Missouri River for navigation and power generation, and the needs of fish and wildlife. Six USACE reservoirs stretch from Nebraska and South Dakota upstream to Montana, and provide up to 86 billion cubic meters (23 trillion gallons or 70 million acre-feet) of storage for flood control, irrigation, navigation, and power generation. It is the largest reservoir system in the US, and includes Fort Peck, Oahe, Garrison, Big Bend, Fort Randall, and Gavins Point. Barge traffic is used as a cheaper alternative to trucks and rail transportation, and moves bulk cement, grain, gravel, liquid asphalt, and other commodities. The cost of transporting by truck is as much as seven times greater than by barge, and the rising cost of fuel makes it much greater.

If the combined storage levels of all six reservoirs fall below 38 billion cubic meters (10 trillion gallons or 31 million acre-feet) by March 15 of any year, the USACE will not commit water releases for navigation for the coming season – April 1 to December 1. The USFWS states that flows that are more natural are needed along the river to improve the health of the river, and to protect fish and wildlife in the region. The National Academy of Sciences (NAS) supports the USFWS position. In its 2002 Report [3], *The Missouri River Ecosystem: Exploring the Prospects of Recovery*, the NAS details the terrible degradation that has occurred along the river, and how seasonal flow changes are vital to restore ecosystem health. It states that the benefits of such changes would be US$87 million, with almost no impact on flood control – the primary purpose of the six upstream dams and reservoirs.

13.3.9 Case Study: Invasive Species

Overview

An invasive species is an organism – such as a microbe, plant, or animal – that enters an ecosystem through natural processes or with human assistance, and poses a threat to public health or the economy. Zebra mussels (Figure 13.16) and salt cedar (tamarisk) (Figure 13.17) are examples of invasive species in the US that have significant negative effects on water resources.

Primary Federal Agencies Involved

US Fish and Wildlife Service (fish and wildlife protection), US National Park Service, US Bureau of Land Management, US Department of Agriculture, and the US Bureau of Reclamation.

Fig 13.16 Zebra mussels (Dreissena polymorpha). These are live mussels: you can see that they are open
(Getty image credit: Ed Reschke)

Fig 13.17 Tamarisk (salt cedar) is an ornamental with vibrant, pink flowers in spring. It is also a highly invasive water user. Choosing to plant non-native species often has negative consequences (Getty image credit: Rolf Nussbaumer)

Other Agency Involvement

State and local agencies, and private conservation groups.

Issues

Currently, 180 non-native species exist in the Great Lakes, with zebra mussels (*Dreissena polymorpha*) causing up to US$200 million in economic damages annually. The Great Lakes (Erie, Huron, Michigan, Ontario, and Superior) provide water to over 25 million people. Zebra mussels are a small mussel, about the size of a human fingernail, and are native to the Caspian Sea in Asia. It is believed the mussels were introduced into the Great Lakes by ballast water that was released from ocean freighters. Zebra mussels were first discovered in Lake St. Clair, near Detroit, Michigan, in 1988. Since then, they have spread rapidly to all the Great Lakes and waterways in several states and provinces, particularly Ontario and Quebec.

Salt cedar (tamarisk, *Tamarix aphylla*) is rapidly spreading in riparian zones along riverbanks in the southwestern US and can consume up to 1.1 cubic meters (300 gallons) of water every day. Salt is released from the plant roots to prevent other plants from growing nearby. It was introduced from Eurasia as an ornamental shrub in the early 1800s. The deep-rooted plant aggressively obtains water from the surrounding soil and groundwater. Tamarisk grows to a height of 3–5 meters (12–15 feet). It is causing considerable problems by using so much water that it is reducing water supplies in federal and local water projects. Tamarisk is also encroaching on lands of national wildlife refuges.

The US National Park Service, US Fish and Wildlife Service, US Department of Agriculture, and several university researchers among others are experimenting with biological controls. Two species of beetle, the Mediterranean tamarisk beetle, Diorhabda elongata and the subtropical tamarisk beetle, Diorhabda sublineata, approved for use in controlling tamarisk have been highly successful in the US Southwest after testing in Texas [4]. The northern tamarisk beetle (*Diorhabda carinulata*) has been used with similar success in Colorado since 2005.

13.3.10 Case Study: Tribal Agency Water Issues

Overview

In 2006, the Native American Rights Fund (NARF) filed a federal lawsuit on behalf of the Kickapoo Tribe in Kansas to enforce construction of the Plum Creek Reservoir Project. In addition, the Tribe is seeking a declaration that the Tribe and its members hold the senior water rights to the Upper Delaware River system in Kansas. The Tribe is also seeking an injunction against state or federal funding for water-related projects on private land, since it is destroying access to a dependable supply of sufficient quality for the tribal lands.

Federal Agencies Involved

Bureau of Indian Affairs, US Department of Agriculture – Natural Resources Conservation Service (NRCS), and the US Environmental Protection Agency.

Other Agency Involvement

Kickapoo Tribe, State of Kansas, Native American Rights Fund, and local watershed agencies.

Issues

According to the Native American Rights Fund litigation, the Kickapoo Tribe has been working with the NRCS since 1983 to construct the multi-purpose Plum Creek Project. They signed a federal contract in 1994 to begin construction. Part of the project required the condemnation of private property, but the local watershed district voted not to pursue any condemnation proceedings. The Tribe does not have the authority to condemn private property. It also does not have access to a safe drinking water supply. (The Tribe's water plant was under a federal USEPA notice of violation of the Safe Drinking Water Act of 1974.)

In 1908, the US Supreme Court established the Winters Doctrine in *Winters v. United States*, which established the concept of a federal reserved water right. In effect, any reservation created by the US government carries with it a water right sufficient to fulfill the purposes of the reservation. The priority date of the federal reserved water right is the same date as the creation of the reservation – for the Kickapoo Tribe that date is October 24, 1832, with the Treaty of Castor Hill.

During the drought of 2003, the Delaware River in Kansas did not flow for over 60 days, and the Kickapoo Tribe was forced to truck over 26,500 cubic meters (7,000,000 gallons) of drinking water to the reservation. Commercial operations were required to reduce water consumption by almost 60 percent. The current water treatment system was constructed in the mid-1970s, and was inadequate to meet drinking water and fire protection needs [5]. Kansas reached an agreement with the Kickapoo Tribe in 2016 establishing the Tribe's senior water rights and requesting federal funds for construction of a reservoir to ensure adequate water to meet the Tribe's needs [6].

More recently, the Dakota Access Pipeline project in North Dakota and other Midwest states has created great controversy regarding the potential for surface and groundwater pollution. The pipeline extends 1,886 km (1,172 mi) from the shale oil fields of the Bakken formation in northwest North Dakota, across South Dakota and Iowa, and ends at an oil terminal near Patoka, Illinois. Water pollution issues related to potential pipeline leaks and breaks have created grave concerns with the Standing Rock Sioux Tribe and others.

13.4 Selected Regional, State, and Local Water Agency Case Studies

13.4.1 Case Study: Regional Water Agency Issues

Watershed Protection

Overview

Lake Tahoe is located on the California/Nevada border, and is one of the deepest (more than 0.4 kilometers (0.25 mile) deep) and most clear alpine lakes in the world. The lake has been degraded by environmental problems such as contaminated storm runoff, erosion, and air pollution. The deep blue color of the lake has been fading – about 0.3 meter (1 foot) of clarity is lost annually due to these pollutants (Figure 13.18).

Regional Water Agencies Involved

Tahoe Regional Planning Agency, and the Lahontan Regional Water Quality Control Board.

Issues

Federal, regional, state, and local agencies are working together to develop a water quality plan to identify sources of pollution. Then, they will designate reductions in sediments and nutrients that are necessary to restore Lake Tahoe's clarity and quality. These reductions, or limits, are called total maximum daily loads (TMDLs). Since 1997, the USEPA has provided more than US$22 million to promote water-quality efforts around the lake, and has devoted a full-time staff person to work with other officials for the long-term protection of the lake.

Development around Lake Tahoe has been closely regulated for over two decades. However, a new issue has created concern for the future protection of the lake – boating. A previous Shorezone Plan would remove a 20-year moratorium on the construction of boat piers on about two-thirds of the lake. This would occur within an area previously considered sensitive habitat to fish. However, decades of scientific study have concluded that boat piers are not detrimental to fish populations. It's been estimated that hundreds of piers and boat slips would be constructed if the restrictions are removed, and boat traffic on the lake would increase by 30 percent in the next two decades, or 60,000 motorized trips. To mitigate potential water pollution, the

Fig 13.18 Lake Tahoe is in danger of losing the legendary blue waters due to water quality degradation (Getty image credit: Valentin Prokopets)

Shorezone Plan includes the nation's first environmental boating plan – called the "Blue Boating Program." Certification for all boats on the lake is required. Certification will ensure that a vessel's engine is properly maintained, and the hull is free of invasive species.

13.4.2 Case Study: State Water Agency Issues

Groundwater

Overview

In 2005, the state of Mississippi sued the Memphis Light, Gas, and Water Division (the city-owned utility) for US$1 billion due to alleged injury from well pumping near the Tennessee/Mississippi border. Mississippi claims that scores of wells operated by Memphis are lowering groundwater levels in the Memphis Sands aquifer across the state line.

State Water Agencies Involved

States of Mississippi and Tennessee.

Other Agency Involvement

Memphis Light, Gas, and Water Division, and the City of Memphis.

Issues

In 2002, a report by the State of Tennessee ironically noted that groundwater pumping near Memphis, Tennessee, could deplete groundwater supplies in Mississippi to the south. Ongoing pumping has created a drawdown in the Memphis Sands (a high-quality aquifer) extending beneath northern Mississippi. Tennessee officials argue that groundwater pumping for use in the city of Memphis is not injurious to its neighbor to the south.

The State of Mississippi is arguing before a federal court that Memphis should be required to obtain its water supplies from the Mississippi River; a prospect that could cost billions of dollars to access and treat. It is a unique lawsuit over groundwater supplies in two neighboring states. (It follows an attempt by a private company to pump groundwater in Mississippi, load it onto ocean freighters, and transport it to Saudi Arabia.) Memphis contends that irrigators on the other side of the Mississippi River, in neighboring Arkansas to the west, are also depleting groundwater supplies that could be injuring the state of Mississippi. City of Memphis officials also argue that most aquifer recharge occurs in Tennessee, and that entitles the City of Memphis to unlimited pumping from "their" aquifer [7].

13.4.3 Case Study: Interstate Water Conflicts

Multi-Use Conflicts

Overview

In 1990, the states of Alabama, Florida, and Georgia began working on a water allocation formula to divide waters of the Apalachicola–Chattahoochee–Flint (ACF) and Alabama–Coosa–Tallapoosa (ACT) River Basins between the three states. The issue has been heard by federal courts numerous times. Teams of negotiators are still working to find a long-term solution to water shortages in the region.

Federal Agencies Involved

US Army Corps of Engineers, US Fish and Wildlife Service.

Other Agency Involvement

States of Alabama, Georgia, and Florida.

Issues

Urban growth in Atlanta, Georgia, has increased demands for public water supply. In addition, environmental needs in Florida have led to disputes between the three basin states of Alabama, Georgia, and Florida. Governors of the three river states have appointed members to negotiating committees to attempt to reach a settlement – called the Tri-State Agreement – for the distribution of waters from the ACF and ACT River Basins.

The USACE operates Lake Lanier, near Atlanta, Georgia, and has added to the controversy. It plans to redistribute water supplies from hydropower generation (a non-consumptive use) to drinking water use. Georgia also plans to construct a new reservoir upstream of Alabama – effectively cutting off a portion of the water supply to the lower river states. In 1997, the US

Congress and the three states agreed to provisions of an interstate water allocation compact. However, it was never ratified by the state governments due to ongoing rivalries regarding its terms. In 2002, a preliminary 30-year agreement was announced between the three states, but it also was not finalized. Deadlines to extend negotiations have been made dozens of times in the past decade. It is now being heard before the US Supreme Court.

13.4.4 Two Case Studies: Local Agency Water Issues

Drinking Water – Case Study 1

Overview

In April 1993, more than 100 people died after drinking contaminated water provided by the city of Milwaukee, Wisconsin. A tiny yet potent parasite (*Cryptosporidium parvum*) (Figure 13.19) infested the city's water supply system and made more than 400,000 sick. This led to an overhaul of the aging water treatment plants in the city. The outbreak was linked to inadequate water treatment of drinking water from Lake Michigan and was the largest water-borne disease outbreak in the more-industrialized world.

Primary Local Agencies Involved

City of Milwaukee, and the Wisconsin Department of Natural Resources.

Other Agency Involvement

Centers for Disease Control and Prevention (identification of disease, source, and prevention).

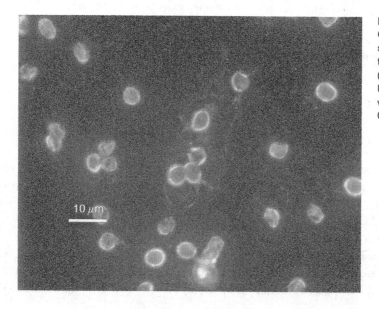

Fig 13.19 The parasite Cryptosporidium parvum can be seen using immunofluorescence techniques
(Photograph by H.D.A. Lindquist, USEPA at http://en.wikipedia.org/wiki/File:Cryptosporidium_parvum_01.jpg)

Issues

An unusually snowy and wet spring created higher than normal surface water runoff in 1993. The runoff probably contributed to the increased presence of the parasite. It is a very tough, one-celled protozoan able to withstand numerous environmental extremes and water treatment practices. *Cryptosporidium* is often found in rivers and lakes contaminated with animal waste or sewage from wastewater treatment plants. Cattle and dairy cows, especially calves, are often a major source of *Cryptosporidium*.

Severe diarrhea, caused by the parasite, struck hundreds of thousands of Milwaukee residents in 1993. A week-long boil order was put in place before the tainted water could be consumed. Tragically, for over 100 people – many with health ailments – the directive came too late. At the time, the USEPA had no drinking water standard for *Cryptosporidium*. The outbreak sickened 403,000, required 44,000 doctor visits, hospitalized 4,400, and caused the loss of US$96 million in wages and medical expenses. The city spent US$90 million to install a new ozone purification system, since chlorination does not kill *Cryptosporidium*. The water intake pipe was also extended an additional 1,280 meters (4,200 feet) into Lake Michigan to draw purer water [8].

Drinking Water – Case Study 2

Overview

The Flint, Michigan, water crisis began in 2014 when lead seeped into tap water when the source of drinking water was changed by city officials. Rust-colored water contained traces of leached lead from old pipes throughout the aging industrial city. The switch in water sources (from treated Detroit Water and Sewerage Department water (obtained from Lake Huron and the Detroit River) to the Flint River was explained as a cost-savings decision by Flint city officials. In the years since, residents, and particularly children, have experienced severe health issues after using and drinking the contaminated water.

Primary Local Agencies Involved

City of Flint, Genesee County, and the Michigan Department of Environmental Quality.

Other Agency Involvement

US Environmental Protection Agency, Federal Emergency Management Agency (FEMA), and the Centers for Disease Control and Prevention.

Issues

Flint, Michigan, has struggled economically since the 1980s when several General Motors automobile assembly plants closed during the 1980s and 1990s. Michigan's governor declared a financial emergency for the city in in 2002, and things did not improve over the decade. In 2011, the governor appointed several unelected managers to run the city. Those managers reported directly to the Michigan State Treasury and not to the citizens of Flint. In 2014, the cost-saving decision was made to switch to the less expensive water source of the Flint River. Issues arose immediately over bacteria-contaminated drinking water and then later over the high levels of lead caused from the corrosive nature of the new water source on household plumbing.

In October 2015, Michigan Governor Rick Snyder created an independent taskforce to review the events leading up to the health crisis. Their report stated that the crisis was a striking example of environmental injustice, since Flint's poor, largely African American population "did not enjoy the same degree of protection from environmental and health hazards as that provided to other communities." The Michigan attorney general brought criminal charges against two Michigan Department of Environmental Quality employees and the Flint city utilities administrator. In June 2016, more than two years after the switch to Flint River water and then eight months after the return to Detroit Water and Sewerage Department supplies, the USEPA declared that water in Flint was once again safe to drink, provided an approved filter was used to remove the remaining traces of lead.

13.4.5 Case Study: Groundwater/Surface Water Conflict

Overview

On May 8, 2006, 440 shallow alluvial groundwater wells were ordered shut down due to inadequate water supplies to protect senior water rights on the South Platte River of Colorado. Most wells were used for irrigation, and crops had already been planted. It is believed to be the first time in the history of the US that irrigation well pumping was curtailed after the start of the crop-growing season.

Primary Local Agency Involved

Central Colorado Water Conservancy District.

Other Agencies Involved

State of Colorado, various cities, and irrigation groups.

Issues

Colorado follows the strict Doctrine of Prior Appropriation, and requires that surface-water users, and tributary groundwater pumpers, follow the same priority system – "first in time, first in right." Pumping from tributary (alluvial) groundwater had been ongoing for over 100 years, but in 1969, the Colorado Legislature approved legislation to require that such wells also follow the "first in time, first in right" rule. Alluvial wells were given a priority equal to the date the well was drilled and put to beneficial use, generally in the 1930s and 1950s. Most senior surface-water rights along the South Platte River, by contrast, have priority dates in the mid- to late-1800s. The well shutdown in 2006 was devastating to well irrigators and the local agricultural economy. Property values of formerly irrigated lands plummeted, and lack of well water forced some irrigators to leave the area. In addition, a few wells that were used for drinking water and irrigation of athletic fields at public schools were shut down and not used for a period of time.

13.5 Privatization of Water Systems

Water, according to *Fortune Magazine* [9], is "One of the world's great business opportunities. It promises to be to the twenty-first century what oil was to the twentieth." According to a report by the Canadian Broadcasting Corporation in 2003, three giant global corporations have quietly taken control of water systems that service almost 300 million people around the world [10]. The results have ranged from mixed to disastrous.

Guest Essay By Dr. Laurel Phoenix

Dr. Phoenix is a professor in the Environmental Studies Department at the University of Wisconsin, Green Bay. Her research interests are water resources management and environmental land use planning, with an emphasis on municipal water and sewer infrastructure and source water development.

Water Privatization: Some International Trends

Who should own or control the water you drink? Should water be considered a human right or just another commodity to be bought and sold? These philosophical questions have important social, economic, political, and environmental consequences depending on how they are answered. This essay will discuss the growing trend of international water privatization, primarily drinking water privatization, some of the concerns associated with this privatization, and highlights from places around the world where privatization has proven to be against the public interest.

   Water privatization is the situation in which a private company controls part or all of a public utility, like drinking water treatment and delivery or sewage collection and treatment. What drives this new push for privatization? First, world population is growing but freshwater supply is limited. In fact, freshwater supply is decreasing because contamination effectively eliminates some water as potential source water. With a limited, non-substitutable, and essential resource like water – versus increasing demand from growing populations – the potential price of the resource goes up. Although some would be willing to pay more for water, the poorest citizens would not be able to translate their need for water into a market demand, since they would have no money to pay for it. Coupled with this is the global problem shared by cities of aging or inadequate water supply systems that need upgrading or expanding – such cities are finding it increasingly difficult to acquire the funding necessary to start such large capital projects. For example, cities such as Baltimore, Washington, DC, St. Louis, and Boston have water pipes as old as 100 years or more, made of hollowed logs or brittle cast iron, which easily cracks under increased pressures to pump more water through the system to meet new demand.

   Responding to this demand are several giant corporations that have realized the potential profits in providing urban water and sewer services. Although there are several smaller engineering firms that supply these services to cities, the three largest multinationals are based in Europe. Many public water and sewage utilities have already been privatized in Europe, whereas 80 percent of Americans are still served by public utilities. These corporations are driven to expand by news that in America, it will cost US$1 trillion to upgrade and expand all public water and wastewater facilities. Analysts predict the American drinking water market would grow an additional US$120–140 billion by 2019. In America and Europe (now including former Eastern European countries), the three largest multinationals were predicted to control 65–75 percent of public utilities by 2020. The World Health Organization (WHO) estimates 1.2 billion persons worldwide lack access to safe drinking water, and 2.5 billion persons lack access to adequate sanitation. The Second World Water Forum meeting in 2000 predicted that the global cost of providing safe drinking water alone would be US$100 billion for the next 25 years [11].

Private corporations argue that they can manage or build water systems more efficiently than public entities, although there are no data yet to support this for the wealthier countries. These companies also argue that a big advantage of giving them a 20-year contract is that they have access to far more investment funds than a city could get to start expensive capital projects. This can be true in some instances, but some companies are growing so fast that their own debt is looking burdensome, and a city wouldn't want to be caught if a rapid investment bubble burst during the life of the contract. Sometimes private companies have lobbied for increased public funding for capital expansion, so they don't have to invest so much of their own money.

Water companies are increasing their donations to politicians from the local to national level, utilizing more lobbyists to influence national laws supporting privatization, and conducting multifaceted public relations campaigns to sway citizens to favor privatization. These strategies help them enter new markets and keep their shares attractive.

The result of increasing water demand and rapid response to fill demand has led to some instances of unsuccessful privatization, where the public interest was not served. From several of these examples, Gleick et al, have created a list of privatization concerns that need addressing in future privatization contracts so that urban citizens of any country can benefit [12]. First, if local governments give up their historical responsibility to provide basic services to all of their citizens through privatization, then they must maintain oversight and regulatory power to ensure that the public interest is still upheld and citizens are served responsibly and equally.

Another concern is that privatization can increase economic inequities if water is priced beyond the ability of the poor to pay. In these cases, some kind of subsidy needs to be created. Similarly, improper pricing will not encourage efficiency and conservation by those who have sufficient funds to pay their water bills.

If contracts do not include contract oversight and public participation, the public interest can be ignored, either through inadequate maintenance, failure to increase access to unserved neighborhoods, or ignoring water-quality problems that could harm people over time. Similarly, dispute resolution processes must be clearly written for cities to address any failures in service provision or corporation requests to increase water bills. A corporation may have even less interest in consequences on downstream communities or local ecosystems than a city would, so actions bearing potential consequences on those outside entities must be protected in contracts.

Privatization also means that any profits made will leave the local community to the parent companies. This is of greater significance in wealthier countries where cities have had long experience with running their own public systems. Privatization should never be absolute and give a corporation the ownership of water or water rights themselves, but only give them a contract to treat and distribute the water. This is where water as a commodity and water as a social good and human right are most clearly at odds.

Public control over water ties in with economic development plans, and water offered to neighborhoods outside of the city is usually conditioned by first allowing their annexation. Private control of water would leave a city no control over how it could use water to attract new investment, or enlarge its political boundaries.

The following events give but a brief glimpse of examples in and outside the US where the growing pains of the water infrastructure industry have been felt.

United States and Canada

In Atlanta, Georgia, a 20-year contract signed in January of 1999 was cancelled in January of 2003 after too many problems with brown water, boil water orders, slow maintenance, billing, meter installation, and a collapsed road. The city has again created its own water department and will have to deal once again with upgrading old and broken pipes while the area continues to grow.

In 1999, the mayor of Stockton, California, wanted to privatize the city's water and wastewater systems. Local citizens protested and collected enough signatures to force a referendum on the issue. An outside auditor said that the city would not save the promised US$175 million over 20 years, but would save over US$1 million per year by *not* privatizing. Despite this, the city council signed a 20-year, $600 million contract two weeks before the March 2003 referendum vote was scheduled. In the fall of 2004, a California Supreme Court judge ruled that the council had abused its discretion by signing a contract before an EIS could be written on the project, as California state law requires.

Inspired by politician Stuart Smith, the city of Hamilton, Ontario, privatized its water and sewage treatment in 1995, experiencing ever-increasing rates, sewage spills, and unpaid environmental fines during the next six years. Sewage bubbled up through manhole covers and into basements, and flowed in Burlington Bay for days. A drastic cutback in employees by the private company had compromised safety, maintenance, and the ability to make repairs quickly. Hamilton now has a different private company but may go back to a publicly run entity in the future.

International Examples

Internationally, multinationals have realized privatization agreements in over 130 countries. In developing countries, these water corporations gain access to a privatization deal when the World Bank or the International Monetary Fund makes debt relief conditional on privatizing a country's utilities. These water corporations then move in, usually with a lucrative guaranteed profit, and start upgrading water treatment or sewer facilities. It is also common that as soon as the privatization is official, water rates are raised and the poorer folks who can't afford the higher rates are cut off. Quite often, the poorest edge communities of these cities that most needed treated water are not serviced at all.

The first and most well-known example is Cochabamba, Bolivia. Water rates rose soon after the contract was signed in 1999, and several subsequent rate hikes tripled or quadrupled monthly bills. Some families were now paying a full 20 percent of their monthly income on water. One year later, civil disobedience, strikes and street marching in 2000 resulted in the Bolivian government sending in troops and restricting civil liberties. After shooting into crowds on several occasions caused seven deaths and numerous injuries, the Bolivian government halted the escalation by canceling the contract and telling the water company that their workers' safety couldn't be guaranteed unless they left town immediately. The multinational is still trying to sue Bolivia for the loss of its anticipated future earnings.

In anticipation of privatizing their water supply, communities in South Africa in 1998 stopped offering subsidies on water bills. This was so that it would look to the incoming companies that the locals could afford to pay their water bills. However, the subsidized poor could not afford the higher rates, and their water service was cut off. When people were driven to use contaminated streams and ponds, a cholera epidemic killed close to 300 persons and sickened over a quarter-million by 2002.

Buenos Aires, Argentina, signed a privatization contract in 1993, despite numerous conditions of enormous benefit to the water company. By 2002, the company hadn't met its performance objectives, had cut off service to many who couldn't pay the 20 percent increase in rates, and never built a sewage plant that was promised by contract, leaving 95 percent of the city's sewage to flow untreated into the river. Argentina is suing the company for this and other broken contracts.

Despite the problems mentioned here, the water industry still has the potential to help build and maintain water and sewer infrastructure around the world, if contracts include safeguards for the public interest and are tightly regulated. The World Bank and similar institutions could guarantee countries this safety through requiring standard contracts that list such conditions. These could include: conditions for yearly guaranteed minimum investments, creating new access to underserved and unserved populations first (e.g., the poorest squatter communities adjacent to the cities), forfeiting payments or other incentives if water quality is not high or repairs are not timely, limits on the degree and frequency of water rate hikes, clear performance benchmarks, transparent accounting so locals can regulate the company, and no layoffs of existing public workers for the first few years.

Clearly, the control of local water should stay in the public sphere. We should no more be restricted from access to water as we are to air, and yet when privatization is not balanced with human rights, we risk this very thing. As the rush to privatize slows down to a more reasoned trickle, well-written contracts, and corporate transparency could allow successful public–private partnerships to occur.

Summary Points

- This chapter centers on recent water use in the United States. Numerous federal agencies play a role in the protection and management of water resources and the environment in the US. The **President's Council on Environmental Quality** coordinates the efforts of federal agency activities.
  - **US Environmental Protection Agency (USEPA)**: agency charged with regulation of the nation's resources. Guided by the 1969 **National Environmental Policy Act (NEPA)** and the **Clean Water Act** of 1972.

- **US Fish and Wildlife Service (USFWS)**: works to conserve, protect, and enhance fish, wildlife, and plants and their habitats. The **Endangered Species Act** of 1973 is within its domain.
- **US Army Corps of Engineers (USACE)**: developed in 1802, plans, designs, builds, and operates water resources and other civil works projects.
- **US Bureau of Reclamation (USBR)**: developed to help settle the arid West, they built dams and reservoirs. Today they work with improved water management and water conservation.
- **US Geological Survey (USGS)**: mission is to provide reliable scientific information to describe and understand the Earth; minimize loss of life from disasters; manage water, biological, energy, and mineral resources; and enhance and protect our quality of life.
- **US Department of Agriculture (USDA) Natural Resources Conservation Service (NRCS)**: charged with helping landowners conserve soil, water, and other natural resources.
- **US Department of Human Health and Services Centers for Disease Control and Prevention (CDC)**: mission is to promote health and quality of life by preventing and controlling disease, injury, and disability.

- The diversity of government agencies – and their sometimes-competing missions – adds to the complexity and slow progress of mitigating some water resource issues.

- Major water issues in the US include:
  - Water quantity: the amount of available water.
  - Water quality: is water quality good enough for the designated use?
  - Flood control: how can floods be controlled without harming the water resource?
  - Water security: is our nation's water supply protected?
  - Threatened and endangered species: is there enough, good quality, water for humans and animals?
  - Wetlands: can wetlands be protected or restored?
  - Global warming: certainly, an international issue, but is the US doing all that it can to deal with this issue?
  - Navigation: how do we protect the water necessary for navigating our waterways?
  - Invasive species: can non-native microbes, plants, or animals be controlled?
  - Tribal water issues: do tribes have water rights that need protection?
  - Watershed protection: who and how should protect the lands that feed water into streams, lakes, rivers, and the oceans?

- States deal with some water issues.
  - Groundwater rights.
  - Interstate conflicts over water.
  - Drinking water problems (fall under federal and state jurisdiction).
  - Groundwater – surface water conflicts.

- Water privatization is a growing trend in the treatment and delivery of municipal water resources in many countries.
  - The success of privatization depends a great deal on the integrity of the companies and the people who back them.

Questions for Analysis

1. Is it necessary for federal agencies to become involved in water resources issues? Would more local control provide a better model for water management?
2. What potential conflicts could develop between the various federal agencies described in this chapter? Why?
3. Of the selected water issues described in the second part of this chapter, which two would you consider the most critical? Why?
4. Several southeastern US states are reconsidering their water laws in the light of increasing demands by agriculture, municipalities, and industries and their recent drought status. What lessons can these states learn from the western US?
5. Are water laws from the US, Canada, or the European countries useful as models for developing countries?
6. Is privatization of municipal water systems a successful business model?

Further Reading

Gore, Al, 2006, *An Inconvenient Truth: The Planetary Emergency of Global Warming and What We Can Do About It*, New York: Rodale Books.
Hardin, Garrett, 1968, "The tragedy of the commons," *Science* 162, 1243–1248.

References

1 Hugh Hammond Bennett, 1959, *The Hugh Bennett Lectures*, Raleigh, N.C.: North Carolina State College.
2 US Geological Survey, http://water.usgs.gov/realtime.html
3 National Academy of Sciences, 2002, *The Missouri River Ecosystem: Exploring the Prospects for Recovery*, Washington, DC: National Academy Press, available at www.nap.edu/catalog/10277/the-missouri-river-ecosystem-exploring-the-prospects-for-recovery#toc (accessed 2021)
4 Allen Knutson, Mark Muegge, and C. Jack DeLoach, 2015, "Biological control of salt cedar," http://lubbock.tamu.edu/files/2015/06/Biological_Control_of_Saltcedar.pdf (accessed 2020).
5 Native American Rights Fund Legal Review, 2007, "Kickapoo Tribe in Kansas files lawsuit in Federal Court to end 30-year era of systematic deprivation of the Tribe's water rights," March, http://www.narf.org/pubs/nlr/nlr31–2.pdf
6 Kansas Department of Agriculture, 2016, "Kansas reaches water rights agreement with Kickapoo Tribe," https://agriculture.ks.gov/news-events/news-releases/2016/09/09/kansas-reaches-water-rights-agreement-with-kickapoo-tribe (accessed 2020).
7 Memphis Commercial Appeal, 2007, "Water fight at MLGW," March http://www.commercialappeal.com/mca/local/article
8 Milwaukee Journal Sentinel Online, 2007, "10 years ago, Crypto gripped the city," March http://www2.jsonline.com/news/metro/apr03/131542.asp
9 Shawn Tully, 2000, "Water, water everywhere," *Fortune Magazine*, May 15.
10 *CBC News*, 2007, "Water for profit," March, http://www.cbc.ca/news/features/water/

11 World Water Council (WWC), Second World Water Forum, 2000, https://www.worldwatercouncil.org/en/hague-2000, accessed March 2021.

12 Peter H. Gleick et al., 2002, *The New Economy of Water: The Risks and Benefits of Globalization and Privatization of Fresh Water*, The Pacific Institute for Studies in Development, Environment, and Security, http://pacinst.org/reports/new_economy.htm

14 Water Conflicts, Solutions, and Our Future

> The blessed work of helping the world forward, happily does not wait to be done by perfect men.
>
> George Eliot (Mary Ann Evans), English novelist (1819–1880) [1]

> We've arranged a civilization in which most crucial elements profoundly depend on science and technology. We have also arranged things so that almost no one understands science and technology. This is a prescription for disaster. We might get away with it for a while, but sooner or later this combustible mixture of ignorance and power is going to blow up in our faces.
>
> Carl Sagan, American Astronomer (1934–1996) [2]

Chapter Outline

14.1 Introduction

This is our concluding "Think About It" chapter. Our core goal for this book was to emphasize the need to not just learn facts, but also to recognize interactions and consequences; and analyze what can happen under a variety of circumstances that you may encounter in your career. There is much gloom in the world, but there is also tremendous hope and optimism. There is hope in students, such as yourselves, who will make a difference by increasing your understanding of our Earth and its inhabitants –from microbes to humans. This book is intended as an introduction to the amazing world of water resources and the issues surrounding it. We hope that it will help start your journey to prepare to accomplish the following, whether as citizens or scientists, in a variety of fields of study:

1. Make informed decisions about environmental issues – in particular, those involving water resources.
2. Understand the basics of the sciences and ecosystems involving water resources.

3. Know that understanding how a system works is essential to making decisions with as many positive outcomes for all resource uses as possible.
4. Realize that some decisions are exceedingly difficult to fix, once made and implemented.
5. Know how different water resources interact within different ecosystems and be able to predict changes in one resulting from changes in another.
6. Understand that ecosystems cannot be managed piecemeal without unintended consequences. It may help to remember Murphy's Law – "anything that can go wrong will go wrong."
7. Understand the competing uses for water and the consequences of trying to meet all expectations for each use.
8. Understand that making decisions is never simple when people, profit, politics, the environment, and water resources are involved.
9. Be prepared, as Henry David Thoreau said, to "Be not simply good; be good for something."

Please debate, argue, disagree, or agree about anything we say here. The point is to develop critical thinking skills. Let's start with the idea of water as a basic requirement of life, and is vital to the economic activities of communities, regions, and nations. Access to clean and adequate freshwater is a basic human right and must not be denied any human being. However, clean water comes with a cost – both economic and environmental. This leads to the question: Is water a social issue, an economic issue, or both? Alternatively, should water be a commodity? Should individuals and private companies be allowed to profit from water development, or should governments control and regulate all water use for the maximum benefit of society and the environment? Can or will governments successfully regulate all water use for the maximum benefit of society and the environment?

Adequate water supplies are an essential component of economic activity – either directly or indirectly. Therefore, water in many cases is an economic good or commodity and is developed and used for economic gain; for example, water privatization efforts. However, does this lead to maximum social good and environmental protection? Or could government control of local and regional water supplies provide better benefits? What are the repercussions when uninformed, profit-driven people (including politicians backed by powerful lobbyist) are making our laws? What incentives would exist for individuals or companies to forgo profit to achieve maximum social benefits if there were no rules or regulations?

A second aspect of water as an economic commodity is the impact on the environment. Water is an ecosystem resource, which, in adequate and dependable supply, maintains ecosystem health. Nevertheless, humans cannot function without water, nor can factories, office buildings, agriculture, or ecosystems. What costs do human uses impose on the natural environment? What is the true value of those uses? Diversion of water from rivers and aquifers too often results in injury to ecosystems and wildlife. Our natural systems can only remain healthy if adequately protected from excess resource extraction. Failure to return appropriate quantity and quality of water to ecosystems will deteriorate, and eventually destroy, many natural systems. These systems are part of our lifeline. They provide vital functions – from the smallest species to humans. Examples are the food webs that ultimately provide sustenance to all forms of life.

Water is a strange resource. Unlike minerals, timber, or other natural resources used by humans, water is mobile. It flows down rivers and is captured in wetlands and other natural and human-made storage features. It crosses state, provincial, and international boundaries, both above and below ground. A single drop of water can spend millennia in groundwater but only hours in the atmosphere. One's use of water prevents others from using the resource, at that moment in time, and may result in less water for others due to consumption, evaporation, or transpiration. Freshwater supplies are temporally and spatially a finite resource. As our world population grows, along with increased per capita consumption, will continue to place severe and competing demands on it throughout the world.

Agriculture, for crop and animal production, is our greatest user of freshwater across the globe. At the same time, the world's food sustainability is dependent on freshwater resources, soils, and climates that optimize food production. The FAO estimates that as world population grows to 9.2 billion by 2050, it will require a 60 percent increase in agricultural production to feed everyone [3]. It is a painful truth that all of us who live in affluent societies consume more food than we need for good nutrition, while in 2018, 113 million people in 53 countries were hungry, many to the point of starvation; 821 million experience chronic hunger, and children are particularly harmed by being micronutrient starved. Micronutrient starvation means that the brain and body do not have the essential elements to develop a healthy body and brain. Stunting, blindness, and mental retardation can result preventing these children from living full productive lives.

It is obvious that quality of food is just as important as quantity. The lack of nutritious food and adequate safe water is a humanitarian situation that has been driven primarily by three factors: (1) conflict, insecurity from local violence to political crisis; (2) climate/weather/natural disasters, including drought, floods, hurricanes, earthquakes, and extended dry or wet spells; and (3) economic issues such as hyperinflation, economic down turn, loss of income, or a mutually reinforcing combinations of two or more of these drivers. Any or all of these situations, which can negatively affect food and safe water supplies, can result in crises [4].

14.2 Tragedy of the Commons

The concept of the "tragedy of the commons" is both a simple, and yet overwhelmingly complex explanation for understanding the degradation of our environment. In 1968, biology professor Garrett Hardin wrote an article for the journal *Science* called "The tragedy of the commons [5]." Professor Hardin argued that individuals focus on personal gain when dealing with "common" resources (owned by no one; therefore, owned by all). His idea was that such common use does not reward conservation or wise-use ethics, because the person who gives in or does not take their share of the resources generally loses it to someone else. The rate of destruction of the common resource is determined by population growth – more people, more resource use – and is not necessarily simply a result of human greed.

We all cherish individual freedom and economic gain; however, unlimited use of common resources leads to increased competition and use. Such freedom of use too often results in the tragedy of the commons, or destruction of the resource. Individual freedom to use a common resource, with no restrictions of use, will not lead toward the greatest public good. Freedom of use of a commons can destroy a commons.

Government plays a vital role in the protection of common resources. If an individual does not see their neighbor giving up a portion of their share of the privileges in a commons, then the individual will probably not be willing to give up any portion of their share, either. Privatizations of property, and the enactment of property laws, have served to create and enforce this necessary mutual coercion (through fines or prison terms). If a field is privately owned, the owner will control the use by the public, thus protecting it from overuse, at least in theory. Private ownership of property, and natural resource laws, can help protect a resource from overuse, or not.

Examples of "common resources" are the air we breathe and ocean fisheries. Air is available for all to use but use without restraints leads to air pollution. In 1625, the Dutch scholar Hugo Grotius (1583–1645) said, "The extent of the ocean is in fact so great that it suffices for any possible use on the part of all peoples for drawing water, for fishing, for sailing." However, increased demands have forced governments to place fishing restrictions to limit the freedom of use of this global commons.

In 1998, Garrett Hardin stated that "Individualism is cherished because it produces freedom, but the gift is conditional: the more the population exceeds the carrying capacity of the environment, the more freedoms must be given up [6]." Critically thinking, we can wonder which and whose freedoms must be given up?

Think About It: Personal Politics and Science

Hardin's essay is widely used, over 40,000 citations, and debated. Adding to the debate is the realization that Garrett Harding, the man, presented himself as a scientist and environmentalist. However, his personal views have been documented as racist, eugenicist, nativist, and Islamophobic [7]. In his opinion piece for the *Scientific American*, the author presented a "think about it" scenario. Environmental sustainability cannot exist without environmental justice. Are we really prepared to follow Hardin and say there are only so many lead pipes we can replace? Only so many bodies that should be protected from cancer-causing pollutants? Only so many children whose futures matter [8]?

Hardin is listed by the Southern Poverty Law Center as a known white nationalist [9]. The Southern Poverty Law Center provides examples, contrasting his scientific writings in which he maintained that he was not racist with his writings for far-right publications such as *The Social Contract and Chronicles*. The Center stated that

> [Hardin] used his position as a famous scientist and environmentalist to provide a veneer of intellectual and moral legitimacy for his underlying nativist agenda, serving on the board of directors of both the anti-immigrant Federation for American Immigration Reform and the white-nationalist Social Contract Press.

> Do you think that a person's political beliefs influence their work as professionals or their legacy? How do we deal with the politics of people, considered important in a particular field, when we learn disturbing aspects of their behaviors and beliefs? Are their personal beliefs relevant, and if so, how, and why?

14.3 Safe Drinking Water

In 2007, the United Nations declared the theme of World Water Day "Coping with water scarcity." UN Secretary-General Ban Ki-Moon of the Republic of Korea stated that 700 million people in

43 countries suffer from water scarcity, and by 2025, the figure could reach 3 billion people. In 2019, the World Health Organization put the number of people without access to a safely managed water source at 2.1 billion; that is 3 in 10 worldwide. Another 844 million people without access to a basic water source; this includes 263 million people who must travel (usually walking) over 30 minutes just to access water that isn't very clean, and another 159 million who rely on wholly untreated surface water sources for drinking, cooking and cleaning [10]. It is not just the world's growing population that is pressuring water resources; excessive use is also evident. The global population tripled in the twentieth century, but the use of water increased six-fold [11]. Between now and 2050, water demands are expected to increase by 400 percent from manufacturing, and by 130 percent from household use [12].

The right to water is a basic human right, but the right to clean, adequate water supplies, within the homes of all humans, is far from realized. We discussed earlier in this text the severe conditions endured by many living in poverty. Some women and girls must walk several kilometers (miles) to fill jugs of water for use in the home. Too often, the water is polluted, causing intestinal and other diseases. Illnesses require women and girls to be the family caregivers. Education suffers if school is skipped to care for the sick, or to stand in line waiting to fill a jug of water. The inevitable human process, leading to a tragedy of the commons, has led to overuse and the pollution of limited water supplies in too many locations on Earth. People are also a resource. When people are not educated, that resource (their opportunities to contribute to society) is minimized. This too is a tragedy.

Privatization of water resources has been attempted in many parts of the world to meet the growing need for safe drinking water. However, too often people in poverty cannot afford the services provided and are forced to spend more of their subsistence income on daily water supplies than they can sustain. Private water providers may choose to develop systems for more prosperous neighborhoods in urban settings. Poor neighborhoods are often left to whatever supplies they can find – sometimes polluted, sometimes located great distances from homes, and sometimes, both.

On a smaller scale, some civic, religious, health, and other nonprofit aid organizations are developing and supplying low-cost water filtering systems to provide families in poverty safer drinking water supplies. The United Methodist Michigan Area Haiti Task Force, for example, is building very low-cost, concrete bio-sand portable water filtration systems to reduce disease and to make safe drinking water more accessible [13]. Treating water as a human right, rather than as a commodity, helps those living in poverty. However, if such programs are not adequately funded and maintained through donations of time and money, they will falter and disappear.

14.4 Surface and Groundwater Conflicts

Streams in many parts of the United States and other regions around the world are drying up. In some cases, drought is accelerating these conditions. However, groundwater pumping and use is often a contributor to the problem. It is estimated that globally there are 592 transboundary aquifers. Transboundary water agreements need to be robust enough to deal with increasingly uncertain environmental and climatic conditions, and the social and demographic changes that will raise global population to 9.7 billion by 2050, and double the number of people who live in urban areas [14].

Many regions are starting to curtail the use of shallow groundwater supplies, through controversial well-pumping restrictions, to preserve surface water flows in adjacent streams. Colorado, Georgia, Idaho, Kansas, Nebraska, New Mexico, Wyoming, and other regions of the world are experiencing surface water shortages and have or are considering groundwater pumping restrictions. It is an issue that will continue to be economically controversial for many years to come.

It is not obvious that pumping restrictions would ever be needed in areas that have over 125 centimeters (50 inches) of rainfall a year. However, portions of the southeastern US are confronted with this problem because groundwater supplies are sometimes not renewed or replenished at the same rate as they are pumped. This can cause aquifer depletions that result in loss of river baseflow and all the ecosystem damage that entails. It is not the total depth of an aquifer that is important to rivers. It is the location of the river bottom, in relation to the groundwater table, that determines the impact on river baseflow. Water users in in the western US, and also in Arkansas, Louisiana, Mississippi, and other locations around the world are dealing with these types of groundwater/surface water conflicts.

An earlier example in the book considered the well shutdown in Colorado in 2006. Other states are considering methods to protect the economic needs of surface water users while retaining some use of groundwater. Should we consider which commodity provides greater economic benefits – surface water or groundwater? This question will be debated in the coming years, and will be complicated by state law, water-quality issues, and environmental protection.

Unlimited resource use will lead to overuse and destruction. The overdevelopment of shallow groundwater resources has led to today's growing conflict between surface and groundwater uses. Governmental intervention is solving the problem by shutting off wells; a process that leads to economic devastation for some, but protection of the resource for use by others. These battles, however, do not always consider the ecosystem because they are economically driven.

Unfortunately, battlegrounds over water in other parts of the world are, too often, deadly confrontations. In a chilling statement, the United Nations Secretary General (2007–16) Ban Ki-Moon stated "Let me be clear: The use of starvation as a weapon of war is a war crime." In 2018 the UN Security Council passed a council resolution which prohibited the use of hunger as a weapon of war. The fact that starvation as a weapon is a centuries-old practice does not diminish the complete lack of humanity in starving, women, children, or adults for any reason. The same is true for water.

Guest Essay By Kath Weston

Professor Kath Weston is a professor of Anthropology at the University of Virginia, and is also a British Academy Global Professor at the University of Edinburgh. She has also offered courses on environmental issues at Harvard University and Tokyo University. Her areas of specialization include water politics, political ecology, gender and sexuality, political economy, historical anthropology, kinship, surveillance, and class relations. She is the author of several publications, including *Gender in Real Time* and *Families We Choose*. Her latest books are *Animate Planet: Making Visceral Sense of Living in a High-Tech Ecologically Damaged World* (Duke University Press, 2017), and *Traveling Light: On the Road with America's Poor* (Beacon, 2008).

Water and War

Water wars are familiar to anyone who follows water politics. Upstream versus downstream, desert city versus coastal city, corporate-run versus state-run distribution systems, senior water rights versus junior water rights, sustainability versus profitability: as the climate changes and growing economies put pressure on resources, the conflicts seem to multiply. However, there is also another, more conventional kind of warfare in which water figures prominently. Millions of people in the world's battle zones owe their lives – or their deaths – to decisions about how to secure water in highly militarized situations.

To illustrate the importance of considering the relationship between water, war, and militarization, let's look briefly at three cases. The first comes from Sri Lanka, where endemic warfare has prompted the relocation of wells, with significant impacts on household economies, gender relations, and sociality. The second comes from Darfur in the Sudan, where United Nations engineers have had to balance decisions about their ability to supply water to peacekeeping forces with the needs of refugee populations. The final case concerns the Andaman Islands, now governed by India, where local people have been exploring innovative ways to rehabilitate water-harvesting structures left behind by a conquering army.

I have chosen these three cases from many possible ones because they push at the boundaries of established topics in water management such as water allocation, privatization, and the revival of what were once thought to be historically superseded technologies. Access to water during and after a military conflict is, of course, critical to survival. At the same time, a better understanding of the relationship between water and war on the part of scholars, water managers, non-governmental organizations (NGOs), engineers, and activists has the potential to inform, unsettle, and reframe contemporary debates about water politics and the environmental consequences of armed combat.

Sri Lanka

Sri Lanka is an island nation split into wet and dry zones, with annual rainfall that ranges from 100 to 500 centimeters (40 to 200 inches) according to region. The tsunami of 2004 infiltrated aquifers in coastal areas and contaminated drinking wells with saltwater [15]. Destruction of mangrove forests in the name of development only exacerbated the problems that faced engineers engaged in the tricky work of purifying these water sources. Long before the tsunami, however, water security had already emerged as an issue in areas of the country most directly disrupted by war [16]. For decades the Sri Lankan government, the Liberation Tigers of Tamil Eelam (LTTE), and paramilitary groups had been locked in on-again, off-again combat that threatened water supplies in a much more deliberate way than the massive wave.

As far back as 1986 at Peruveli in Trincomalee District, the Sri Lankan military and "home guards" attacked a refugee camp and dumped bodies into local wells, rendering the water unpotable. In 2004, controversy erupted near Batticaloa when the city attempted to tap an underground spring on reserved state land, only to be stopped by residents who threatened to poison the supply wells if the state tried to evict them. According to the chief water works engineer for Batticaloa, residents claimed that paramilitaries had sold them the land under wartime conditions between 1990 and 1994. In 2005, the Sri Lankan government posted troops to guard drinking water sources at the Vavuniya army camp after allegedly receiving reports that the LTTE had attempted to poison the water used by security forces. Near Jaffna in the north, rumors fly periodically about the poisoning of village wells by both sides in the combat. Abandoned wells have also been used as hiding places by young people hoping to avoid conscription by the Tigers.

Some of these incidents are better documented than others, but even threats and rumors can have profound effects when taken seriously. In the north, reports of well poisoning have motivated people to relocate wells from village commons to the interiors of household compounds. This shift represents a kind of privatization seldom mentioned in the literature on water politics, yet the ripple effects on community life can be significant. In these communities, women and children no longer gathered to exchange information, trade stories, and forge relationships when they went to fetch their daily water from a shared source. The relocation of wells speaks to the potentially far-reaching impacts of wartime decisions that emerge not from market operations or policy prescriptions but from strategies initiated at the household level. What may begin with a concern for everyday survival can end up reconfiguring social relations for years to come.

Darfur

Water poisoning has also been used as a military tactic in the genocidal conflict in Darfur. In this desert landscape, both the Sudanese army and the Janjaweed militia have reportedly thrown bodies into wells to make established villages uninhabitable. Water managers have also been targeted in this conflict. In 2006, the killing and abduction of hydraulic engineers in West Darfur led to a temporary suspension of water supply and sanitation for several refugee camps. Even without such disruptions, water is so scarce that the standard ration in many camps amounts to a mere 5 liters per day for all needs – cooking, drinking, and washing – in a region where temperatures routinely reach 50 °C [17].

As in Sri Lanka, the escalating militarization in Darfur affected household divisions of labor. Securing water was women's work in many communities, but as the countryside became increasingly unsafe, men began to haul water. In refugee camps where NGOs built pipelines, women often resumed the task of water supply, standing in line for up to six hours to draw their water rations. In some instances, this was longer than it had taken them to walk for river water back home, but the centralized wells and pipelines combined with the long waits to create an opportunity to build relationships in the disorienting world of the camps.

By 2004, the Darfur conflict had displaced almost 1.5 million people [18]. Before new refugee camps could be founded to relieve overcrowding and minimize the spread of waterborne diseases, water sources had to be located. Gathering so many people into camps had upset the balance of an ecologically and hydrologically sensitive region. Scientists used satellite radar to identify promising sites for drilling and NGOs funded new wells, but water supply could not keep up with the magnitude of the crisis. When the United Nations had to make decisions about sending in peacekeeping troops, they faced a dilemma [19]. Even if the engineers attached to a UN force successfully drilled for water, how would they handle the pressure to share it with thirsty refugees and local residents? This sort of water politics can materially affect the prospects for ending a war.

Andaman Islands

Where water is concerned, not all the legacies of warfare are so grim. After Japan invaded the Andaman Islands during World War II, the Japanese army dug 191 wells to supply its troops. When Japan withdrew from the Andamans during the later stages of the war, they abandoned these wells along with an airstrip, bunkers, and other structures designed to service the military.

Because these islands in the Bay of Bengal experience two monsoons per year, residents have relied on rainfall to meet most of the archipelago's needs for drinking water. By 2007, erratic rainfall plus the small size of the reservoir that supplied the capital city of Port Blair had translated into a water supply that arrived in homes only once every three days. In an effort to augment the water that the city provides, residents and authorities began to restore the old Japanese-built wells, which had fallen into disrepair [20].

The Andamans case proves even more interesting if considered in light of recent work on what could be called technological revivalism. Technological revivalism seeks to meld historic technologies that are often more environmentally appropriate with contemporary materials and changing circumstances. Historically these technologies were often associated with warfare and the spread of empire. In Syria, for example, residents have cleared, cleaned, and capped Roman-built water tunnels with modern cement to keep their villages alive. In India, people have rehabilitated cisterns that the colonial government once ordered closed for fear that "terrorists" (i.e., opponents to British rule) would hide in them.

Architects have studied the water-harvesting, cooling, and recycling systems that fed gardens in the Mughal Empire to inspire contemporary designs [21]. In this small but important sense, the long-term environmental consequences of war are not limited to highly publicized and devastating impacts such as deforestation or chemical contamination. With creativity and motivation, even the most horrific conflicts can someday paradoxically end up sustaining life.

14.5 Environmental Restoration

Environmental restoration is being attempted in several locations around the world to avert the destruction of shared resources. In 2007, Prime Minister John Howard of Australia announced that the Australian Government Water Fund would be used to provide AUS$220 million for watershed improvements in the Commonwealth. A significant component of the plan was the retirement of water rights (with compensation) to increase environmental water flows in rivers and streams. In California, water from the Owens River Valley – formerly used to supply drinking water to the Los Angeles metropolitan area – is being returned to its basin of origin to help restore local ecosystems.

Dam removal is another sign of efforts to return streams to conditions that are more natural. Efforts in Wisconsin, Maine, and California are evidence of growing concern to protect common resources. Generally, however, dam removal is successful only if the present owners of the dam are no longer receiving economic benefits from the structure or have been ordered to make repairs that exceed economic benefits. Unfortunately, lawsuits are generally required to force the owners and operators of water systems to re-evaluate present water uses and opportunities to return some supplies for environmental restoration programs.

14.6 Global Climate Change

Globally, glaciers are disappearing. The glaciers of Peru are melting. The Cordillera de Vilcanota (Quelccaya Ice Cap), for example, has been the traditional source of water for residents of Lima, Peru, and is the largest single glacier in the Peruvian Andes Mountains. The Alps of Europe also have glaciers that have shrunk to half their size, while in Africa the glacier on Mount Kenya has shrunk by 40 percent since 1963. Permafrost is disappearing in the coldest parts of the globe with new maximum temperature records being broken across the globe. Storms are becoming more fierce and more frequent, droughts are destroying food crops and causing wildfires. The hydrologic cycle is off-kilter. We have provided many examples of the influence of global climate change on the environment throughout this text.

The Kyoto Protocol was signed in Kyoto, Japan, by over 100 governments around the world pledging to reduce combined emissions of greenhouse gases by 5 percent by 2012. The US government did not sign the treaty, although many local and state governments have officially expressed support for it. Other countries decided that the protocol would not work without the support of two of the world's major resource users, the United States and China. The US Environmental Protection Agency (USEPA) established a Water Program Climate Change Workgroup which is now, once again, active. The United States joined in the Paris Climate Agreement in which 197 nations have agreed to cut carbon emissions to try to stop the Earth's temperature from rising 2 degrees above the preindustrial level. The best-case scenario was 1.5 degrees. The likelihood of achieving this goal was, in a sad twist of politics, less likely when the United States removed itself from the accords on November 4, 2020. However, that action was rescinded in February 2021 under the leadership of President Joe Biden.

Climate change is real, and it is serious. In just a 2 °C hotter world, according to an analysis of 70 peer-reviewed studies by Carbon Brief [22]:

- Seas could rise an average of 56 cm, or nearly 2 feet.
- 30 million people in coastal areas could be flooded each year by 2055.
- 37 percent of the population could face a severe heatwave at least every five years.
- 388 million people could be exposed to water scarcity and 195 million will be exposed to severe drought.
- Maize crop yields could fall 9 percent by 2100.
- The global per-capita GDP could fall 13 percent by 2100.

Global climate change is a tragedy in progress.

14.7 Values

People often express their feelings when discussing water issues, and the need to protect their personal "values." For some, maximum use of a resource is the same as achieving a maximum public good. However, *maximum use* of a water resource does not always equate to *maximum good* for everyone, including the environment. Instead, *optimum use* may be preferred, but is something less than maximum possible use. Each individual, striving only for personal gain and profit – is not necessarily thinking of the public good. In addition, individual intent does not necessarily lead toward the greatest public good. Instead, it generally leads to increased use of a resource. A change in personal and societal values are needed to shift the current paradigm of maximum use of water resources for only human and economic needs.

The development, use, and protection of freshwater supplies often creates polarized viewpoints. Social values toward ecosystem protection may be severely impacted by the economic values placed on resource development. What kind of framework can be created to find a middle ground, somewhere between all or none, or between big and small? In 1992, the United Nations Conference on Environment and Development (UNCED) meeting in Rio de Janeiro recognized the multiple needs of water: "Integrated water resources management is based on the perception of water as an integral part of the ecosystem, a natural resource, and a social and economic good [23]."

Personal values will drive future activities – whether they are based upon social, economic, or environmental needs. The future use and protection of water resources and the environment are in a critical state around the world. How will you contribute to the solutions needed to preserve our environment while providing the gift of safe, adequate water supplies to people around the world? We must balance our efforts. We must protect the basic human right to water while also meeting the economic, social, and environmental needs of our world. A truly daunting and yet exciting challenge awaits all of us.

Further Reading

Gore, Al, 2006, *An Inconvenient Truth: The Planetary Emergency of Global Warming and What We Can Do about It*, New York: Rodale Books.

Hardin, Garrett, 1968, "The tragedy of the commons," *Science* 162, 1243–1248.

Jahren, Hope, 2020, *The Story of More, How We Got to Climate Change and Where to Go from Here*, New York: Vintage Books, a Division of Penguin Random House LLC.

Rich, Nathaniel, 2019, *Losing Earth, A Recent History*, New York: Farrar, Straus, and Giroux.

References

1 George Eliot, 1858, *Scenes from Clerical Life*, Edinburgh: William Blackwood.

2 Carl Sagan, 1996, *The Demon Haunted World: Science as a Candle in the Dark*, A Ballantine Book published by Random House Publishing, first published as Carl Sagan, 1995, *The Demon-Haunted World: Science as a Candle in the Dark*, New York: Random House.

3 Food and Agriculture Association of the United Nations (FAO), 2011, "Save and grow: A policymaker's guide to the sustainable intensification of smallholder crop production," Rome. www.fao.org/zhc/hunger-facts/en/

4 Food Security Information Network, 2019, "Third Annual Report on Food Crises (GRFC)," www.fsinplatform.org/sites/default/files/resources/files/GRFC_2019-Full_Report.pdf (accessed and downloaded 2020).

5 Garrett Hardin, 1968, "The tragedy of the commons," *Science* 162, 1243–1248 https://science.sciencemag.org/content/162/3859/1243 (accessed 2020).

6 Garrett Hardin, 1998, "Extensions of the tragedy of the commons," *Science* 280, 682 https://science.sciencemag.org/content/280/5364/682.full (accessed 2020).

7 Mildenberger, Matto, 2018, "The tragedy of the 'tragedy of the commons'," *Scientific American*, https://getpocket.com/explore/item/the-tragedy-of-the-tragedy-of-the-commons?utm_source=pocket-newtab (accessed 2020).

8 Ibid.

9 Southern Poverty Law Center, "About Garrett Hardin," www.splcenter.org/fighting-hate/extremist-files/individual/garrett-hardin

10 World Health Organization, 2019, "Drinking water," www.who.int/en/news-room/fact-sheets/detail/drinking-water

11 FAO (Food and Agriculture Organization of the United Nations), 2009, "How to feed the world in 2050," Rome: FAO. www.fao.org/wsfs/forum2050/wsfs-background-documents/wsfs-expert-papers/en/ (accessed 2020).

12 Organisation for Economic Co-operation and Development (OECD), 2020, "Environmental outlook to 2050: the consequences of inaction," www.oecd.org/env/indicators-modelling-outlooks/oecdenvironmentaloutlookto2050theconsequencesofinaction-keyfactsandfigures.htm (accessed 2020).

13 The United Methodist Church, "Michigan group works for clean drinking water in Haiti," www.umc.org/site/apps/nl/content3.asp?c=lwL4KnN1LtH&b=2429867&ct=3708797 (accessed 2020).

14 IGRAC (International Groundwater Resources Assessment Centre), 2015, "Transboundary aquifers of the world map," www.un-igrac.org/who-we-are

15 CBC News, 2006, "Wells in Sri Lanka still contaminated by tsunami," May 8, www.cbc.ca/news/technology/wells-in-sri-lanka-still-contaminated-by-tsunami-1.626118 (accessed 2020).

16 Chandrasekara Dissanayake, 2005, "Of stones and health: medical geology in Sri Lanka," *Science* 309, 883–885.

17 Office for the Coordination of Humanitarian Affairs (OCHA), Relief Web, 2020 (original source: Oxfam, "Water shortages add to daily misery of camp life," February 11, 2005), https://reliefweb.int/report/sudan/sudan-water-shortages-add-daily-misery-camp-life (accessed 2020).

18 UN News Centre, 2004, "UN agencies find number of Sudanese displaced from Darfur jumps to 1.45 million," September 22, www.un.org/apps/news/story.asp?NewsID=12007&Cr=sudan&Cr1= (accessed 2020).

19 The Economist, 2007, "Call the Blue Helmets: Can the UN cope with increasing demands for its soldiers?" www.economist.com/briefing/2007/01/04/call-the-blue-helmets (accessed 2020).

20 Sanjib Kumar Roy, 2007, "Dry days: Andaman faces water shortage, turns to World War II wells," *Down to Earth: Science and Environment Fortnightly*, March 15.

21 Anil Agarwal and Sunita Narain, eds., 1997, *Dying Wisdom: Rise, Fall, and Potential of India's Traditional Water Harvesting Systems*, New Delhi: Centre for Science and Environment.

22 Carbon Brief, 2020, "The impacts of climate change at 1.5C, 2C and beyond," https://interactive.carbonbrief.org/impacts-climate-change-one-point-five-degrees-two-degrees/?utm_source=web&utm_campaign=Redirect (accessed 2020).

23 United Nations Conference on Environment and Development (UNCED), 1992, "Earth Summit," www.un.org/en/conferences/environment/rio1992 (accessed March 2021).

Index

Printed in the United States
by Baker & Taylor Publisher Services

Printed in the United States
by Baker & Taylor Publisher Services